AF757100

Annual Report on Operation of China's Natural Gas Industry: 2023

Liehui Zhang · Chaohong Gao · Xiaoqiang Zheng · Xianfeng Liu

Annual Report on Operation of China's Natural Gas Industry: 2023

石油工业出版社有限公司
PETROLEUM INDUSTRY PRESS

Springer

Liehui Zhang
Southwest Petroleum University
Chengdu, China

Chaohong Gao
China Petroleum Enterprise Association
Beijing, China

Xiaoqiang Zheng
Southwest Petroleum University
Chengdu, China

Xianfeng Liu
Southwest Petroleum University
Chengdu, China

ISBN 978-981-95-2370-2 ISBN 978-981-95-2371-9 (eBook)
https://doi.org/10.1007/978-981-95-2371-9

Jointly published with Petroleum Industry Press.
The print edition is not for sale in China (Mainland). Customers from China (Mainland) please order the print book from: Petroleum Industry Press.
ISBN of the Co-Publisher's edition: 978-751-83-6640-8

The original submitted manuscript has been translated into English. The translation was done using artificial intelligence. A subsequent revision was performed by the author(s) to further refine the work and to ensure that the translation is appropriate concerning content and scientific correctness. It may, however, read stylistically different from a conventional translation.

© Petroleum Industry Press 2025

This work is subject to copyright. All rights are solely and exclusively licensed by the Publisher, whether the whole or part of the material is concerned, specifically the rights of reprinting, reuse of illustrations, recitation, broadcasting, reproduction on microfilms or in any other physical way, and transmission or information storage and retrieval, electronic adaptation, computer software, or by similar or dissimilar methodology now known or hereafter developed.
The use of general descriptive names, registered names, trademarks, service marks, etc. in this publication does not imply, even in the absence of a specific statement, that such names are exempt from the relevant protective laws and regulations and therefore free for general use.
The publishers, the authors, and the editors are safe to assume that the advice and information in this book are believed to be true and accurate at the date of publication. Neither the publishers nor the authors or the editors give a warranty, express or implied, with respect to the material contained herein or for any errors or omissions that may have been made. The publishers remain neutral with regard to jurisdictional claims in published maps and institutional affiliations.

This Springer imprint is published by the registered company Springer Nature Singapore Pte Ltd.
The registered company address is: 152 Beach Road, #21-01/04 Gateway East, Singapore 189721, Singapore

If disposing of this product, please recycle the paper.

Editorial Committee

Chairpersons

Zhang Liehui, Gao Chaohong, Zheng Xiaoqiang, Liu Xianfeng

Vice Chairpersons

Ning Minjing, Shen Xilin, Yang Xiaoxi, Pang Jian, Zhao Chunlan, Yu Cuiting, Liu Qin, Deng Tianjuan

Members

Chen Yinan, Duan Binyang, Duan Linli, Duan Zhaofang, Gao Jun, Gao Yun, Guo Zhenyu, He Sha, He Runmin, He Yiping, Jiang Tongwen, Lang Dongxiaom Li Rou, Li Ting, Liu Hui, Liu Yucheng, Luo Shuanghan, Qi Weiyi, Tang Dalin, Wang Bin, Yan Jiantao, Yang Xinping, Yin Tao, Yu Xiaozhong, Zang Qingying, Zhang Huifang, Zhao Yulong, Zhou Juan

Preface

The world today is in a period of great change, one that has not been seen in a century. The international situation is becoming increasingly complex, amid the intensification of geopolitical risks, the frequent occurrence of extreme weather events, and the "instrumentalization" of energy. The political, economic, and social changes in the post-pandemic era pose real challenges to the development of world energy and have impacted the current world energy structure. For instance, some Western countries have abandoned their emission reduction commitments and begun to turn to high-carbon fossil fuels to make up for the energy gap. Indecisive energy policy further intensifies the uncertainty of the world's green and low-carbon energy transition, negatively affecting the restructuring of the world energy industry. In 2023, China's energy industry fully implemented the "14th Five-Year Plan" and "dual carbon" goals, steadily improved its energy supply capacity, enhanced its role in securing social and economic development, deepened structural reforms on the energy supply side, and demonstrated a trend toward diversified, clean, and intelligent development.

Natural gas, as a clean, low-carbon, and efficient fossil fuel, is considered one of the most important types of energy and plays a significant role in achieving China's "dual carbon" goals. In recent years, China's natural gas industry has been making comprehensive efforts in system reform, supply security, and efficient utilization. It has been continuously facilitating the strategy of "comprehensive layout, technological innovation, multi-energy complementarity, multi-element integration, flexible efficiency, and optimization and upgrading." Meanwhile, it has accelerated the pace of natural gas infrastructure construction and promoted the complementary, coordinated, and integrated development of natural gas and new energy. Furthermore, continuous improvements have been made to natural gas-related laws and regulations, and a comprehensive production, supply, storage, and sales system has been gradually established. Essentially, a multi-gas supply pattern has been formed, which mainly consists of domestic conventional and unconventional natural gas, onshore pipeline imported natural gas, and offshore imported liquefied natural gas (LNG). Moreover, major consumption fields, such as urban gas, industrial fuel gas, chemical gas, and power generation gas have been expanded.

In 2023, China's natural gas production hit a record high of 232.4 billion cubic meters, achieving a growth rate of 5.58%. Under the influence of multiple factors such as stable economic recovery, the fall in natural gas prices, and the insufficient supply of hydropower, China's natural gas consumption growth rebounded from the bottom, achieving rapid growth, with total consumption reaching 394.5 billion cubic meters, an increase of 7.6%. Natural gas imports reached 119.97 million tons, a year-on-year increase of 9.9%, and the degree of foreign dependence reached 42.3%. Throughout the year, China's natural gas industry developed steadily and healthily, making breakthroughs in exploration and development, pipeline construction, and market-oriented reforms, demonstrating strong supply elasticity and consumption resilience. Looking forward, China's natural gas industry will continue to make efforts in ensuring the safety of the natural gas supply, promoting the integrated development of natural gas and new energy, fostering the green and low-carbon transformation of energy, and accelerating the digital and intelligent development of the natural gas industry. The industry will further implement the decisions and deployments of the CPC Central Committee and the State Council on the construction of the natural gas production, supply, storage, and sales system, and the requirements for ensuring supply and stabilizing prices. It will promote the implementation of the "14th Five-Year Plan," play a greater role in the low-carbon transformation of energy, and contribute to the national energy security and the accelerated construction of an energy power.

This book features narrative and factual descriptions, data analyses and discussions, development forecasts, and in-depth studies of hot topics. It is divided into five chapters: General Overview, Policy, Industry, Outlook, and Special Topics, covering the upstream, midstream, and downstream of the industry. It provides a systematic analysis of the annual operation of the natural gas industry. The General Overview part introduces the development of global and Chinese natural gas and summarizes the characteristics of their development. The Policy part covers national and local natural gas policies, with analysis and interpretation of major policies. The Industry part focuses on important links of China's natural gas industry chain, including exploration and development, storage and transportation, import and export, and consumption, to show the overall operation of China's natural gas industry. The Outlook part forecasts the future price trends of gaseous and liquefied natural gas in China through a multi-factor price prediction method. The Special Topics Chapter focuses on themes such as "Analysis of China's Natural Gas Industry Prosperity Index," "Natural Gas Supply and Consumption in the Sichuan-Chongqing Region in 2023 and Outlook for 2024," "Digital Transformation Research and Practice in the Natural Gas Industry," and "Full Life Cycle Planning of Natural Gas Development Projects." It explores the industrial practice and future development of China's natural gas industry and reviews the major events in China's natural gas exploration and development in 2023 to clarify the significant breakthroughs in the industry.

Zhang Liehui, Zheng Xiaoqiang, and Liu Xianfeng are the chief architects of this book and were responsible for organizing the compilation of the entire work. Shen Xilin, Zheng Xiaoqiang, Liu Xianfeng, Yang Xiaoxi, Pang Jian, Ning Minjing, Li

Ting, Liu Qin, Zhao Chunlan, Deng Tianjuan, Yu Cuiting from the China Energy Index Research Center of Southwest Petroleum University, and He Runmin and Zhou Juan from the Economic Research Institute of Southwest Oil and Gas Field Branch of PetroChina Co., Ltd. (hereinafter referred to as "PetroChina") were responsible for the final edition of this book; Chen Yinan, Zhao Yulong and others participated in the revision of the manuscript. The specific writing assignments for each chapter are as follows: Yang Xiaoxi, Lin Anqi, Qiu Yufei for the General Overview part, Zheng Xiaoqiang, Li Ting, Li Wentao, Yuan Yanru for the Policy part, Liu Xianfeng, Deng Tianjuan, Pang Jian, Ning Minjing, Xu Yihan, Zhao Yaru, Long Xuemei, Shi Yuefeng, Li Ling for the Industry part, Zhao Chunlan, Yu Cuiting, Wu Xiang for the Outlook part, Shen Xilin, He Runmin, Zhou Juan, Gao Yun, Liu Qin, Liu Xianfeng, Xu Yihan, Li Yingying for the part on Special Topics, and Li Ting for the Appendix. During the writing of this book, we referred to the statistical data and information from the National Development and Reform Commission, the National Bureau of Statistics, the General Administration of Customs of China, and other relevant research institutions. We have also drawn on research achievements of experts and scholars and absorbed relevant ideas and viewpoints from the reference literature. We would like to express our heartfelt thanks to all of them! We hope that this book can provide a basis for the government to formulate natural gas development policies, serve as a reference for oil and gas companies to develop strategies and investment decisions, and provide basic research materials for relevant researchers. Due to the limited level of the authors, there may be some shortcomings in the report, and we sincerely ask for the readers' corrections!

Chengdu, China — Liehui Zhang
Beijing, China — Chaohong Gao
Chengdu, China — Xiaoqiang Zheng
Chengdu, China — Xianfeng Liu
March 2024

Contents

General Overview 1
Overview of World Natural Gas Development 2
Global Energy Production and Consumption 2
Global Natural Gas Production 8
Global Natural Gas Consumption 12
Summary 16
The Current Situation of Natural Gas Development in China 18
Energy Production and Consumption in China 18
Natural Gas Production in China 24
Natural Gas Consumption in China 30
Summary 37

Natural Gas Policy 39
National Natural Gas Policy 40
Green Development 40
Energy Governance 47
Energy Utilization 57
Natural Gas Prices 62
Local Natural Gas Policy 64
Local Natural Gas Prices 72
Natural Gas Pipeline Construction 81

The Natural Gas Industry in China 87
China's Natural Gas Exploration 88
The Overall Situation of Oil and Gas Exploration 88
Natural Gas Reserves in Provinces, Autonomous Regions, and Municipalities 90
Natural Gas Development in China 96
The Overall Situation of Natural Gas Development 96
Natural Gas Output of Provinces, Autonomous Regions, and Municipalities 96

The Development of Major Oil and Gas Fields 100
The Development Situation by Product Type 102
Natural Gas Imports and Exports in China 108
Natural Gas Imports 108
Natural Gas Exports 129
Analysis of Natural Gas Import and Export Trends 132
China's Natural Gas Consumption 135
Natural Gas Consumption 135
Natural Gas Prices 140
Development Trend of Natural Gas in the Later Period of the "14th Five-Year Plan" 155

Outlook 159
Natural Gas Price Outlook 159
Introduction to the Outlook Model 159
Gaseous Natural Gas Price Outlook 163
Liquefied Natural Gas (LNG) Price Outlook 170
Summary 189

Special Topics in the Natural Gas Industry 191
Special Topic 1: Analysis of the Prosperity Index of China's Natural Gas Industry 192
Analysis of the Prosperity Index of China's Natural Gas Industry 192
Analysis of the Prosperity Index of Natural Gas Production Companies 193
Analysis of the Prosperity Index of Natural Gas Sales Companies 198
Conclusion 200
Special Topic 2: Natural Gas Supply and Consumption in Sichuan and Chongqing in 2023 and Outlook for 2024 201
Natural Gas Market Supply in Sichuan and Chongqing in 2023 202
Natural Gas Market Consumption in Sichuan and Chongqing in 2023 203
Natural Gas Market Outlook in Sichuan and Chongqing in 2024 207
Special Topic 3: Research and Practice on the Digital Transformation of the Natural Gas Industry 210
Background and Significance of Digital Transformation in the Natural Gas Industry 210
Current Status and Trends of Digital Development in the Natural Gas Industry 215
Design of Digital Integration Scheme for the Natural Gas Industry 224
Practice and Achievements of Digital Transformation in the Natural Gas Industry 227

Suggestions for Digital Transformation in the Natural Gas Industry 233
Special Topic 4: Full Life Cycle Planning of Natural Gas Development Projects 236
Overview of Full Life Cycle Management of Natural Gas Development Projects 236
Planning Framework for the Full Life Cycle Management of Natural Gas Development Projects 241
Full Life Cycle Management CSSQ Planning of Natural Gas Development Project 247

Appendix: Major Events in China's Natural Gas Exploration and Development in 2023 259

Bibliography 277

About the Authors

Liehui Zhang is a Professor and Doctoral Supervisor currently serving as the Party Secretary of Southwest Petroleum University. He is a distinguished recipient of several prestigious awards, including the Ministry of Education's University Young Teachers Award, the Sun Yueqi Young Science and Technology Award, the Fok Ying Tung Young Teachers Award, and the Sichuan Province Young Science and Technology Award. He holds significant positions within key professional organizations, serving as Deputy Director for both the Sub-Committee on Natural Gas Development of the National Natural Gas Standardization Technical Committee and the Petroleum Engineering Professional Committee of the China Petroleum Association. He is also a Member of the Ninth Natural Gas Committee of the China Petroleum Society and Head of its Natural Gas Development Group, as well as a Member of the Society of Petroleum Engineers (SPE). Furthermore, he serves on the editorial boards of leading academic journals, including *Hydrodynamics, Special Oil and Gas Reservoirs, Petroleum Geology and Oilfield Development in Daqing, Reservoir Evaluation and Development,* and *Natural Gas Industry.* His long-term research and teaching efforts are focused on the seepage mechanics of complex oil and gas reservoirs, well testing, and numerical simulation.

Chaohong Gao serves as Chief Editor and is the former full-time Vice President of the China Petroleum Enterprise Association. With a long-term focus on management within the petroleum and chemical industries, he has authored numerous articles for the journal *China Petroleum Enterprise.* He has spearheaded the compilation of several influential publications, including the book *Petroleum Spirit* and a series of annual reports such as the "Annual Report on the Development Analysis and Outlook of China's Oil and Gas Industry," the "Report on Development of China's Low-Carbon Economy," the "Report on Development of Refined Oil and New Energy," and the "Annual Report on Operation of China's Natural Gas Industry." His extensive research on the operation and management of the oil and gas industry has yielded significant achievements, for which he has received numerous honors and awards.

Xiaoqiang Zheng is a Professor and Doctoral Supervisor who currently serves as the Dean of the School of Economics and Management at Southwest Petroleum University. He has been recognized as a leader of academic and technological talents and as the leader of a high-level philosophy and social sciences team in Sichuan Province. He holds multiple esteemed positions within national academic societies, including Member of the Energy and Environmental Management Division of the Chinese Academy of Management Sciences; Standing Director of the Energy Economy and Management Research Division of the Chinese "Double Law" Research Society; Standing Director of the Energy Systems Engineering Division of the Chinese Academy of Systems Engineering; Member of the Neural Economic Management Committee of the Chinese Academy of Technological Economics; Member of the Youth Committee of the Chinese Petroleum Society; and Member of the Carbon Neutrality Professional Committee of the China Petroleum Enterprise Association. His primary research focuses on energy economics and management, energy systems engineering, and energy security and sustainable development.

Xianfeng Liu is a Professor and Master's Supervisor. He has been selected as a reserve candidate for Academic and Technical Leadership in Sichuan Province and serves as the Deputy Director of the China Energy Index Research Center at Southwest Petroleum University. His primary research focuses on energy economics and management, energy systems engineering, and energy security and sustainable development.

General Overview

As countries around the world gradually enter the process of modernization, energy mainly centered around oil and natural gas has become the foundation of national security and economic development, and the situation will remain for some period into the future. The continuous existence of uncertainties in the international environment is also leading to new and profound changes in the development of world energy. In 2023, with the overall recovery of domestic economic growth, the growth rate of energy consumption has accelerated. The transformation of China's energy consumption structure has been further advanced. Energy enterprises have adhered to innovation-driven approaches to enhance the modernization level of the industry, and the capacity and quality of energy supply and assurance have continued to improve.

This section adopts a macro perspective. It follows a train of thought that moves from the international level to the domestic level, from the stage of production to that of consumption, and from looking at the overall quantity to analyzing the internal structure. Through this approach, it unravels the general situation regarding energy production and consumption over the past ten years. It then zeroes in on the natural gas industry, presenting in-depth analyses across five specific aspects. Firstly, an overview of the global energy production and consumption totals. While all types of energy generally showed a growth trend, the production and consumption of non-fossil energy increased at a faster rate than that of fossil energy. Secondly, the global production, consumption, and interregional imports and exports of natural gas have been affected by a variety of factors such as extreme weather, energy transformation, and geopolitics. As a result, the supply and demand relationship in the global natural gas market has fluctuated. In 2022, the total production of natural gas remained essentially the same as in 2021, while consumption decreased by 3.09%. The proportion of natural gas production and consumption in fossil energy both declined. Thirdly, an overview of China's energy production and consumption totals. China is the largest energy producer and consumer in the world. In 2023, both energy production and consumption in China continued to grow. The total energy

© Petroleum Industry Press 2025

L. Zhang et al., *Annual Report on Operation of China's Natural Gas Industry: 2023*,
https://doi.org/10.1007/978-981-95-2371-9_1

production reached 4.83 billion tons of standard coal, an increase of 3.65% year-on-year. The total energy consumption reached 5.72 billion tons of standard coal, up by 5.73% year-on-year. Fourthly, the total production of natural gas in China and its proportion in energy production. In 2023, the natural gas output reached 232.4 billion cubic meters, up by 5.58% year-on-year. Its proportion in the total energy production continued to rise compared with 2022. Fifthly, the total consumption of natural gas in China and its proportion in energy consumption. In 2023, the consumption of natural gas was 394.5 billion cubic meters, up by 7.60% year-on-year. Its proportion in the total energy consumption remained relatively stable. In the face of the new era and the new journey, the internal and external environment is undergoing profound changes. The energy industry in China will further accelerate the construction of a new type of energy system to achieve higher-quality development of energy.

Overview of World Natural Gas Development

Natural gas is a combustible gas which is formed as a result of ancient organisms buried underground being subjected to high temperature and high pressure for billions of years. It boasts a high calorific value, enjoys stable combustion, and is clean and environmentally friendly. Moreover, it is the cleanest type of energy among the fossil fuels. As a result, natural gas, characterized by attributes such as clean and low-carbon features, high calorific value, and stable combustion, serves as an important bridge linking high-carbon fossil energy with zero-carbon new energy. It also emerges as the optimal partner for the continuous development of the new energy industry. With the technological progress of the natural gas industry, especially the continuous advancement of shale gas development, the proportion of natural gas in the world's energy consumption structure is rising rapidly. Moreover, it is playing an indispensable role in optimizing the national energy consumption structures.

Global Energy Production and Consumption

In 2022, global energy demand continued to recover. However, the ever-changing international situation and the frequent occurrence of extreme weather events constantly exerted impacts on the global energy industry. Overall, primary energy consumption in 2022 increased by 1% when compared to 2021 and about 3% when compared to 2019. In the context of global energy's transition to green and low-carbon, the energy demand structure has changed, with the proportion of renewable energy constantly increasing and the degree of electrification of terminal energy gradually rising. Fossil fuels accounts for about 82% of primary energy consumption, while non-fossil energy accounts for about 18%, among which renewable energy has grown rapidly, reaching a proportion of 7.5%.

(I) Overview of global energy production

1. Global fossil fuel production

Fossil fuel refers to the carbon-containing energy sources represented by coal, oil, and natural gas. It constitutes an essential material basis for the development and prosperity of modern civilization. Although clean energy technologies are growing rapidly, the world's energy supply still relies mainly on fossil energy, whose dominant position is irreplaceable. The total output and growth rate of global fossil energy in the past decade are shown in Table 1 and Fig. 1.

During the period from 2013 to 2022, the total output of global fossil energy experienced two major declines, in 2016 and 2020. This was especially the case

Table 1 Total global fossil fuel production and growth rate (2013–2022)

Year	Total output (million tons of oil equivalent)	Growth rate (%)
2013	10,974.2	0.94
2014	11,121.5	1.34
2015	11,215.9	0.85
2016	11,049.5	−1.48
2017	11,249.0	1.81
2018	11,684.2	3.87
2019	11,812.0	1.09
2020	11,230.8	−4.92
2021	11,601.0	3.30
2022	12,058.6	3.94

Data source "World Energy Statistics Yearbook 2023"

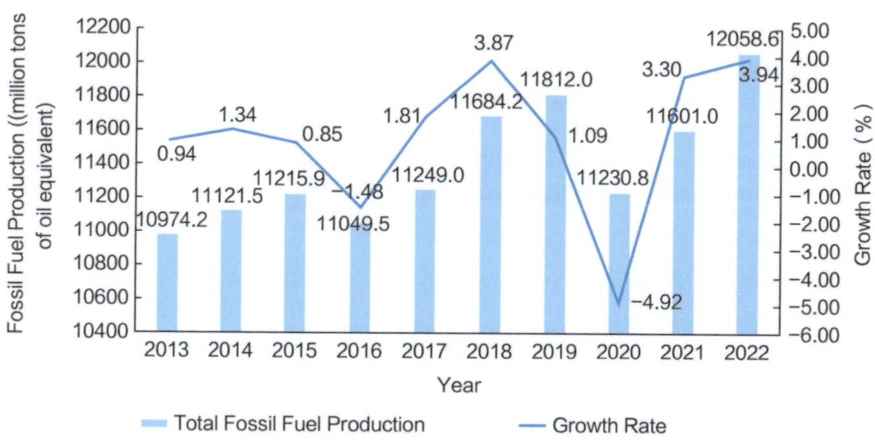

Fig. 1 Total global fossil fuel production and growth rate (2013–2022). *Data source* "World Energy Statistics Yearbook 2023"

Table 2 Total global non-fossil energy power generation by type and growth rate (2013–2022)

Year	Nuclear (TW·h)	Hydropower (TW·h)	Renewable energy (Excluding Hydropower) (TW·h)	Total non-fossil energy power generation (TW·h)	Growth rate (%)
2013	2,490.3	3,788.1	1,238.5	7,516.9	4.77
2014	2,541.1	3,889.0	1,406.0	7,836.2	4.25
2015	2,575.3	3,879.5	1,631.4	8,086.2	3.19
2016	2,613.7	4,014.4	1,843.5	8,471.6	4.77
2017	2,637.0	4,071.2	2,175.8	8,884.0	4.87
2018	2,700.2	4,189.8	2,481.5	9,371.5	5.49
2019	2,795.8	4,243.3	2,793.5	9,832.5	4.92
2020	2,689.0	4,359.0	3,151.3	10,199.4	3.73
2021	2,802.5	4,288.8	3,664.6	10,755.9	5.46
2022	2,679.0	4,334.2	4,204.3	11,217.5	4.29

Data source "World Energy Statistics Yearbook 2023"

in 2020, when the decline in fossil energy production reached its peak. After that, fossil energy production rebounded in 2021 and continued to rise in 2022, increasing from 11,230.8 million tons of oil equivalent in 2020 to 12,058.6 million tons of oil equivalent in 2022.

2. Global non-fossil energy production

According to the data statistics in the "World Energy Statistics Yearbook 2023,"[1] the changes in the total power generation and growth rate of global non-fossil energy by type from 2013 to 2022 are shown in Table 2 and Fig. 2.

In 2022, the total global non-fossil energy production was about 11,217.5 TW·h, which represented an increase of 4.29% from 2021. Nuclear energy generation was 2,679.0 TW·h, accounting for 23.88% of the total global non-fossil energy generation, a slight decrease from 2021. Hydropower energy generation was 4,334.2 TW·h, with the largest proportion in global non-fossil energy generation, accounting for 38.64%. Renewable energy generation (excluding hydropower) was 4,204.3 TW·h, accounting for 37.48% of the total global non-fossil energy generation.

(II) Overview of global energy consumption

Despite the rapid development of renewable energy, fossil fuels (oil, natural gas, and coal) still dominated global energy consumption in 2022. In terms of consumption growth rates, the growth rate of primary energy demand consumption slowed in 2022,

[1] The global non-fossil energy production is calculated based on the equivalent amount of fossil fuels that would be required to produce a specific quantity of electricity in a standard thermal power plant, with the unit being terawatt-hours (TW·h). To maintain data consistency, this book also adopts the same unit.

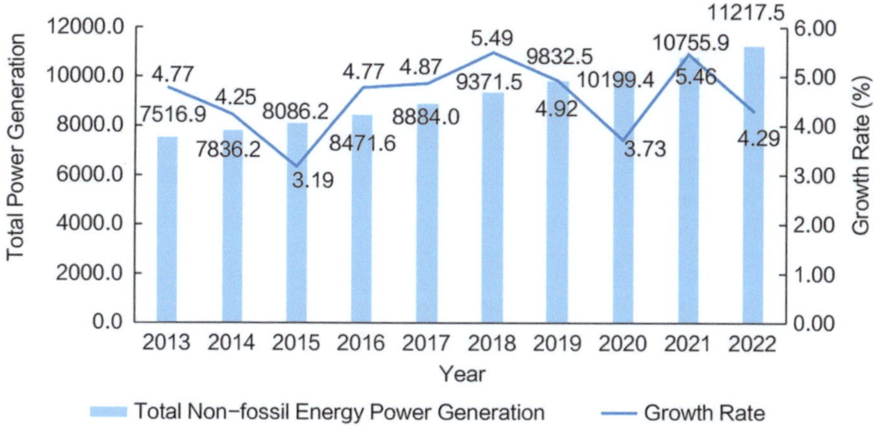

Fig. 2 Total global non-fossil energy power generation and growth rate (2013–2022). *Data source* "World Energy Statistics Yearbook 2023"

registering an increase of just 1.1%. This is somewhat lower than the average level of the past decade and considerably lower than the growth rate of 5.46% in 2021. However, it is still about 3% higher than the figure in 2019. The significant increase in the total global energy consumption in 2021 was largely due to the sharp drop in energy demand in 2020 caused by factors such as the global economic recession and sharp reduction in demand during the COVID-19 pandemic. In 2022, it returned to normal growth rates, and the primary energy consumption in all regions except Europe and the Commonwealth of Independent States (CIS) countries has increased.

1. Total global energy consumption

By sorting out the data in the "World Energy Statistics Yearbook 2023," we can obtain the total global energy consumption as well as its growth rate from 2013 to 2022, which are as shown in Table 3 and Fig. 3.

From 2013 to 2019, the total energy consumption maintained a relatively stable upward trend, with an average annual growth rate of 1.54%. In 2020, the global energy growth rate fell into negative territory due to the impact of the COVID-19 pandemic and the global economic downturn. After 2021, with the gradual recovery of the global economy, global energy consumption resumed its upward trend. The primary energy consumption in 2022 was 14,436.5 million tons of oil equivalent, which represented an increase of 1.11% from 2021.

2. Global fossil and non-fossil energy consumption

By referring to the data in the "World Energy Statistics Yearbook 2023," we can obtain both the total global fossil energy consumption and growth rate from 2013 to 2022 (as shown in Table 4 and Fig. 4) and the total global non-fossil energy consumption and growth rate from 2013 to 2022 (as shown in Table 5 and Fig. 5).

Global fossil fuel consumption showed a steady upward trend before 2019 and a significant decline for the first time in 2020. However, in 2021, global fossil fuel

Table 3 Total global energy consumption and growth rate (2013–2022)

Year	Consumption (million tons of oil equivalent)	Growth rate (%)
2013	12,838.9	1.75
2014	12,974.7	1.06
2015	13,082.7	0.83
2016	13,240.2	1.20
2017	13,517.8	2.10
2018	13,887.1	2.73
2019	14,038.6	1.09
2020	13,539.1	−3.56
2021	14,278.0	5.46
2022	14,436.5	1.11

Data source "World Energy Statistics Yearbook 2023"

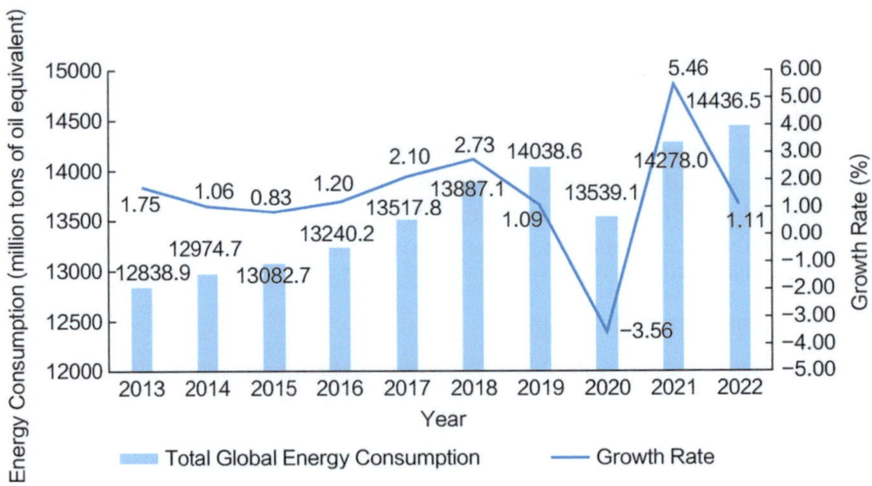

Fig. 3 Total global energy consumption and growth rate (2013–2022). *Data source* "World Energy Statistics Yearbook 2023"

consumption returned nearly to the levels before 2020 and continued to show a slight upward trend in 2022, rising from 11,751.8 million tons of oil equivalent in 2021 to 11,807.8 million tons of oil equivalent in 2022.

From 2013 to 2022, global non-fossil energy consumption showed a steady upward trend. It rose from 1,825.6 million tons of oil equivalent to 2,628.6 million tons of oil equivalent, an increase of 44% during the ten-year period. Compared to 2021, the growth rate of non-fossil energy consumption in 2022 was 4.05%. According to the "World Energy Statistics Yearbook 2023," the fastest growth was in renewable energy (excluding hydropower). Renewable energy (excluding

Table 4 Total global fossil energy consumption and growth rate (2013–2022)

Year	Total fossil fuel consumption (million tons of oil equivalent)	Growth rate (%)
2013	11,013.4	1.32
2014	11,079.4	0.60
2015	11,135.5	0.51
2016	11,211.0	0.68
2017	11,401.8	1.70
2018	11,663.3	2.29
2019	11,710.5	0.41
2020	11,137.6	−4.89
2021	11,751.8	5.51
2022	11,807.8	0.48

Data source "World Energy Statistics Yearbook 2023"

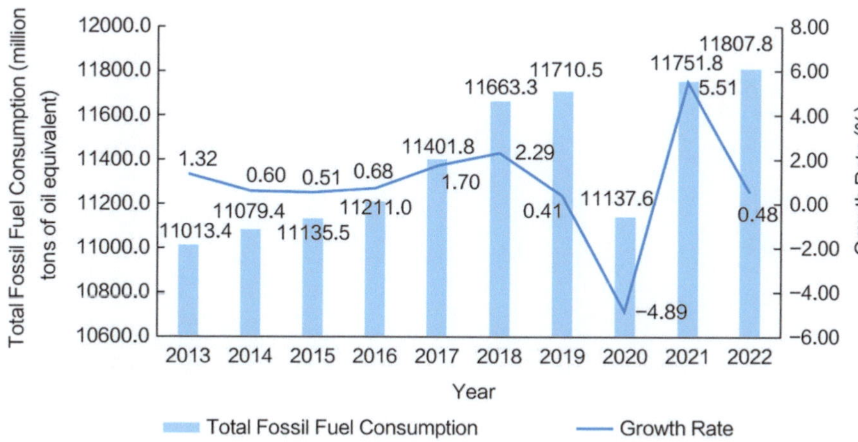

Fig. 4 Total global fossil energy consumption and growth rate (2013–2022). *Data source* "World Energy Statistics Yearbook 2023"

hydropower) also accounted for the largest proportion in the global non-fossil energy in 2022, while the proportions of nuclear energy and hydropower have been relatively stable in the past decade.

Table 5 Total consumption and growth rate of global non-fossil energy (2013–2022)

Year	Non-fossil energy consumption (million tons of oil equivalent)	Growth rate (%)
2013	1,825.6	4.43
2014	1,895.3	3.82
2015	1,947.3	2.74
2016	2,029.2	4.21
2017	2,116.0	4.28
2018	2,223.8	5.10
2019	2,328.0	4.69
2020	2,401.5	3.16
2021	2,526.2	5.19
2022	2,628.6	4.05

Data source "World Energy Statistics Yearbook 2023"

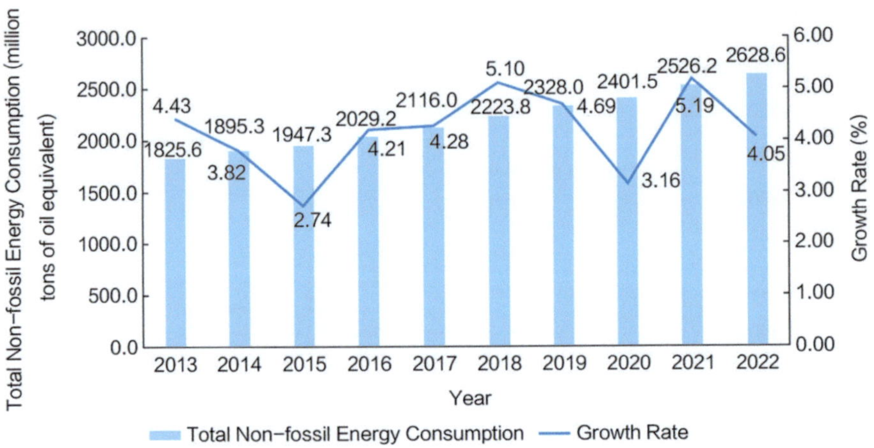

Fig. 5 Total consumption and growth rate of global non-fossil energy (2013–2022). *Data source* "World Energy Statistics Yearbook 2023"

Global Natural Gas Production

Natural gas, being an efficient and clean fossil fuel that has a high calorific value, plays an important role in human survival and life. Compared with other fossil fuels such as oil and coal, natural gas emits less carbon dioxide. At the same time, it possesses a solid industrial development foundation and huge development potential. In terms of both supply side and consumption, natural gas enjoys unique comparative advantages. It can play a significant role in smoothing the peak load of the power grid in the context of an influx of new energy, advancing low-carbon development in the transportation sector, decreasing emissions in the industrial sector, and improving the urban environment. Besides, it can also buy time for disruptive technological breakthroughs in areas like energy storage, high penetration green hydrogen, carbon capture-use-storage, and the smooth transition of the energy system.

(I) Global natural gas production

In recent years, to address climate issues and promote energy transition, the production of natural gas has been given attention by countries around the world, and its output has generally shown an upward trend over the past decade. The total global natural gas production and growth rate from 2013 to 2022 are shown in Table 6 and Fig. 6.

From 2013 to 2019, global natural gas production showed a trend of steady growth, with an average growth rate of 2.56%. However, in 2020, due to factors such as the global economic downturn, there was a decline for the first time in nearly a decade. Subsequently, natural gas production increased significantly in 2021 as the global economy recovered. In 2022, natural gas production remained relatively stable, decreasing slightly by 0.24%.

(II) The proportion of natural gas in global fossil fuel production

Natural gas is considered to be a clean alternative to coal and oil and plays an important role in promoting the energy transition. Over the ten years from 2013 to 2022, the proportion of natural gas in fossil fuel production increased by 2.46%. The proportion of natural gas in fossil fuel production in 2022 and the structure of global fossil energy production from 2013 to 2022 are shown in Figs. 7 and 8 respectively.

From 2013 to 2022, the proportion of natural gas in total fossil fuel production was always the lowest. Over the past decade, the proportion of natural gas in fossil

Table 6 Total global natural gas production and growth rate (2013–2022)

Year	Total production (million tons of oil equivalent)	Growth rate (%)
2013	2,895.8	1.19
2014	2,956.0	2.08
2015	3,018.1	2.10
2016	3,049.7	1.04
2017	3,159.3	3.59
2018	3,317.5	5.01
2019	3,414.4	2.92
2020	3,321.7	−2.72
2021	3,487.6	4.99
2022	3,479.3	−0.24

Data source "World Energy Statistics Yearbook 2023"

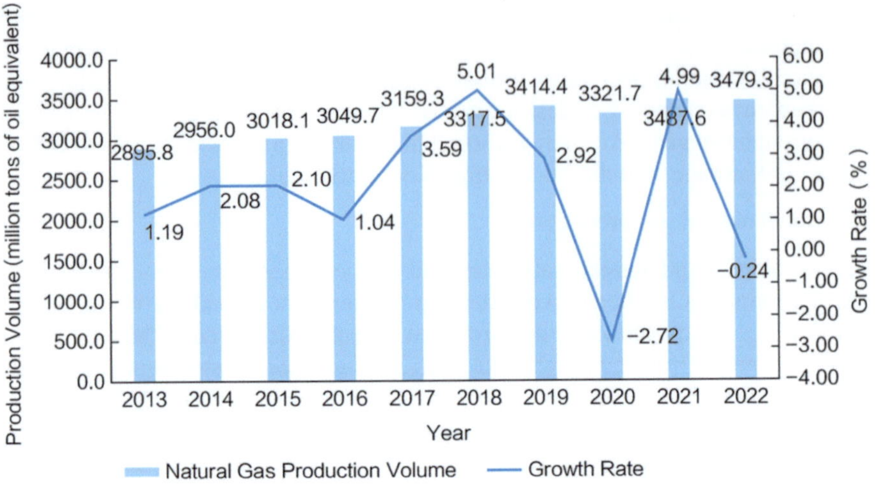

Fig. 6 Total global natural gas production and growth rate (2013–2022). *Data source* "World Energy Statistics Yearbook 2023"

Fig. 7 The proportion of natural gas in global fossil fuel production in 2022. *Data source* "World Energy Statistics Yearbook 2023"

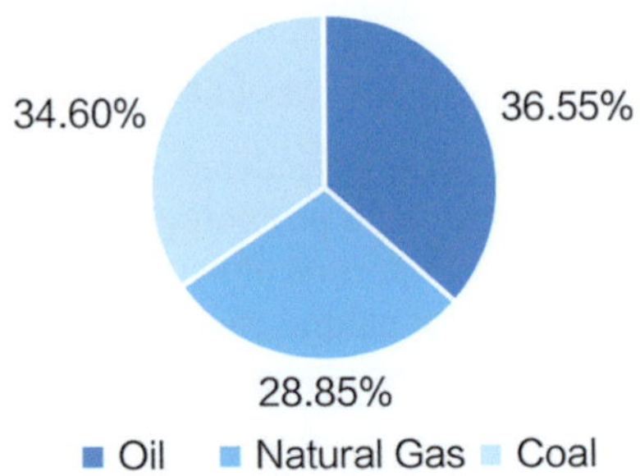

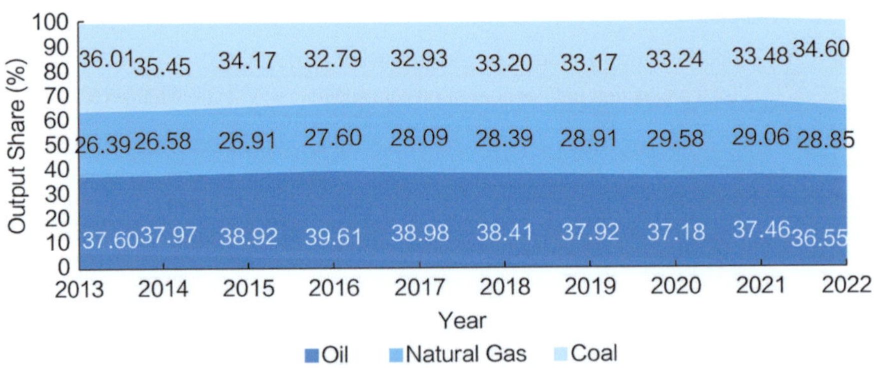

Fig. 8 Structure of global fossil energy production (2013–2022). *Data source* "World Energy Statistics Yearbook 2023"

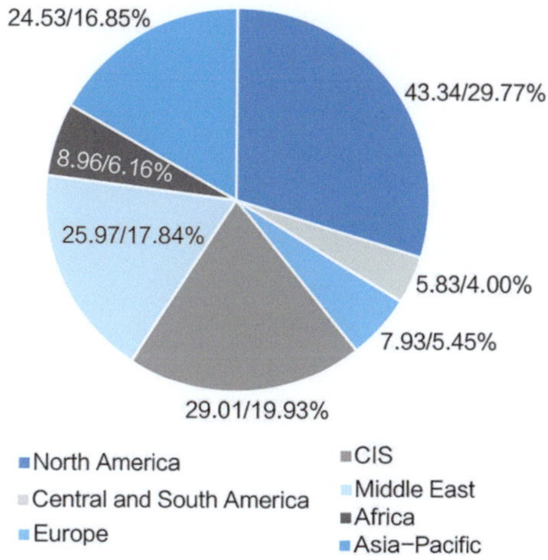

Fig. 9 Global natural gas production and proportion by region in 2022. *Data source* "World Energy Statistics Yearbook 2023"

energy production has generally increased, rising from 26.39% in 2013 to 28.85% in 2022. Although the total production of natural gas is growing, the proportion of natural gas in fossil fuel production declined in 2022, decreasing by 1.21% compared to 2021. Complex geopolitical events, changes in the global energy market structure, and the slowdown in global economic growth are all affecting the changing direction of the energy structure.

(III) Global natural gas production and export by region

1. Global natural gas production by region

By analyzing the data in the "World Energy Statistics Yearbook 2023," the global natural gas regional production (unit: exajoules) and its proportion of total fossil fuel production in 2022 can be obtained (as shown in Fig. 9).

In 2022, North America produced the most natural gas in the world, accounting for as much as 29.77%, with the United States being the world's largest producer of natural gas. The Commonwealth of Independent States ranked second, accounting for 19.93%. This was followed by Middle Eastern countries, accounting for 17.84%. The Asia–Pacific region ranked fourth, accounting for 16.85%. Africa, Europe, and Central and South America followed closely, accounting for 6.16%, 5.45%, and 4.00% respectively.

2. Global Natural Gas Export by Region

Based on the data in the "World Energy Statistics Yearbook 2023", the natural gas export figures are obtained by adding up the total amounts of pipeline exports and liquefied natural gas exports and are used to obtain the proportion of global natural gas exports by region in 2022 (as shown in Fig. 10).

In terms of natural gas exports, the top three regions in 2022 were North America (led by the United States), the Commonwealth of Independent States (led by Russia), and the Asia–Pacific region, accounting for 27.79%, 24.36%, and 19.97%

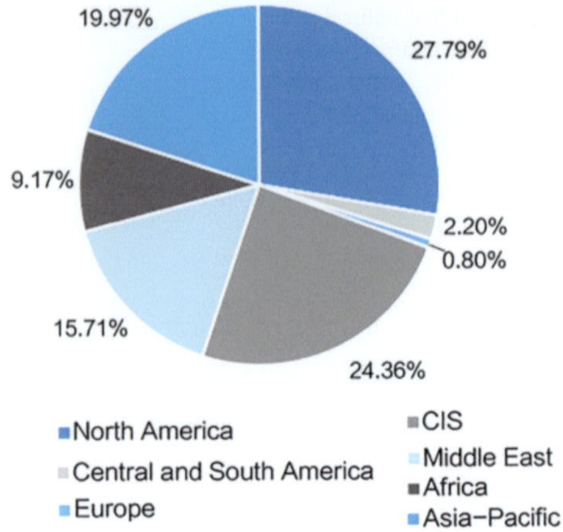

Fig. 10 Global natural gas export share by region. *Data source* "World Energy Statistics Yearbook 2023"

of total global exports respectively. The Middle East ranked fourth in global exports, accounting for 15.71%. This was followed by Africa, accounting for 9.17%. The export volumes of Central and South America and Europe accounted for less than 3% each.

Global Natural Gas Consumption

In the transition period from the fossil fuel era to the non-fossil fuel era, natural gas, as a representative of clean fossil fuel, plays an important role. Over the past 20 years, global natural gas consumption has increased significantly, but its proportion in energy consumption has only slightly increased. From the current data on the proportion of natural gas in global energy demand, the proportion of natural gas is still lower than that of oil and coal, and it will take a relatively long time for the world to enter the natural gas era.

(I) Total global natural gas consumption

Since 2022, factors such as geopolitical influences, high spot prices for LNG, and a relatively warm climate have put pressures on the growth of natural gas demand, significantly slowing the growth rate of global natural gas market consumption. In 2022, global natural gas demand fell by 3%, slightly below the 4 trillion cubic meters mark first set in 2021, with the sharp decline in demand in Europe and Asia offset by strong growth in North America.

Table 7 and Fig. 11 reflect the changes in the total consumption and growth rate of global natural gas from 2013 to 2022.

Global natural gas consumption showed a steady upward trend from 2013 to 2019, with an average growth rate of 2.35%, but it declined for the first time in 2020. After the rebound in consumption in 2021, the total consumption of natural gas declined

Table 7 Global natural gas consumption and growth rate (2013–2022)

Year	Consumption (million tons of oil equivalent)	Growth rate (%)
2013	2,903.2	1.62
2014	2,923.4	0.70
2015	2,993.2	2.39
2016	3,062.4	2.31
2017	3,142.3	2.61
2018	3,300.1	5.02
2019	3,360.6	1.83
2020	3,321.4	−1.64
2021	3,499.3	5.36
2022	3,391.1	−3.09

Data source World Energy Statistics Yearbook 2023

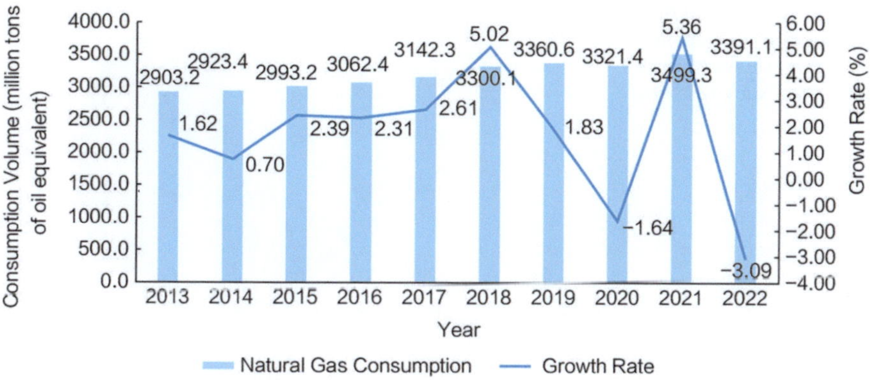

Fig. 11 Global natural gas consumption and growth rate (2013–2022). *Data source* World Energy Statistics Yearbook 2023

again in 2022 due to the Ukraine Crisis and other factors, falling from 3,499.3 million tons of oil equivalent in 2021 to 3,391.1 million tons of oil equivalent in 2022.

(II) The global consumption share of natural gas

With the introduction of the "dual carbon" goals, the global energy sector is shifting from high-carbon to low-carbon sources. Some typical countries around the world have expanded the use of natural gas and continuously adjusted and transformed their energy structures, completing the transition of their main energy source from "coal" to "oil and gas." Overall, the proportion of natural gas has been increasing over the past decade. However, in 2022, due to factors such as weak global economic recovery, abnormally high international gas prices, and increased use of alternative energy sources, the proportion of natural gas declined slightly compared to 2021.

1. The proportion of natural gas in global fossil fuel consumption

By analyzing the data in the "World Energy Statistics Yearbook 2023," we can get the proportion of natural gas in fossil fuel consumption in 2022 and the consumption structure of global fossil fuel from 2013 to 2022 (as shown in Figs. 12 and 13).

Between 2013 and 2022, natural gas had the lowest share in the production of fossil fuels. However, overall, its share in fossil fuels increased during this decade, rising from 26.36% in 2013 to 28.72% in 2022. Focusing solely on 2022, the global consumption share of natural gas experienced a slight decline, dropping from 29.78% in 2021 to 28.72%.

2. The proportion of natural gas in total global energy consumption

With the world's transition to low-carbon energy, the proportions of natural gas and new energy in global energy consumption has generally been on the rise over the past decade. The proportion of natural gas in global energy consumption in 2022 and

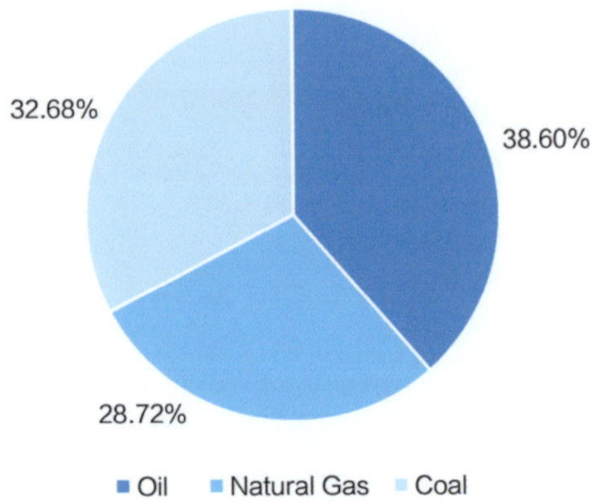

Fig. 12 Global fossil fuel consumption structure in 2022. *Data source* "World Energy Statistics Yearbook 2023"

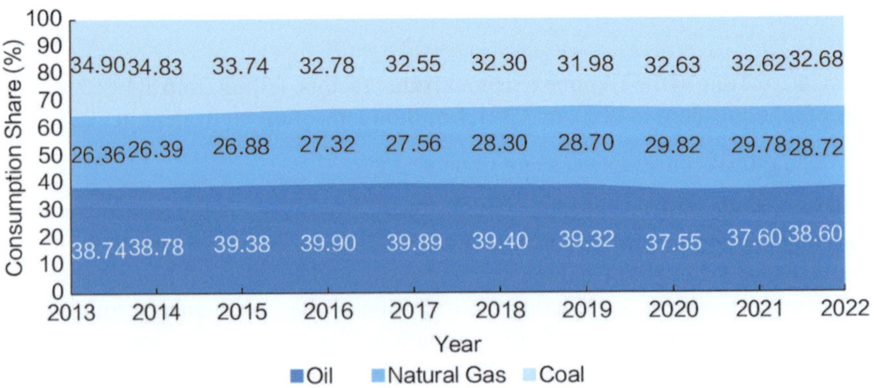

Fig. 13 Changes in global fossil fuel consumption structure (2013–2022). *Data source* "World Energy Statistics Yearbook 2023"

the structure of global energy consumption from 2013 to 2022 are shown in Figs. 14 and 15.

From 2013 to 2021, the share of natural gas in primary energy generally showed a steady upward trend, but in 2022, the share of natural gas in primary energy slightly declined, falling from 24.51% in 2021 to 23.49%.

(III) Global natural gas consumption and import by region

1. Global natural gas consumption by region

By analyzing the data in the "World Energy Statistics Yearbook 2023," we obtain the volume (in exajoules) and share of natural gas consumption by region in 2022, as shown in Fig. 16.

Fig. 14 The share of natural gas in global energy consumption in 2022. *Data source* "World Energy Statistics Yearbook 2023"

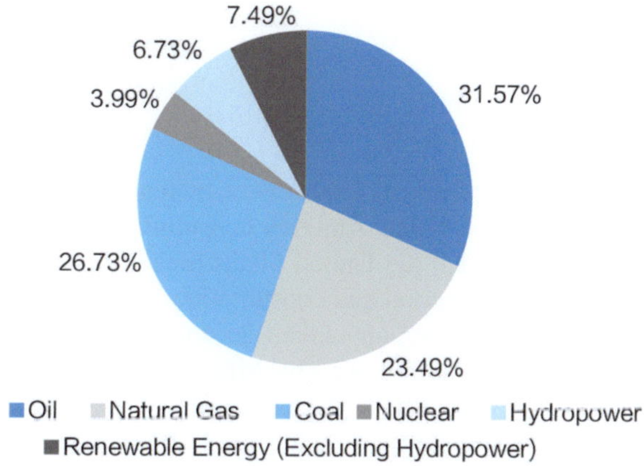

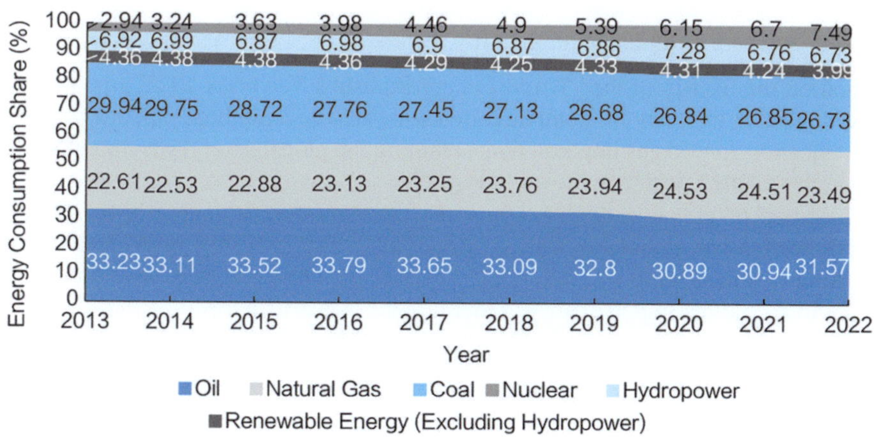

Fig. 15 Changes in global energy consumption structure (2013–2022). *Data source* "World Energy Statistics Yearbook 2023"

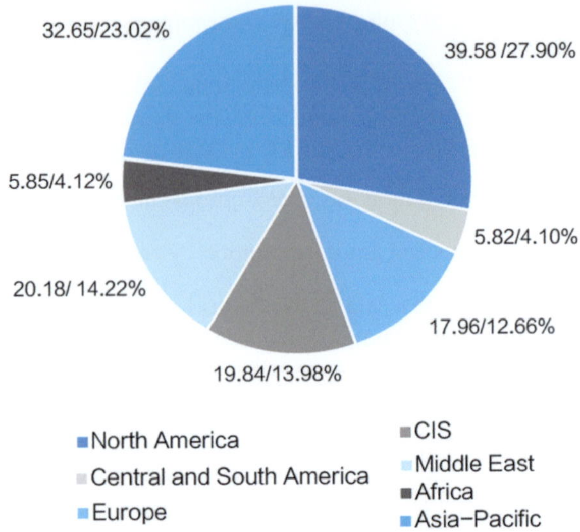

Fig. 16 Global natural gas consumption volume and share by region in 2022. *Data source* World Energy Statistics Yearbook 2023

In 2022, the consumption volume by region was similar to the production volume by region, with North America consuming the most, accounting for as much as 27.90%, of which the United States ranked first among all countries in the world in terms of its natural gas consumption. The consumption volume of countries in the Asia–Pacific region ranked second in regional consumption, accounting for 23.02%. This was followed by Middle Eastern countries, accounting for 14.22%. The Commonwealth of Independent States (CIS) ranked fourth, accounting for 13.98%, and Europe accounted for 12.66%, ranking fifth. Africa and Central and South America followed closely, with roughly the same shares, at 4.12% and 4.10% respectively.

2. Import share of global natural gas by region

Based on the data from the "World Energy Statistics Yearbook 2023", the natural gas imports in this part are combined by taking into account both pipeline imports and liquefied natural gas imports, thus obtaining the global natural gas import share by region in 2022, which is shown in Fig. 17.

In terms of global natural gas imports, the Asia–Pacific region is the largest importer of global natural gas, with an import share of up to 42.45% in 2022; followed by Europe, with an import share of 33.15%; then North America, with an import share of 17.15%, and the import share of other regions is less than 5%.

Summary

According to the "Paris Agreement," humanity will enter an era dominated by non-fossil energy by the middle of the twenty-first century. However, judging by the global

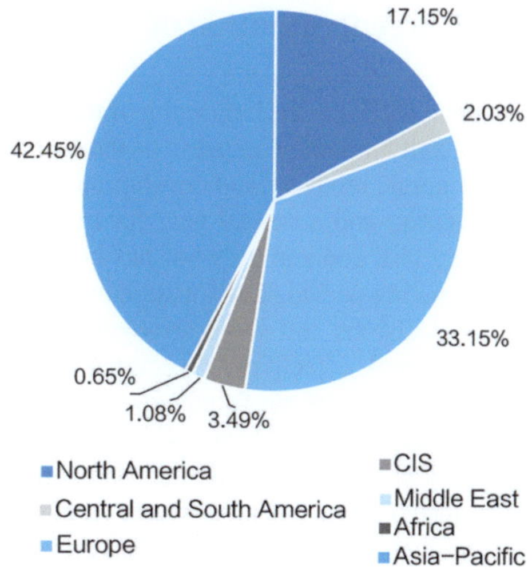

Fig. 17 Global natural gas import share by region. *Data source* "World Energy Statistics Yearbook 2023"

energy production and consumption situation, affected by the COVID-19 pandemic and the international situation, the global economic situation generally continues to progress in a manner of "soft landing". Meanwhile, the energy market also demonstrates volatile characteristics. The "World Energy Statistics Yearbook 2023" shows that the total global energy consumption in 2022 reached 1.631 billion tons of oil equivalent, an increase of about 2% year-on-year. Among them, oil accounted for 31.57%, natural gas accounted for 23.49%, and coal accounted for 26.73%. The three fossil fuels accounted for 81.79% in total, showing that traditional fossil fuels still dominate the global energy structure, and we are currently in the oil era of the fossil fuel era.

With the advancement of energy transformation and the "dual carbon" goals, natural gas, serving as the link between high-carbon fossil fuels such as coal and oil and carbon-free new energy sources such as solar energy, wind energy, and geothermal energy, is the bridge for the transition of the energy structure from high-carbon to carbon-free. Moreover, most countries also believe that the development of natural gas is a necessary transitional measure to achieve the "dual carbon" goals. At the same time, the high calorific value of natural gas has promoted the large-scale application of natural gas, driven the substitution of oil and natural gas for coal, and contributed to the realization of the second energy transformation. However, as of 2022, the proportion of natural gas is still lower than that of oil and coal, indicating that it will take some time for the world to enter the natural gas era, so that its proportion can first exceed that of coal and then that of oil.

The Current Situation of Natural Gas Development in China

The year 2023 was the beginning of the full implementation of the guiding principles of the 20th National Congress of the Communist Party of China. It was also a year for economic recovery and development after the transition in COVID-19 prevention and control, and a crucial year for implementing the 14th Five-Year Plan, serving as a key link connecting what has been achieved before and what will be done in the following period. The Chinese energy industry has thoroughly implemented the policy decisions and work arrangements of the Party Central Committee and the State Council, effectively responded to multiple challenges such as a weak global economic recovery, profound adjustments in the energy market, and continued large fluctuations in energy commodity prices. Moreover, on the basis of ensuring a stable energy supply, it has accelerated the construction of a new energy system, actively participated in global governance to address climate change, and the high-quality development of energy has reached a new level.

Since 2023, the Chinese economy has continued to recover and improve, which has driven the continuous growth of energy consumption. Looking at the various types of energy, electricity consumption has grown rapidly, raw coal consumption has been stable, refined oil consumption has recovered, and natural gas consumption has significantly warmed up. In addition, the ability to secure energy supply has steadily improved, the green and low-carbon transformation has been deeply advanced, investment in the energy industry has grown rapidly, and the structural reform of the energy supply side has continued to deepen.

Energy Production and Consumption in China

As a major global natural gas consumer, China holds that increasing the proportion of natural gas in the primary energy consumption structure is the inevitable path to building a clean, low-carbon, efficient and secure energy system. Guided by the "dual carbon" goals, the efforts regarding energy conservation, carbon reduction and efficiency improvement have been steadily pushed forward. Thanks to the steady increase in urbanization rate, the consistent growth in natural gas consumption in the transportation sector, and the substantial growth of natural gas power generation, domestic natural gas production will continue to grow, and the natural gas industry as a whole is currently in a stage of steady development.

(I) Overview of China's energy production

During the "14th Five-Year Plan" period, energy development has entered a period of adaptation in the context of reform. At this stage, various policies and measures will place emphasis on coordination and stability. Meanwhile, efforts will be made to strengthen the restraining, guiding and supporting functions of the "dual carbon" goals in relation to the high-quality development strategy and the energy development and transformation strategy. Thanks to the continuous implementation of policies and measures aimed at ensuring energy supply and stabilizing prices, China's energy supply has witnessed a steady increase while its capacity to secure energy supply has been continuously strengthened, and energy production has achieved a stable growth.

1. China's total primary energy production
2. China is the world's largest energy producer. The strong support of policies, innovation in technology, and the continuous growth in market demand have enabled China's total energy production to maintain steady growth for a long time. During the "14th Five-Year Plan" period, the government fully recognized the prominent role of fossil fuels in ensuring the stability of China's energy system, vigorously developed the natural gas industry, and fully utilized the advantages of natural gas's low-carbon effect and the flexibility of gas-fired power for peak load regulation, thus leaving ample development time for the renewable energy-based power system. Table 8 and Fig. 18 show the changes and growth rate of China's total primary energy production over the ten years from 2014 to 2023.

In 2022, China's total energy production reached 46.6 billion tons of standard coal, an increase of 7.62% year-on-year; in 2023, China's total energy production was 48.3 billion tons of standard coal, an increase of 3.65% year-on-year. From 2014 to 2023, the total energy production increased significantly, and the year-on-year growth rate fluctuated relatively stably. The growth rate of total energy production in 2023 slowed slightly compared to 2022, but it was still rising steadily.

2. Total production of energy by type in China

As the world's largest energy producer, China plays a pivotal and unshakable role in the production of fossil fuels. In the long run, the supply elasticity of "coal, oil, gas, and non-fossil fuel" will be strengthened, and the energy security capacity will

Table 8 Total primary energy production and growth rate in China (2014–2023)

Year	Total production (10 kilotons of standard coal)	Growth rate (%)
2014	362,212	0.96
2015	362,193	−0.01
2016	345,954	−4.48
2017	358,867	3.73
2018	378,859	5.57
2019	397,317	4.87
2020	408,000	2.69
2021	433,000	6.13
2022	466,000	7.62
2023	483,000	3.65

Data sources "China Statistical Yearbook 2023" and "Statistical Bulletin of National Economic and Social Development of the People's Republic of China 2023"

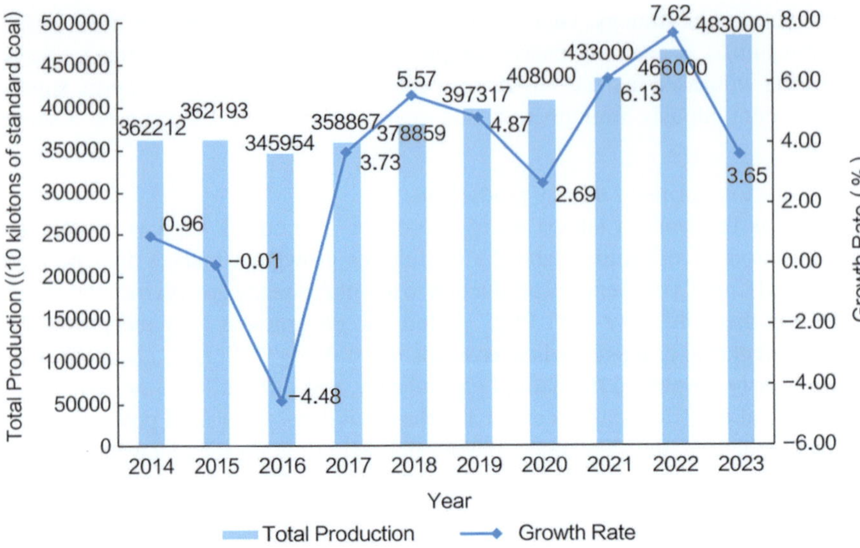

Fig. 18 Total energy production and growth rate in China (2014–2023). *Data sources* "China Statistical Yearbook 2023" and "Statistical Bulletin of National Economic and Social Development of the People's Republic of China 2023"

be significantly improved. China remains committed to energy reform, continuously intensifying the reform of the energy institutional framework and mechanism while optimizing the energy structure. Table 9 and Fig. 19 show the production of different types of energy in China from 2014 to 2023.

Based on aggregate statistics and unit standardization calculations, it has been found that in 2023, compared with 2022, the production of raw coal in China increased by 3.4%. Meanwhile, the production of crude oil increased by 2.1%, the production of natural gas increased by 5.6%, and the total production of primary electricity and other energy sources increased by 6.9%. Overall, raw coal continues to be the dominant energy source in China's fossil energy supply, while crude oil and natural gas account for relatively smaller proportions.

(II) Overview of China's energy consumption

In 2023, under the impact of the Israel-Hamas conflict, a shipping crisis involving oil and gas transportation emerged in the Red Sea. As a result, the prices of the three major energy types, namely crude oil, natural gas, and raw coal, all fell significantly. Moreover, the declines the prices of natural gas and raw coal were much steeper than that of crude oil. In the face of diverse challenges, China's energy industry has vigorously promoted the energy revolution, accelerated the construction of a new energy system, and continued to optimize the energy consumption structure.

Table 9 Energy production by type in China (2014–2023)

Year	Raw coal	Crude oil	Natural gas	Primary electricity and other energy
2014	266,226	30,064	17,024	48,899
2015	261,503	30,786	17,385	52,518
2016	241,476	28,714	17,990	57,774
2017	249,771	27,274	19,379	62,443
2018	262,170	27,278	20,458	68,952
2019	272,162	27,415	22,250	75,490
2020	275,400	27,744	24,480	80,376
2021	288,811	29,011	25,980	89,198
2022	314,084	29,358	27,494	95,064
2023	324,763	29,975	29,034	101,623

Unit 10 kilotons of Standard Coal
Note The 2023 data has been calculated on the basis of the "China Statistical Yearbook 2023" and the "Statistical Bulletin of National Economic and Social Development of the People's Republic of China 2023"
Data sources "China Statistical Yearbook 2023" and "Statistical Bulletin of National Economic and Social Development of the People's Republic of China 2023"

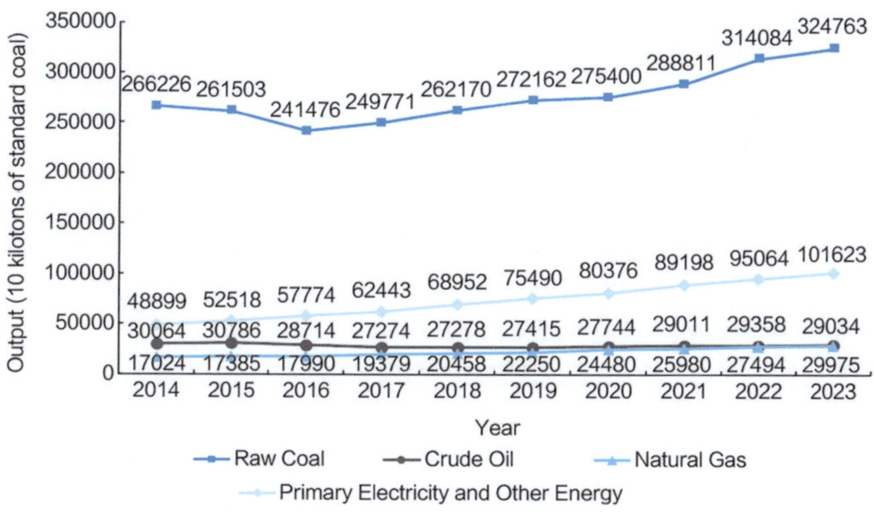

Fig. 19 Total energy production by type in China (2014–2023). *Data sources* "China Statistical Yearbook 2023" and "Statistical Bulletin of National Economic and Social Development of the People's Republic of China 2023"

1. Total energy consumption in China

The total energy consumption can reflect a country's economic strength and the basic living standards of its people. Facing the tests of the external environment, China's total energy consumption has maintained stable growth. The stable increase in China's total energy consumption indicates the stability of social development and the stable growth of China's economic level. The total energy consumption and growth rate in China from 2014 to 2023 are shown in Table 10 and Fig. 20.

In 2022, China's total energy consumption was 54.1 billion tons of standard coal, while in 2023, it was 57.2 billion tons of standard coal. Consequently, during this period, China's total energy consumption increased by 3.1 billion tons of standard coal, with a year-on-year growth rate of 5.73%. From 2014 to 2023, China's total energy consumption maintained a steady growth trend, increasing from 42.83 billion tons of standard coal in 2014 to 57.2 billion tons of standard coal in 2023, a significant increase of 33.54%.

2. Energy consumption by type in China

Under the strategy of carbon peaking and carbon neutrality, the next 15 to 20 years will remain a period of rapid growth in natural gas consumption. The government is vigorously developing natural gas power generation with the aim of reducing carbon emissions from the power system and laying the foundation for the development and large-scale application of renewable energy. At the same time, it ensures the steady realization of the "dual carbon" goals in both domestic development and international cooperation. Table 11 and Fig. 21 present the consumption of different types of energy in China from 2014 to 2023.

According to the "Statistical Bulletin of National Economic and Social Development of the People's Republic of China 2023," in 2023, the total consumption of raw coal was 32.1 billion tons of standard coal, an increase of 5.6% compared to 2022.

Table 10 Total energy consumption and growth rate in China (2014–2023)

Year	Total consumption (10 kilotons of standard coal)	Growth rate (%)
2014	428,334	2.74
2015	434,113	1.35
2016	441,492	1.70
2017	455,827	3.25
2018	471,925	3.53
2019	487,488	3.30
2020	498,000	2.16
2021	524,000	5.22
2022	541,000	3.24
2023	572,000	5.73

Data sources "China Statistical Yearbook 2023" and "Statistical Bulletin of National Economic and Social Development of the People's Republic of China 2023"

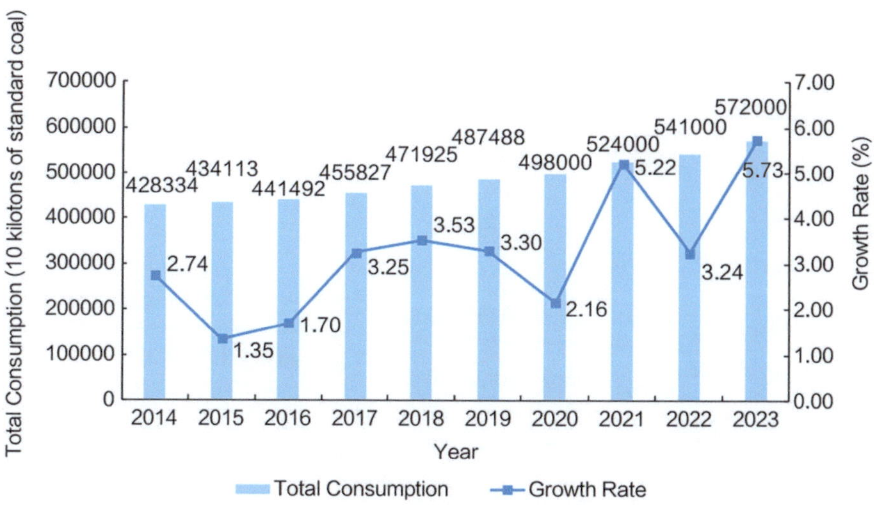

Fig. 20 Total energy consumption and growth rate in China (2014–2023). *Data sources* "China Statistical Yearbook 2023," "People's Republic of China 2023 National Economic and Social Development Statistics Bulletin"

Table 11 Energy consumption by type in China (2014 to 2023)

Year	Raw coal	Crude oil	Natural gas	Primary electricity and other energy
2014	281,844	74,102	23,987	48,402
2015	276,964	79,877	25,179	52,094
2016	274,608	82,559	26,931	57,394
2017	276,231	86,151	31,452	61,992
2018	278,436	89,194	35,866	68,429
2019	281,281	92,623	38,999	74,586
2020	283,362	93,624	41,832	79,182
2021	292,916	97,464	46,112	87,508
2022	304,042	96,839	45,444	94,675
2023	321,068	105,651	48,716	101,018

Unit 10 kilotons of Standard Coal

Note The 2023 data has been calculated on the basis of the "China Statistical Yearbook 2023" and the "Statistical Bulletin of National Economic and Social Development of the People's Republic of China 2023"

Data sources "China Statistical Yearbook 2023" and "Statistical Bulletin of National Economic and Social Development of the People's Republic of China 2023"

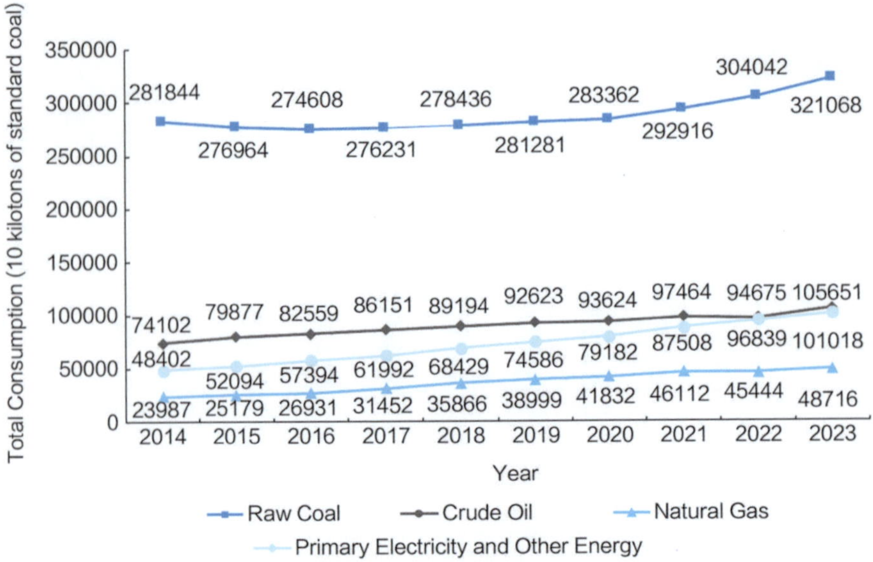

Fig. 21 Energy consumption by type in China (2014 to 2023). *Data sources* "China Statistical Yearbook 2023," "People's Republic of China 2023 National Economic and Social Development Statistics Bulletin"

The total consumption of crude oil was 10.57 billion tons of standard coal, up 9.1% from 2022. The total consumption of natural gas was 4.87 billion tons of standard coal, a rise of 7.2% from 2022. Moreover, the total consumption of primary electricity and other energy hit 10.1 billion tons of standard coal, registering an increase of 6.7% compared to 2022. The transformation of China's energy consumption structure is being deeply advanced, with the total consumption of non-fossil energy continuing to grow at a relatively fast pace. In 2023, natural gas consumption maintained stable growth, while raw coal consumption also saw a relatively significant increase, which may be due to the recovery of industrial production and the increase in industrial energy demand driving the continuous growth of raw coal consumption.

Natural Gas Production in China

With the global focus on reducing greenhouse gas emissions, the natural gas industry is becoming increasingly important in the energy structure. Although natural gas is also a fossil fuel, it can mitigate the global greenhouse effect and improve environmental quality due to its relatively lower carbon emissions and pollution. Moreover, it is the most reliable choice for replacing traditional fossil fuels and rapidly reducing carbon emissions. It serves as a bridge and support during the transition from fossil fuels to non-fossil energy, and it is of great significance for China to achieve energy independence and ensure energy security. Under the energy development arrangements of the "14th Five-Year Plan," while continuously increasing the production of natural gas and other clean energy sources, the energy consumption structure is also being continuously optimized.

(I) Total production of natural gas in China

China's natural gas is in a period of rapid development, and the proven recoverable reserves of natural gas have grown rapidly in the past decade. With the progress of shale gas development and LNG transportation technology, the availability and economy of natural gas have been enhanced. Therefore, as a high-quality and efficient clean energy, the proportion of natural gas in China's total energy production is continuously increasing. Table 12 and Fig. 22 show the total production and growth rate of natural gas in China from 2014 to 2023.

In 2023, China's natural gas production was 2,324 billion cubic meters, with a year-on-year growth rate of 5.58%, maintaining steady growth in a challenging environment. From 2014 to 2023, the production of natural gas has made a qualitative leap, with the production steadily increasing year by year. In 2023, the national natural gas production increased by 1,022 billion cubic meters compared to that in 2014, an increase of 78.49% over the past 10 years. In the future, given the important position of natural gas in energy, the production of natural gas will continue to maintain a high growth trend.

(II) Proportion of natural gas production in China

1. The proportion of natural gas in China's fossil fuel production

As the world's leading energy-producing country, China occupies an unshakable position in the production of fossil fuels. During the global energy crisis triggered by

Table 12 Total production and growth rate of natural gas in China (2014–2023)

Year	Natural gas production (billion cubic meters)	Growth rate (%)
2014	130.2	7.69
2015	134.6	3.38
2016	136.87	1.69
2017	148.03	8.15
2018	160.27	8.27
2019	176.17	9.92
2020	192.5	9.27
2021	207.58	7.83
2022	220.10	6.03
2023	232.4	5.58

Data sources The data from 2014 to 2015 are sourced from "Zhongjing Data," while the data from 2016 to 2023 are obtained from the monthly production data of the National Bureau of Statistics for each year

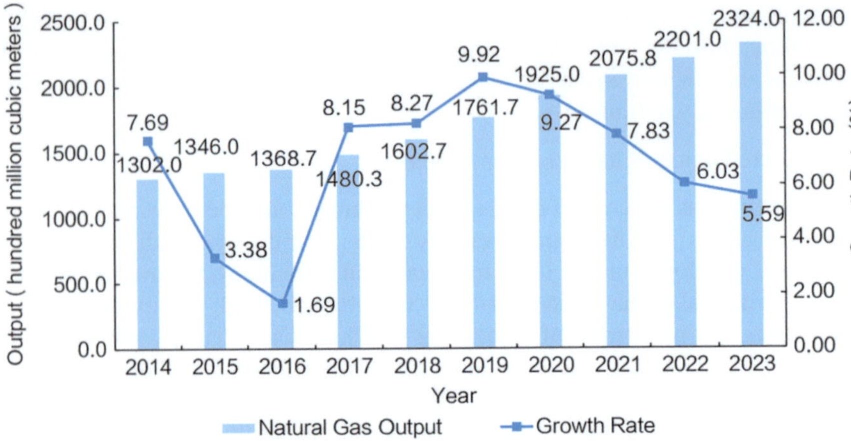

Fig. 22 Total production and growth rate of natural gas in China (2014–2023). *Data sources* The data from 2014 to 2015 are sourced from "Zhongjing Data," while the data from 2016 to 2023 are obtained from the monthly production data of the National Bureau of Statistics for each year

the COVID-19 pandemic, coal, being the most traditional type of fossil fuel, played a crucial role in alleviating the difficulties faced by the world. Moreover, considering the growth in the consumption of oil and natural gas in 2023, it is evident that under the wave of energy transformation, all three traditional fossil energies are shouldering the significant responsibility of meeting the energy demands of human society. In the long run, fossil energy will continue to serve as the mainstay of energy supply. At present, coal production still occupies a dominant position in China's fossil fuel production. At the same time, natural gas, as a clean fossil fuel, has constantly maintained a stable growth trend. The proportion of production of fossil fuels by type in China from 2014 to 2023 are shown in Table 13 and Figs. 23, 24 and 25.

Firstly, it is evident from the horizontal comparison of Table 13 that natural gas has a relatively low share in the total production of fossil fuels, but its production growth potential is huge. With the improvement of production technology and the transition to renewable energy, it is believed that the production of natural gas will continue to grow in the future. Looking at Table 13 vertically, the proportion of natural gas production has grown the most significantly, while the proportion of coal and oil production has shown a certain downward trend. The expansion of natural gas reserves and the enhancement of its production remains robust. The China National Petroleum Corporation, China Petroleum & Chemical Corporation (hereinafter referred to as "Sinopec"), and China National Offshore Oil Corporation (hereinafter referred to as "CNOOC") continue to ramp up their exploration and development efforts, and the construction of production capacity is being efficiently advanced. The three major basins of Sichuan, Ordos, and Tarim continue to increase reserves and production capacity, and both conventional and unconventional gas are developed simultaneously to jointly boost the output of domestic gas.

Table 13 Structure of fossil fuel production by type in China (2014–2023)

Year	Coal (%)	Oil (%)	Natural gas (%)
2014	84.97	9.60	5.43
2015	84.44	9.94	5.62
2016	83.79	9.96	6.25
2017	84.26	9.20	6.54
2018	84.60	8.80	6.60
2019	84.57	8.52	6.91
2020	84.06	8.47	7.47
2021	84.01	8.44	7.55
2022	84.67	7.91	7.42
2023	84.62	7.81	7.57

Data sources "China Statistical Yearbook 2023" and "Statistical Bulletin of National Economic and Social Development of the People's Republic of China 2023"

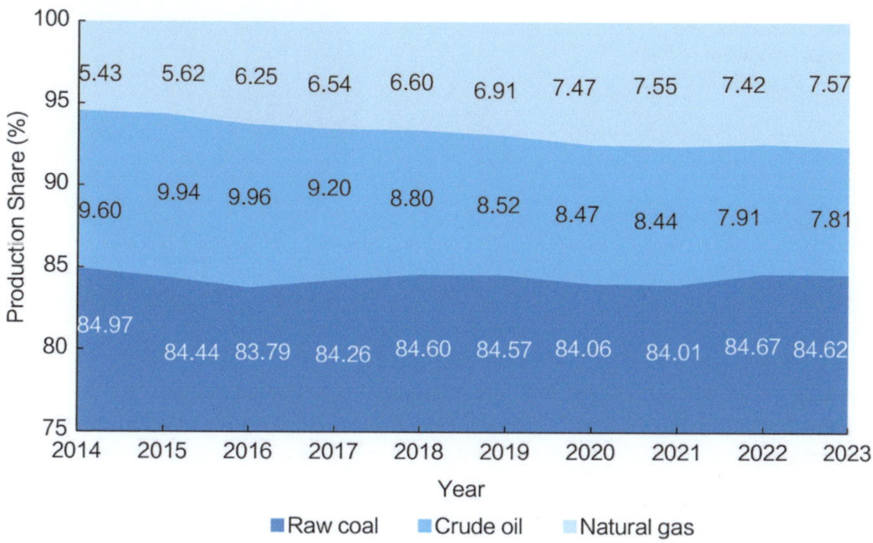

Fig. 23 Structure of fossil fuel production by type in China (2014–2023). *Data sources* "China Statistical Yearbook 2023" and "Statistical Bulletin of National Economic and Social Development of the People's Republic of China 2023"

Figures 24 and 25 display the proportion of fossil fuel production in 2022 and 2023. In 2022, natural gas accounted for 7.42% of fossil energy production, coal accounted for 84.67%, and oil accounted for 7.91%. In the production of fossil fuels, coal production has maintained a dominant position. In 2023, there were new developments in this regard. Specifically, the proportion of natural gas climbed to 7.57%, while that of coal declined to 84.62% and the proportion of oil stood at 7.81%.

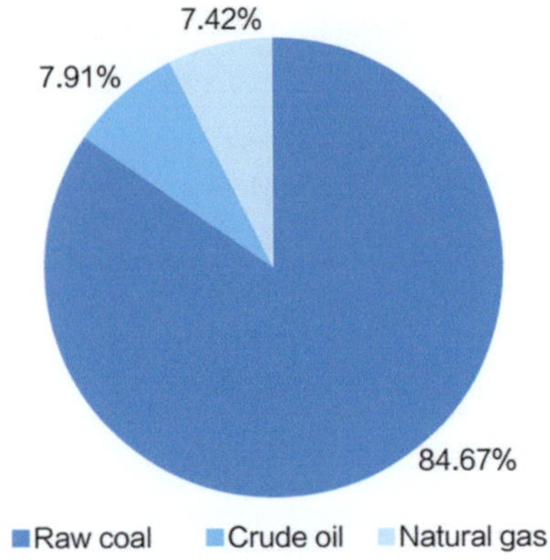

Fig. 24 Structure of fossil fuel production by type in China in 2022. *Data source* "Statistical Bulletin of National Economic and Social Development of the People's Republic of China 2022"

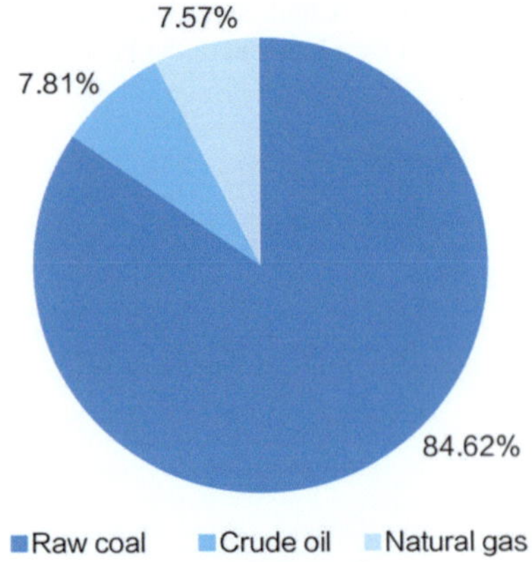

Fig. 25 Structure of fossil fuel production by type in China in 2023. *Data source* "Statistical Bulletin of National Economic and Social Development of the People's Republic of China 2023"

Through a comparison of the two sets of figures, it becomes evident that the share of natural gas in China's fossil energy production in 2023 regained its former stable upward trend.

2. The proportion of natural gas in China's primary energy production

For more than a decade, China has emerged as the world's largest and fastest-growing producer of renewable energy. This has clearly demonstrated the crucial supportive role that natural gas plays in China's "dual carbon" initiative, thereby guiding the orderly and high-quality development of the natural gas industry. The continuous enhancement of natural gas infrastructure construction, coupled with the new achievements made in market-oriented reforms, has further diversified the main players in the natural gas business. All these factors have jointly promoted the steady and sustainable development of the natural gas industry. Table 14 and Fig. 26 show the changes in the proportion of China's primary energy production by type during the ten-year period from 2014 to 2023.

Table 14 Production share of primary energy sources by type in China (2014–2023)

Year	Coal (%)	Oil (%)	Natural gas (%)	Primary electricity and other energy (%)
2014	73.50	8.30	4.70	13.50
2015	72.20	8.50	4.80	14.50
2016	69.80	8.30	5.20	16.70
2017	69.60	7.60	5.40	17.40
2018	69.20	7.20	5.40	18.20
2019	68.50	6.90	5.60	19.00
2020	67.50	6.80	6.00	19.70
2021	66.70	6.70	6.00	20.60
2022	67.40	6.30	5.90	20.40
2023	66.91	6.17	5.98	20.94

Note The data for 2023 has been calculated on the basis of the "China Statistical Yearbook 2023" and the "Statistical Bulletin of National Economic and Social Development of the People's Republic of China 2023"

Data sources "China Statistical Yearbook 2023" and "Statistical Bulletin of National Economic and Social Development of the People's Republic of China 2023

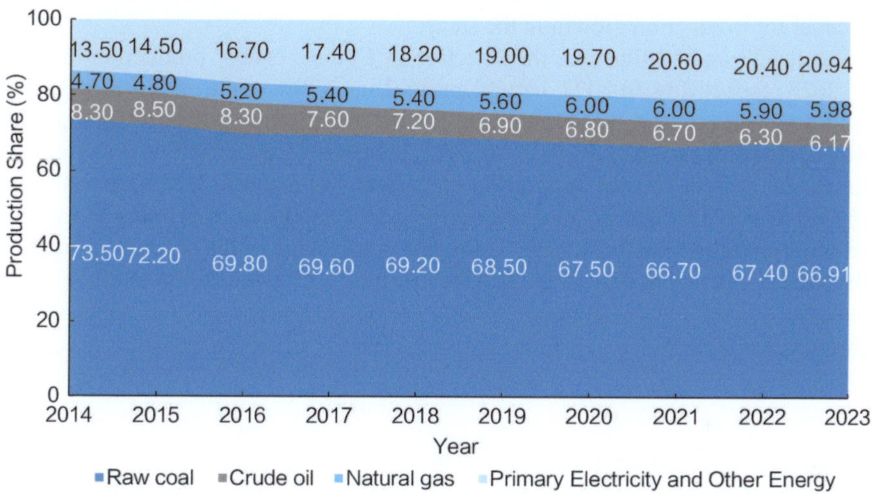

Fig. 26 Production share of primary energy sources by type in China (2014–2023). *Data sources* "China Statistical Yearbook 2023" and "Statistical Bulletin of National Economic and Social Development of the People's Republic of China 2023"

Drawing on the data regarding the proportion of China's primary energy production by type over the past decade, a vertical comparison of Table 14 and Fig. 26 reveals the following trends. Coal, characterized by high pollution and high emissions, has exhibited a downward trend in its share of production. Similarly, oil has seen a slight dip in its production proportion as well. In contrast, primary electricity

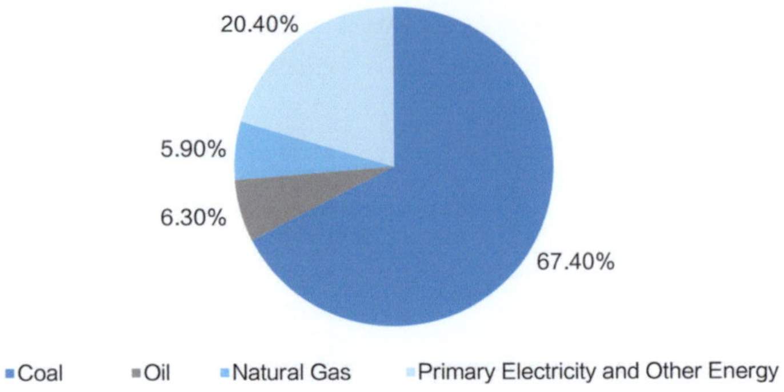

Fig. 27 Structure of China's energy production in 2022. *Data source* "Statistical Bulletin of National Economic and Social Development of the People's Republic of China 2022"

and other forms of energy have shown an upward tendency within China's energy production landscape. It is worth noting that natural gas, as a clean and stable fossil fuel, has witnessed a continuous annual increase in its total output. Moreover, its proportion in primary energy production has remained largely stable. When making a horizontal comparison, it becomes evident that natural gas still holds substantial potential for production growth in the future.

Figures 27 and 28 illustrate the structure of China's energy production in 2022 and 2023 respectively. In 2022, natural gas took up 5.90% of the energy production, coal accounted for 67.40%, oil made up 6.30%, and primary electricity along with other energies constituted 20.40%. In 2023, there were new developments in these figures. To be specific, the proportion of natural gas reached 5.98%, that of coal dropped to 66.91%, that of oil decreased to 6.17%, and that of primary electricity and other energies rose to 20.94%. Through a comparison of the two sets of figures, it becomes evident that despite the fact that the production volume of natural gas has continued to increase, its share in the total energy production has not witnessed an upward trend.

Natural Gas Consumption in China

Natural gas is a clean and environmentally friendly high-quality energy source, which contains almost no sulfur, dust, and other harmful substances. It produces less carbon dioxide when burned than other fossil fuels, resulting in a lower greenhouse effect, and can fundamentally improve the quality of the environment. China insists on increasing energy conservation and emission reduction efforts, strengthening the clean and efficient use of coal. Meanwhile, the energy industry is intensely promoting the work related to reaching the carbon peak. Energy conservation, efficiency enhancement, and green low-carbon transformation are all progressing in tandem. Under the guidance of the "dual carbon" goals, China's energy supply and

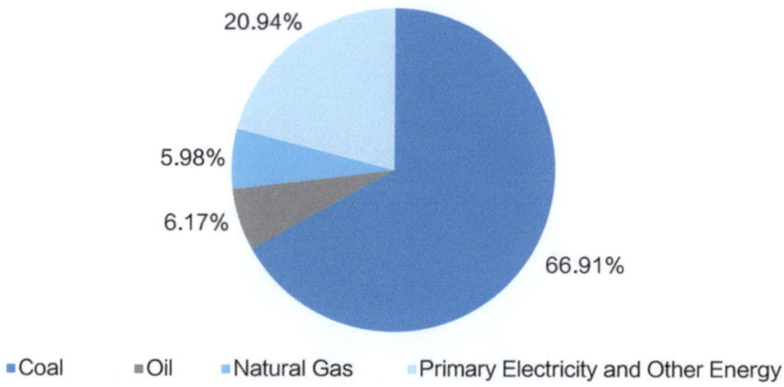

Fig. 28 Structure of China's energy production in 2023. *Data source* "Statistical Bulletin of National Economic and Social Development of the People's Republic of China 2023"

demand structure is undergoing profound changes, with natural gas playing a key role in transitioning the energy consumption structure from high-carbon and polluting sources to low-carbon and green alternatives.

(I) Total consumption of natural gas in China

Guided by the "dual carbon" goals, natural gas, as a clean energy source, plays an important role in the development of the energy consumption structure and has been greatly developed and applied. Influenced by multiple factors such as the "coal-to-gas" policy, supply-side structural reform, and the growth of societal electricity demand, the demand for natural gas in China has grown significantly. Table 15 and Fig. 29 show the total consumption and growth rate of natural gas in China from 2014 to 2023.

Table 15 Total natural gas consumption and growth rate in China (2014–2023)

Year	Natural gas consumption (billion cubic meters)	Growth rate (%)
2014	178.6	5.6
2015	193.2	5.7
2016	205.8	6.6
2017	237.3	15.3
2018	280.3	18.1
2019	306.7	9.4
2020	324.0	5.6
2021	372.6	12.7
2022	366.3	−1.7
2023	394.5	7.6

Data source National Development and Reform Commission, "National Natural Gas Operation Express Report"

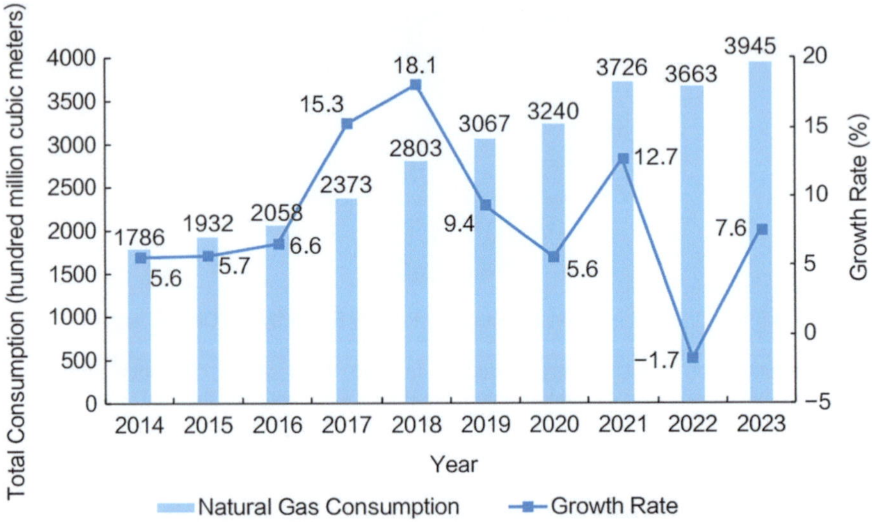

Fig. 29 Total natural gas consumption and growth rate in China (2014–2023). *Data source* National Development and Reform Commission, "National Natural Gas Operation Express Report"

In 2023, the total consumption of natural gas was 394.5 billion cubic meters, with a year-on-year growth rate of 7.6%, reversing the negative growth trend for the first time after 2022 and returning to the growth track. The main reason lies in the fact that under the influence of market laws, China's natural gas consumption experienced negative growth for the first time in 2022. In 2023, both the upstream and downstream sectors of the industry readjusted market boundaries, and the atmosphere of reducing prices to increase the volume gradually became stronger. However, due to the increasing competition from new energy, the momentum for the continued growth of natural gas consumption in China has weakened, and the industry's growth has switched from "high speed" to "medium and low speed", which may pose risks to its development. However, it still takes time for new energy to completely replace traditional fossil fuels. Natural gas is one of the best choices for the transition from fossil fuels to renewable energy, and its bridging role cannot be ignored. Therefore, natural gas will still maintain a relatively stable growth trend in the next decade.

(II) Natural gas consumption share in China

1. The proportion of natural gas in China's consumption of fossil fuels

Although the development momentum of clean energy in China is strong, fossil fuel still dominates the energy sector. Thanks to the frequent introduction of major policies in the natural gas industry and the active and prudent promotion of the reform of the natural gas system, along with the economic recovery and growth, the structure of energy consumption continues to be optimized. The proportion of fossil energy consumption in China from 2014 to 2023 is shown in Table 16, Figs. 30, 31 and 32.

Table 16 Consumption share of fossil fuels by Type in China (2014–2023)

Year	Coal (%)	Oil (%)	Natural gas (%)
2014	74.18	19.51	6.31
2015	72.5	20.91	6.59
2016	71.49	21.50	7.01
2017	70.14	21.88	7.98
2018	69.01	22.10	8.89
2019	68.12	22.43	9.45
2020	67.66	22.35	9.99
2021	67.11	22.33	10.56
2022	68.12	21.70	10.18
2023	67.53	22.22	10.25

Data sources "China Statistical Yearbook 2023" and "Statistical Bulletin of National Economic and Social Development of the People's Republic of China 2023"

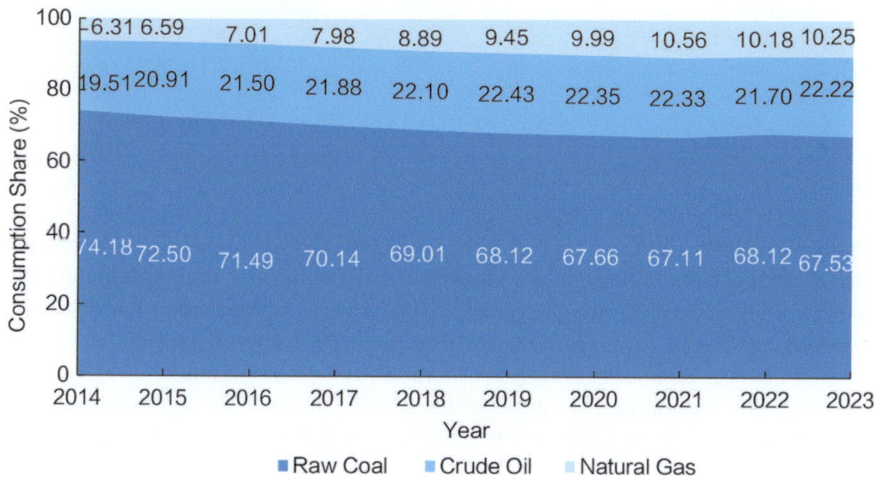

Fig. 30 Consumption share of fossil fuels by type in China (2014–2023). *Data sources* "China Statistical Yearbook 2023" and "Statistical Bulletin of National Economic and Social Development of the People's Republic of China 2023"

Firstly, when looking at Table 16 and Fig. 30 horizontally, it can be seen that the proportion of natural gas in the total consumption of fossil fuels is still relatively low, but its consumption growth potential is huge. Influenced by factors such as the continuous recovery of the economy, the decline in natural gas prices, the insufficient output of hydropower, and the low base in the same period of the previous year, China's natural gas consumption growth rate has bottomed out and rebounded, achieving rapid growth at present. With the continuous advancement of energy transformation, the structure of energy consumption is also constantly changing, and natural gas,

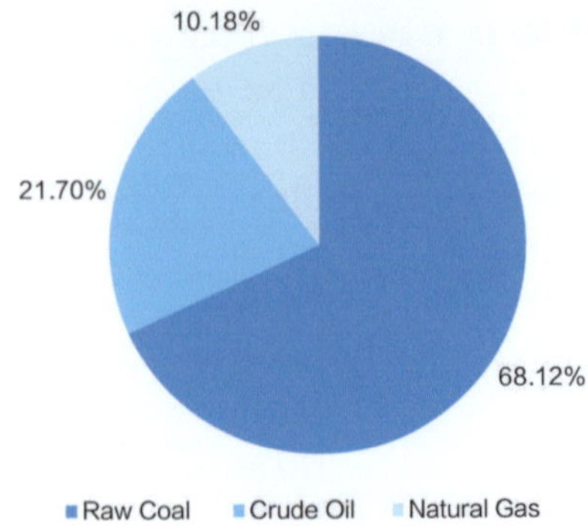

Fig. 31 Structure of fossil fuel consumption in China in 2022. *Data source* "Statistical Bulletin of National Economic and Social Development of the People's Republic of China 2022"

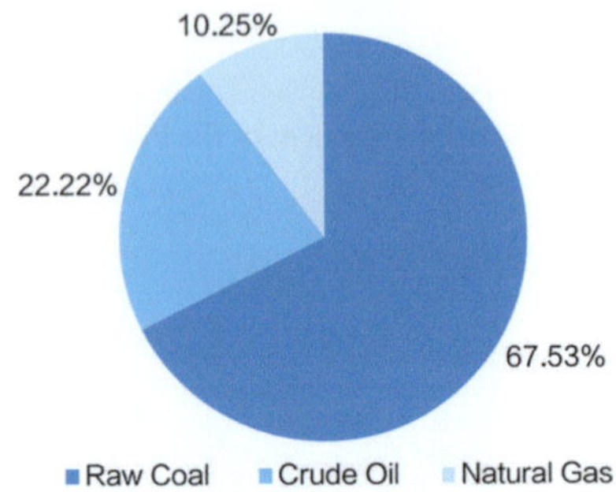

Fig. 32 Structure of fossil fuel consumption in China in 2023. *Data source* "Statistical Bulletin of National Economic and Social Development of the People's Republic of China 2023"

as a clean energy source, plays a significant role in replacing oil and coal. From a vertical perspective, in the face of continuous conflicts in some regions and the slow recovery of the economy, the proportion of natural gas consumption has declined. However, as a bridge for China's energy structure to transition from fossil energy to non-fossil energy to achieve energy conservation and emission reduction in the near and long term, natural gas has a guaranteeing role in accelerating China's energy transformation, so the proportion of natural gas consumption will continue to rise in the future.

Figures 31 and 32 show the proportion of fossil fuel consumption in 2022 and 2023. In 2022, the proportion of natural gas in fossil fuel consumption reached 10.18%, coal accounted for 68.12%, and oil accounted for 21.70%. These figures changed in 2023, with natural gas accounting for 10.25%, coal 67.53%, and oil 22.22%. Comparing the two sets of figures, it can be seen that due to factors such as geopolitics and the economic environment, the proportion of natural gas consumption in fossil energy consumption declined in 2023, while the proportion of oil consumption increased.

2. The proportion of natural gas in China's total energy consumption

The natural gas industry maintains a relatively high level of prosperity. Given the continuous and steady growth of the economy, the overall demand for natural gas is expected to continue to rise. Furthermore, since the supply and demand pattern within the natural gas market has basically stabilized and the likelihood of drastic price fluctuations has decreased, China's natural gas consumption is expected to witness further growth. Table 17 and Fig. 33 show the changing trends of natural gas in the energy consumption structure during the period from 2014 to 2023.

Table 17 Structure of energy consumption in China (2014–2023)

Year	Coal (%)	Oil (%)	Natural gas (%)	Primary electricity and other energy (%)
2014	65.80	17.30	5.60	11.30
2015	63.80	18.40	5.80	12.00
2016	62.20	18.70	6.10	13.00
2017	60.60	18.90	6.90	13.60
2018	59.00	18.90	7.60	14.50
2019	57.70	19.00	8.00	15.30
2020	56.90	18.80	8.40	15.90
2021	55.90	18.60	8.80	16.70
2022	56.20	17.90	8.40	17.50
2023	55.70	18.33	8.45	17.52

Note The data for 2023 has been calculated on the basis of the "China Statistical Yearbook 2023" and the "Statistical Bulletin of National Economic and Social Development of the People's Republic of China 2023"

Data sources "China Statistical Yearbook 2023" and "Statistical Bulletin of National Economic and Social Development of the People's Republic of China 2023"

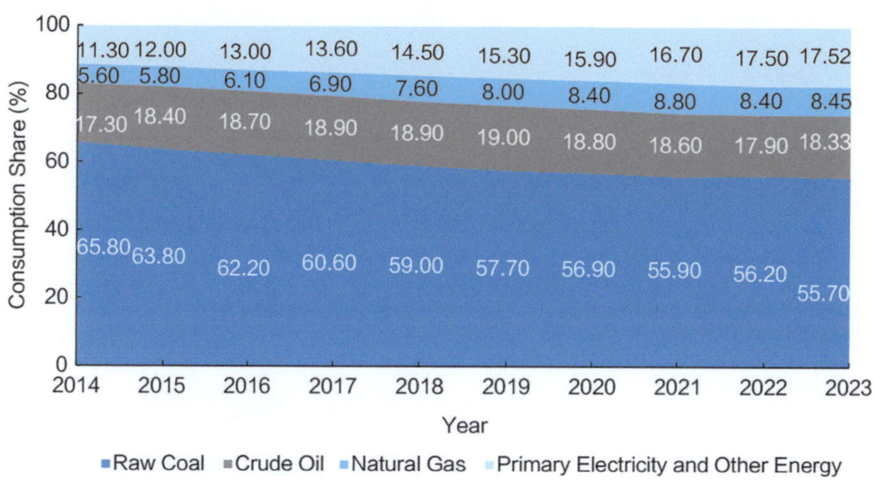

Fig. 33 Structure of energy consumption in China (2014–2023). *Data sources* "China Statistical Yearbook 2023" and "Statistical Bulletin of National Economic and Social Development of the People's Republic of China 2023"

A horizontal analysis of Table 17 and Fig. 33 shows that the proportion of natural gas in China's total energy consumption is relatively small, but its consumption growth potential is huge. With the improvement of consumption technologies and the rapid development of clean energy, the consumption volume of natural gas will continue to grow in the future. As the energy transformation continues to advance, the structure of energy consumption is also constantly changing, and natural gas, as a type of clean energy, has played a great role.

Figures 34 and 35 show the proportion of energy consumption in 2022 and 2023. In 2022, the proportion of natural gas in energy consumption reached 8.40%, coal accounted for 56.20%, oil accounted for 17.90%, and primary electricity and other energy accounted for 17.50%. These figures changed in 2023, with natural gas accounting for 8.45%, coal accounting for 55.70%, oil accounting for 18.33%, and primary electricity and other energy accounting for 17.52%. Although the proportion of natural gas consumption in China's energy consumption rose in 2023, reaching a new historical high, the proportion of natural gas in China's energy consumption is still less than 9%, while the international level is 24%. Natural gas bears the dual mission of ensuring safe supply and achieving green and low-carbon. Vigorously developing natural gas is the inevitable choice for creating a clean, low-carbon, intelligent, efficient, and economically secure energy system. There is still a lot of room for growth in China's natural gas consumption.

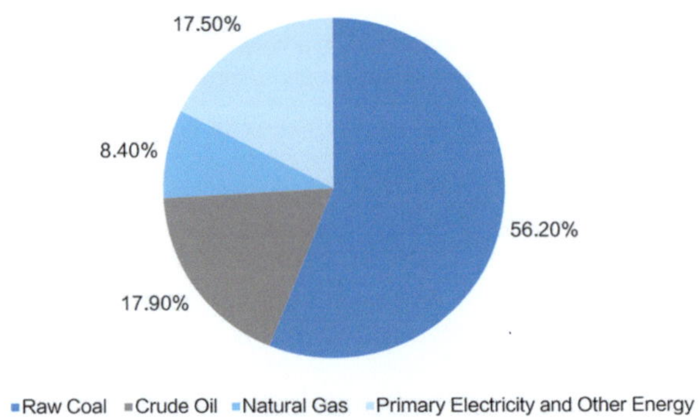

Fig. 34 Structure of energy consumption in China in 2022. *Data source* "Statistical Bulletin of National Economic and Social Development of the People's Republic of China 2022"

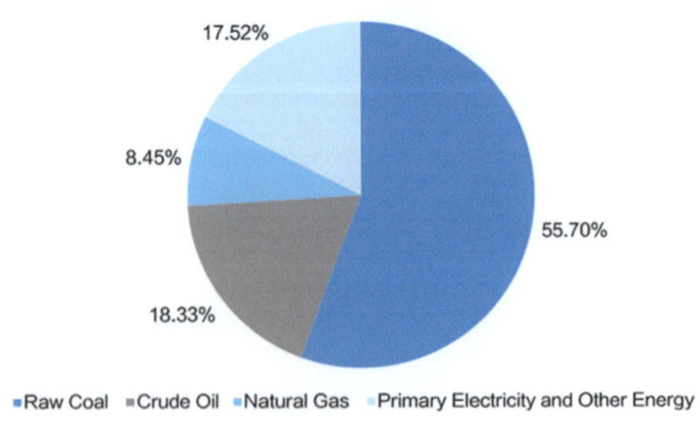

Fig. 35 Structure of energy consumption in China in 2023. *Data source* "Statistical Bulletin of National Economic and Social Development of the People's Republic of China 2023"

Summary

In 2023, China's energy industry fully implemented the "14th Five-Year Plan" and "dual carbon" goals, steadily improved its energy supply capacity, continuously enhanced its basic guarantee role, and continued to deepen the structural reform of the energy supply side, accelerating high-quality development. With the continuous exploration and large-scale development of natural gas resources in large oil and gas basins such as the Ordos Basin, Sichuan Basin, and Tarim Basin, China's natural gas production has entered a period of rapid growth. Affected by factors such as the continuous recovery of the economy, the fall in natural gas prices, insufficient hydropower output, and the low base in the same period of the previous year, China's natural gas consumption growth rate rebounded from the bottom and achieved rapid growth.

At the beginning of 2024, the international environment still remains complex and challenging. Under the influence of multiple factors, including the persistent risk of geopolitical policy conflicts, the slowdown in global economic growth, and the continuous evolution of the restructuring of the international trade system, the restructuring of the energy industry will continue to progress. Looking at the domestic market, China's energy is expected to show a trend of diversification, cleanliness, and intelligence. With the continuous optimization of the energy structure, it is expected that natural gas consumption will maintain steady growth. China's energy industry will continue to actively promote the energy revolution, making energy safer, greener, and more efficient. It will meet the demand for clean and low-carbon energy from new industrialization, provide a solid energy guarantee for high-quality economic and social development, and contribute to the new journey of Chinese-style modernization with the power of energy.

Natural Gas Policy

Natural gas, as a clean, reliable low-carbon fossil fuel, plays an important role in helping China to achieve its "dual carbon" goals. The natural gas industry must resolutely implement the various decisions and arrangements made by the Party Central Committee and the State Council on the construction of the natural gas production, supply, storage, and sales system, as well as the requirements for stable supply and price work. It should steadily advance the implementation of the "14th Five-Year Plan", play a greater role in China's economic development and energy low-carbon transformation, and contribute to ensuring national energy security and accelerate the construction of an energy powerhouse. Therefore, it is necessary to clarify China's natural gas development policy, analyze the development direction of the natural gas industry based on the development patterns of the natural gas market and China's national conditions, propose policy recommendations, and promote the high-quality development of the natural gas industry.

This chapter organizes the policies, regulations, development plans and other documents issued by the national and local governments. Based on these policy documents, the discussion of national natural gas policies are divided into five categories: green development, energy governance, energy utilization, scientific and technological innovation, and natural gas pricing. The discussion of policies regarding local natural gas development are divided into two categories: natural gas pricing policy and natural gas pipeline construction policy. Moreover, this chapter will offer policy recommendations on natural gas supply security, the green and low-carbon development of natural gas, and natural gas pricing policy. In 2023, "supply security and price stability", "integrated development", "green development", and "digitalization" are the focus of China's natural gas policy, that is, to ensure the safety of the natural gas supply, promote the integrated development of natural gas and new energy, promote the green and low-carbon transformation of energy, and accelerate the digital and intelligent development of natural gas. In the foreseeable future, these keywords will continue to be the focus of China's natural gas policy, and the relevant supporting policies will be further improved. At the same time, the resilience, safety, green,

© Petroleum Industry Press 2025

L. Zhang et al., *Annual Report on Operation of China's Natural Gas Industry: 2023*,
https://doi.org/10.1007/978-981-95-2371-9_2

intelligent and modernization level of China's natural gas industry and supply chain will also be gradually improved.

National Natural Gas Policy

In 2023, the National Development and Reform Commission and other relevant departments issued a total of 12 natural gas-related policies, as shown in Table 1. These policies have planned and guided the national energy development in terms of green development, energy governance, energy utilization and natural gas pricing. To gain a comprehensive understanding and effective control of the policy content, we employed the NVivo12 software to conduct a thorough content analysis of China's 2023 national natural gas policies. This analysis generated a word cloud diagram, which visually emphasizes the focal points of the policy content, as illustrated in Fig. 1.

Green Development

(I) "Notice of the National Energy Administration on Printing and Distributing the Action Plan for Accelerating the Integrated Development of Oil and Gas Exploration and Development and New Energy (2023–2025)"

1. Main policy contents

Integrate and coordinate efforts to promote both onshore oil and gas exploration and development and wind and solar power generation. To fully utilize the advantages of onshore oil and gas fields, which have abundant wind and solar energy resources, superior construction conditions, and the potential for large-scale development, efforts should be made to enhance the local consumption capacity of new energy sources. The focus should be on promoting the centralized development of wind power and photovoltaic power generation in major oil fields such as Daqing, Changqing, Shengli, Tarim, Xinjiang, and North China. In and around oil and gas mining areas, the distributed development of wind power and photovoltaic power generation should be actively promoted, with an emphasis on applying low-wind-speed wind power technology and encouraging the local development of wind power. Additionally, in regions like Xinjiang, Qinghai, and Gansu, comprehensive utilization demonstration projects should be constructed to facilitate the synchronous development of oil and gas with solar energy.

Coordinate the advancement of offshore oil and gas exploration and development with offshore wind power construction. By developing offshore wind power, provide green electricity for oil and gas platforms, replace dispersed gas or oil power generation, improve energy use efficiency, reduce carbon emissions, and form a

Table 1 National natural gas policies in 2023

Serial number	Theme	Document name	Date of issuance
1	Green development	"Notice of the National Energy Administration on Issuing the Action Plan for Accelerating the Integration of Oil and Gas Exploration and Development and New Energy Development (2023–2025)"	2023–02-27
2		"Notice of the National Development and Reform Commission and the State Administration for Market Regulation on Further Strengthening the Updating and Implementation of Energy Conservation Standards"	2023–03-08
3		Notice of the National Development and Reform Commission and other departments on issuing the "Implementation Plan for Green Low-Carbon Advanced Technology Demonstration Projects"	2023–08-22
4		Notice of the State Council on Issuing the "Action Plan for Continuous Improvement of Air Quality"	2023–12-07
5	Energy governance	Notice of the National Energy Administration on Issuing the "Key Points of Energy Supervision Work in 2023"	2023–01-04
6		"Several Opinions of the National Energy Administration on Accelerating the Digitalization and Intelligent Development of Energy"	2023–03-28
7		Notice of the National Energy Administration on Issuing the "Guidelines on Energy Work in 2023"	2023–04-14
8		"Opinions of the Ministry of Natural Resources on Several Issues Concerning Deepening the Reform of Mineral Resources Management"	2023–07-26
9		"Notice of the Ministry of Housing and Urban–Rural Development on Issuing the Criteria for Determining Major Safety Hazards in Urban Gas Operation"	2023–09-21
10	Energy utilization	"Natural Gas Utilization Policy (Draft for Comments)"	2023–02-20
11		"Guiding Opinions of the National Development and Reform Commission and Other Departments on Promoting Green Innovation and High-Quality Development of the Refining Industry"	2023–10-25
12	Natural Gas prices	"Notice of the National Development and Reform Commission on the Approval of Cross-province Natural Gas Pipeline Transportation Prices"	2023–12-05

Data sources National Development and Reform Commission, National Energy Administration, Official Website of the Chinese Government

Fig. 1 Word cloud of national natural gas policies

complementary power supply mode for offshore wind power and oil and gas field regional power systems. Prioritize new energy construction in surrounding areas by oil and gas development companies. Fully rely on shore power to ensure the integrated development of the oil and gas industry and offshore wind power. Actively and orderly develop floating wind power to deliver green electricity to deep-sea oil and gas platforms.

Accelerate the enhancement of new energy storage and consumption capacity in the upstream of the oil and gas industry. Promote the large-scale application of new types of energy storage in the upstream of oil and gas. Leverage the multifunctional roles of energy storage in peak shaving, frequency modulation, emergency backup, capacity support, etc., to promote the application of energy storage in multiple scenarios on both the power supply side and the oil and gas exploration and development user side, and orderly advance the coordinated development of energy storage and new energy. On land, reasonably layout natural gas peaking power stations in areas rich in wind and solar resources, and build a comprehensive energy model focusing on the integrated development of wind power and natural gas power generation at sea.

Actively promote the construction of green oil and gas fields as demonstration projects. In oil and gas fields where new energy is abundant and institutional mechanism innovation is being pioneered, a series of low-carbon or zero-carbon oil and gas field construction demonstration projects should be established. These projects aim to promote innovation in the integration of oil and gas exploration and development with new energy technologies, application methods, and institutional mechanisms. The goals include further reducing the self-use of natural gas, expanding the scale of distributed energy access and application, innovating the cooperation model for the full industry chain development and utilization of new energy, and completing the construction and demonstration leadership of low-carbon oil fields.

Refine the systems and mechanisms, and enhance policy support. This will facilitate the low-carbon energy transition and fully leverage the decisive role of the market in resource allocation. Energy authorities at all levels should increase their support, prioritizing new energy projects such as solar, wind, hydrogen, and geothermal energy in energy development plans at all levels. These projects serve as clean alternatives for oil and gas exploration and development.

Improve the efficiency of filing for oil and gas exploration and development projects as well as new energy integration projects. Relying on the national online approval and supervision platform for investment projects, handle the filing procedures as a whole for multi-energy complementary systems such as wind and solar power generation, hydrogen energy, geothermal energy, source-network-load-storage, and micro-grid facilities used for oil and gas exploration and development. These new energy facilities, serving as supporting facilities for oil and gas development project capacity construction, should be integrated into the filing process.

Increase the intensity of technological innovation in the integration and development of oil and gas exploration and development with new energy. Focus on promoting low-cost solar thermal utilization, energy storage technologies (electricity and heat) for oil and gas fields, distributed microgrids, and comprehensive energy smart control in the supporting projects of oil and gas production capacity construction.

2. Policy interpretation

The "Action Plan for Accelerating the Integration and Development of Oil and Gas Exploration and Development and New Energy (2023–2025)" proposes to vigorously promote the integration and development of oil and gas exploration and development and new energy, and actively expand the scale of green electricity development and utilization by oil and gas companies. Moreover, it stipulates that the industry should adhere to the principles of prioritizing ecological conservation, adapting strategies to local circumstances, and fostering diversified and integrated development. Initially, focus should be given to local consumption, the vigorous promotion of wind power and photovoltaic power generation in onshore oil and gas mining and surrounding areas, the promotion of offshore wind power and oil and gas exploration and development, the accelerated improvement of new energy development and utilization and storage capacity in the upstream of oil and gas, and the active promotion of the construction of green oil and gas fields. Energy authorities at all levels should increase their support, and new energy projects such as solar energy, wind energy, hydrogen energy, and geothermal energy that serve as clean alternatives for oil and gas exploration and development should be given priority in energy development planning at all levels.

At the same time, increase the efforts in technological innovation and breakthroughs for the integrated development of oil and gas exploration and development and new energy. Focus on promoting low-cost solar thermal utilization, energy storage technologies (electricity and heat) for oil and gas fields, distributed micro-grids, and comprehensive energy smart management in projects that support oil and gas production capacity construction. Improve the accuracy of wind and solar resource forecasts and the precision of wind power and photovoltaic power generation predictions. Enhance the ability of wind power and photovoltaic power generation to adapt to power system disturbances, thereby supporting the stable operation of oil and gas production.

(II) "Notice of the National Development and Reform Commission and the State Administration for Market Regulation on Further Strengthening the Updating and Implementation of Energy Conservation Standards"

1. Main policy contents

Energy conservation standards form the foundation of the national energy conservation system and provide crucial support for promoting energy conservation and carbon reduction across society. All regions, relevant departments, and industries should place great emphasis on energy conservation standardization work. They should accelerate the updating and upgrading of energy conservation standards in accordance with the law, strengthen the implementation of these standards, continuously consolidate the foundation of energy conservation efforts, and provide robust support for actively and prudently advancing carbon peaking and carbon neutrality, as well as accelerating the green transformation of development methods. In the energy sector, there should be an accelerated development of energy conservation technology standards related to the clean and efficient use of coal, the utilization of new energy and renewable energy sources, the storage and transportation of oil and natural gas, pipeline transportation, and key equipment for power transmission and distribution.

2. Policy interpretation

In order to continuously promote the updating and implementation of energy conservation standards, and to accelerate the research and development as well as the promotion of advanced energy conservation and carbon reduction technologies, the National Development and Reform Commission issued the "Notice on Further Strengthening the Updating and Implementation of Energy Conservation Standards." This notice aims to accelerate the updating and upgrading of energy conservation standards in four aspects. First, the formulation and revision of energy conservation standards in key areas shall be accelerated. Second, the energy consumption limit requirements in key energy-consuming industries will be steadily increased. Third, the energy efficiency level of key energy-consuming products and equipment will be continuously improved. Lastly, the coverage of energy conservation standards will be continuously expanded to ensure their advancement, scientificity, and effectiveness. This notice points out that it is necessary to promote the development of energy conservation technology standards related to oil and natural gas storage and transportation. This will promote the standardization of energy conservation in the energy field, strengthen the implementation of energy conservation standards, continuously consolidate the foundation of energy conservation work, and provide strong support for actively and prudently promoting carbon peaking and carbon neutrality and accelerating the green transformation of development methods.

(III) Notice of the National Development and Reform Commission and other departments on issuing the "Implementation Plan for Green and Low-Carbon Advanced Technology Demonstration Projects.

1. Main policy contents

By 2025, through the implementation of demonstration projects for advanced green and low-carbon technologies, a number of demonstration projects will be carried out, a batch of advanced and applicable green low-carbon technology achievements will be transformed and applied, and several support policies, business models, and regulatory mechanisms conducive to the promotion and application of green and low-carbon technologies will be gradually improved. By 2030, the research and development (R&D), demonstration, and promotion models for advanced and applicable green low-carbon technologies will be essentially mature. Related support policies, business models, and regulatory mechanisms will be further refined, and the international competitive advantage of green low-carbon technologies and industries will be significantly enhanced.

The key directions are as follows:

(1) Carbon reduction at the source. For example, demonstration projects for the clean and efficient development and utilization of fossil fuels, including large-scale coal-fired power units coupled with biomass and low-carbon fuel co-firing power generation, efficient extraction and utilization of coal mine gas, energy-saving and carbon-reducing oil and gas pipelines, and low-carbon (near-zero carbon) oil and gas field demonstrations.
(2) Process carbon reduction. For example, demonstration projects in the industrial field, including advanced low-carbon petrochemicals, modern coal chemicals, green bio-chemicals, the coupling of renewable energy and petrochemical production systems, industrial green microgrids, digital green chemical collaborative carbon reduction, "Industrial Internet + Green Low Carbon", green (zero carbon, near-zero carbon) data centers.
(3) Terminal carbon sequestration. For example, full-process scaled CCUS (Carbon Capture, Utilization, and Storage) demonstration projects, focusing on industries such as petrochemicals, coal chemicals, coal power, steel, non-ferrous metals, building materials, and oil extraction. Select locations with high industrial agglomeration and good geological conditions to construct several full-process scaled CCUS demonstration projects. For the petrochemical, coal chemical, and oil extraction industries, projects with an annual capture, utilization, and storage volume of no less than 500,000 tons should, in principle, be equipped with carbon dioxide transportation pipelines.

2. Policy interpretation

The "Implementation Plan for Green and Low-Carbon Advanced Technology Demonstration Projects" proposes the overall approach, main goals, key directions, and safeguard measures for carrying out demonstrations of green and low-carbon advanced technologies. It plays an important role in accelerating the application and promotion of green low-carbon advanced and applicable technologies, improving the policy environment for supporting the development of new industries and new business formats, and promoting the formation of new competitive advantages in related industries. The "Implementation Plan for Green and Low-Carbon Advanced Technology Demonstration Projects" first clarifies the significance of carrying out demonstrations for green and low-carbon advanced technologies. In light of their importance, the document also points out the key directions and safeguard measures for demonstration projects for green and low-carbon advanced technologies, which revolve around the tasks determined by the "1 + N" policy system for carbon peaking and carbon neutrality. It proposes technological demonstrations not only in the energy and industrial sectors but also in the construction and transportation sectors. It focuses not only on the demonstration and application of advanced and applicable technologies for carbon reduction at the source and process carbon emission reduction but also on the demonstration of terminal carbon sequestration technologies. Demonstration projects related to natural gas include the clean and efficient development and utilization of fossil fuels, industrial field demonstration projects, and full-process scaled CCUS demonstration projects. With the continuous advancement and implementation of these demonstration projects and safeguard measures, the related enterprises will gain confidence for carrying out green low-carbon advanced applicable technology demonstrations. Moreover, they will help promote the industrialization and large-scale low-cost application of research and development achievements, create business models and a policy environment conducive to the development of new green and low-carbon industries in the energy field, and provide strong support for achieving the "dual carbon" goals.

(IV) Notice of the State Council on Issuing the "Action Plan for the Continuous Improvement of Air Quality"

1. Main policy contents

Vigorously develop new energy and clean energy. By 2025, the proportion of non-fossil energy consumption will reach about 20%, and the proportion of electricity in terminal energy consumption will reach about 30%. Continue to increase the production and supply of natural gas, with the newly added natural gas prioritizing the needs of residents' lives and clean heating.

Implement clean energy substitution for industrial furnaces and kilns. Orderly promote electricity substitution for coal, and actively and prudently promote gas substitution for coal. In key areas, no new fuel gasifiers will be added. New, renovated, and expanded furnaces—such as heating furnaces, heat treatment furnaces, drying furnaces, and melting furnaces—will, in principle, use clean and low-carbon energy sources. Industrial kilns currently using highly polluting fuels will be safely and steadily converted to use industrial waste heat, electricity, natural gas, and so on.

Continue to promote clean heating in northern regions. Increase the intensity of replacing loose coal for civil and agricultural use, with a goal of basically eliminating loose coal in the plain areas of key regions. Moreover, gradually promote the clean energy substitution of loose coal in mountainous regions. Cities included in the scope of central government financial support for clean heating in northern regions must complete their transformation tasks with high quality and quantity. For the "coal-to-gas" conversion, it is essential to secure gas supply sources and to implement the conversion based on the available supply.

Improve the price and tax incentive and restraint mechanisms. Reduce the level of urban gas transmission and distribution, reasonably set and strictly regulate transmission and distribution prices, establish and improve the linkage mechanism between end-user sales prices and procurement prices, and implement gas pricing policies for clean heating.

2. Policy interpretation

The introduction of the "Action Plan for the Continuous Improvement of Air Quality" fully reflects the firm determination of the Party Central Committee to continue the war on air pollution, and reflects China's adherence to the development concept of putting the people at the center and promoting the harmonious coexistence between man and nature. The action plan emphasizes that the improvement of air quality is critical. It calls for the solid promotion of the green and low-carbon transformation of industries, energy, and transportation, and for accelerating the formation of green and low-carbon production and lifestyle. This will lead to multiple wins in environmental, economic, and social benefits. The overall approach of the action plan in the energy field is to promote clean energy substitution and reasonably control the total coal consumption. The measures related to natural gas mainly focus on two aspects. One is to promote clean heating in winter, requiring all regions to continuously increase the supply of natural gas, and the newly added natural gas will prioritize the needs of the residents' lives and clean heating, and implement gas sources and supply-oriented changes for 'coal-to-gas'. The other is to implement clean energy substitution for industrial furnaces and kilns, and actively and prudently promote gas substitution for coal. No new fuel gasifiers will be added in key areas.

Energy Governance

(I) Notice of the National Energy Administration on Issuing the "Key Points of Energy Supervision Work in 2023"

1. Main policy contents

Ensure energy security and promote a reliable supply. Conscientiously implement the decisions and arrangements of the Party Central Committee and the State Council on energy work, take energy security and supply as the primary tasks of energy supervision, and urge the relevant departments of local governments and energy companies to fulfill their main responsibilities of ensuring supply and implementing supply policies. It is necessary to fully leverage the functions of assigned institutions as "probes" and "sentries", strengthen the monitoring, analysis and early warning of the supply and demand situation of energy sources such as coal, electricity, and natural gas, and deal with the "three-key issues" that affect energy supply, and report the situation to the party offices of the relevant departments and local governments. Conduct research on the mechanisms of normalized and standardized energy supply monitoring, early warning and supervision to form a joint force for ensuring supply.

Strengthen the supervision of the implementation of energy planning policies. Carry out the mid-term evaluation of the "14th Five-Year Plan" energy plan and supervise the implementation of the plan. Deepen the supervision of the implementation of key energy policies. The assigned institutions are required to supervise how local governments use the powers delegated by the National Energy Administration and where the relevant departments of local governments have not caught up and have not managed the situation well. The assigned institutions are entitled to point out the issues, obliged to report them, and have the responsibility to correct deviations.

Strengthen the supervision over the implementation of major energy projects. Continue to track the progress of major projects such as cross-provincial and cross-regional power transmission channels, oil and gas pipelines, large wind and photovoltaic bases, hydropower stations, integrated power "Source-Network-Load-Storage", and multi-energy complementarity, gradually establish a mechanism for normalized supervision, promptly discover problems in project progression, and put forward supervision opinions and suggestions.

Deepen the fair and open supervision of oil and gas pipeline facilities. Improve the verification list and the regulatory work process for the fair and open supervision of oil and gas pipeline facilities, fully understand the situation of oil and gas companies operating pipeline facilities in implementing fair access, information reporting, independent accounting, service contract signing and performance, and fair and open services. Guide the high-quality and fair opening of LNG receiving stations to ensure the stable supply and pricing of natural gas. Accelerate the introduction of access rules for shippers of natural gas pipeline network facilities, and guide pipeline operation enterprises to optimize and improve the operation methods for capacity allocation of pipeline facilities. Carry out special verification of the natural gas consumption structure in typical provinces and cities.

2. Policy interpretation

The issuance of the "2023 Energy Supervision Work Key Points" has determined the key work of energy supervision in 2023, among which the key supervision tasks for natural gas are to ensure energy security and promote a reliable supply, strengthen the supervision of the implementation of energy planning policies, supervise the implementation of major energy projects, and supervise the fair and open operation of oil and gas pipeline facilities. The "2023 Energy Supervision Work Key Points" clarifies the concept of "industry supervision", further strengthens the supervision of the implementation of major national energy plans, policies, and projects. In terms of the coordination of the energy supply, it urges local government departments and energy enterprises to fulfill their main responsibilities for supply, strengthens the monitoring, analysis and early warning of the supply and demand situation of energy sources such as coal, electricity, and natural gas and handle and report problems in a timely manner. In terms of the implementation of major energy planning policies, it calls for doing a good job in the mid-term evaluation of the "14th Five-Year Plan" energy plan. In terms of supervising the implementation of major energy projects, it demands the continuous tracking of the progress of major energy projects such as cross-provincial and cross-regional oil and gas pipelines approved by the National Energy Administration, gradually establishing a normalized supervision mechanism so as to timely correct deviations and errors and ensure the implementation and orderly progress of key plans and projects. As for the supervision of the natural monopoly sectors, following the strengthening of the fair and open supervision of oil and gas pipeline network facilities last year, continuous efforts will be made this year to carry out various tasks related to the access rules for shippers of natural gas pipeline network facilities. Through this, the fair access to natural gas pipeline network facilities will be promoted, the use of these facilities and the management of shippers will be standardized, and the market order of fair competition will be maintained.

(II) "Several Opinions of the National Energy Administration on Accelerating the Digitalization and Intelligent Development of Energy.

1. Main policy contents

With demand as the driving force, in line with the demands for the digital and intelligent transformation and development of industries including power, coal, and oil and gas, apply digital and intelligent technologies in an integrated manner. Follow the approach of dealing with urgent matters first and starting with the simpler tasks. Subsequently, identify and fill the gaps in the transformation and development across different industries, links, and stages, thus providing solid support for the high-quality development of the energy sector. Empowered by digitalization, leverage the potential of the smart grid to extend and expand the energy network. Promote the formation of an intelligent energy regulation and control system. Enhance the level of precise and efficient resource allocation. Apply digital and intelligent technologies across the entire chain and in all links of the coal, oil, and gas production, supply, storage, and marketing systems. Improve overall energy efficiency, work safety, and the green and low-carbon levels of these industries.

Use digital and intelligent technologies to facilitate the green and low-carbon development and utilization of oil and gas. Accelerate the research and development of professional software for oil and gas exploration and development, promote the construction of digital basins, and drive the establishment of databases, model libraries, and sample libraries for oil and gas exploration and development. Promote the construction of intelligent well logging and intelligent node seismic acquisition systems, and advance the deployment and remote control operations of intelligent drilling and completion, intelligent injection and production, and intelligent fracturing systems. Expand the application of carbon dioxide flooding technology. Promote the integrated intelligent regulation and control level of "Source-Network-Load-Storage", and strengthen the replacement of traditional energy with new energy in production. Promote the information transformation and digital upgrade of oil and gas pipeline networks, promote the construction of intelligent pipelines and intelligent gas storage facilities, and enhance the safe and efficient operation level of oil and gas pipeline network facilities as well as their gas storage and peak shaving capabilities.

Promote pilot demonstrations of diverse application scenarios. Focus on key areas, crucial links, and common demands. Rely on energy projects to explore and expand the application of digitalization and intelligence based on local conditions. Emphasize organizing demonstration projects in application scenarios such as intelligent coal mines, intelligent oil and gas fields, intelligent pipelines, intelligent refineries, integrated energy services, and integrated service platforms to undertake systematic digital and intelligent pilot tasks. Conduct in-depth exploration and take the lead in testing in aspects such as technological innovation, operational models, and development formats.

Enhance the cybersecurity capabilities of the energy system. Promote the development of a dynamic defense system in coal mines that encompasses the entire business life cycle, featuring "early warning, monitoring, and response." Improve the active defense capabilities of industrial hosts in oil and gas fields and strengthen cybersecurity protection for industrial control systems in power plants.

2. Policy interpretation

The "Several Opinions of the National Energy Administration on Accelerating the Development of Energy Digitalization and Intelligence" aims to promote the deep integration of digital technology and energy industry development, strengthen the construction of new infrastructure that combines traditional energy with digital and intelligent technologies, and unleash the potential value of energy data elements. The "Several Opinions of the National Energy Administration on Accelerating the Development of Energy Digitalization and Intelligence" puts forward the basic principles of taking business digitalization as the demand driver, enabling digital empowerment through digital businessization, promoting data resources as a new production factor to facilitate coordinated and efficient development, and making full use of new information technologies for integrated innovation. It provides comprehensive guidance from five aspects: including accelerating the industry's transformation and upgrading, promoting application pilot demonstrations, facilitating breakthroughs in common technologies, improving the development support system, and strengthening organizational guarantees. Regarding natural gas, the "Several Opinions of the National Energy Administration on Accelerating the Development of Energy Digitalization and Intelligence" mainly emphasizes two aspects. First, it is necessary to strengthen network and information security guarantees to effectively enhance the level of energy digital and intelligent development. The improvement of the digital and intelligent development level can only be effectively achieved on the basis of network and information security guarantees, which is specifically reflected in the enhancement of the active defense capabilities of industrial hosts in oil and gas fields. Second, leverage digital and intelligent technologies to assist in the green and low-carbon development and utilization of oil and gas, providing strong support for actively and steadily promoting the achievement of carbon peaking and carbon neutrality. As the energy industry is an important source of carbon emissions, its green and low-carbon transformation is an inevitable trend.

(III) Notice of the National Energy Administration on Issuing the "Guidelines on Energy Work in 2023"

1. Main policy contents

Strengthen the foundation of fossil fuel production and supply. Accelerate the construction of important natural gas production capacity projects such as Shanbei, Chuannan, and Bozhi-Dabei.

Focus firmly on increasing oil and gas production in the key basins of the "five oils and three gases" as well as in relevant sea areas. Promote stable production in existing oil and gas fields, and strive for new breakthroughs in areas such as the exploration and development of oil and gas in deep onshore strata, deep water, and shale oil and gas. Additionally, promote the green and low-carbon development of crude oil through CCUS (Carbon Capture, Utilization, and Storage). Strengthen the construction of energy reserve capacity.

Improve the regulation capacity of the energy system. With underground gas storage facilities as the mainstay and coastal LNG tanks as the auxiliary, promote the intensive layout of gas storage facilities and accelerate the construction of underground gas storage facilities such as Shengping in Daqing, Tongluoxia and Huangcaoxia in Chongqing, Pingdingshan in Henan, and Huaian in Jiangsu.

Strengthen the clean and efficient exploitation and utilization of fossil fuels. Accelerate the integrated development of oil and gas exploration and development with new energy sources, and promote the intelligent and green development of the upstream oil and gas sector. Conduct research and revise natural gas utilization policies. Issue relevant guiding opinions to promote the low-carbon and efficient development of the refining industry, and advance the optimization of the refining production capacity structure and layout.

Accelerate the digital and intelligent upgrading of the energy industry. Promote the in-depth integration of the energy industry with the digital industry, and issue the "Several Opinions on Accelerating the Digital and Intelligent Development of Energy." Actively carry out intelligent construction of equipment, facilities, and technological processes in fields such as power plants, power grids, oil and gas fields, oil and gas pipelines, and end-use energy. Enhance the flexibility and responsiveness of the energy system, and improve its capacity for efficient production and operation. and promote interaction among power sources, power grids, and power loads, as well as the coordinated and complementary use of multiple energy sources.

Enhance the optimal allocation of regional energy resources. Improve the "national unified network" of oil and gas, and focus on the construction of major projects such as the southern section of the China-Russia East Route, the middle section of the West–East Gas Transmission Line Three, the West–East Gas Transmission Line Four, the Second Sichuan-East Gas Transmission Pipeline, and the Hulin-Changchun Natural Gas Pipeline.

Improve the energy regulatory and policy framework. Accelerate the legislative process of the Energy Law of the People's Republic of China, and carry out the formulation and revision of the Renewable Energy Law of the People's Republic of China, the Coal Law of the People's Republic of China, the National Oil Reserve Regulations, and the Petroleum and Natural Gas Pipeline Protection Law of the People's Republic of China. Conduct research and drafting of the Energy Supervision Regulations. Establish and improve the management system for energy data, and strengthen data security governance. Enhance research on standard systems for new power systems, energy storage, and CCUS (Carbon Capture, Utilization, and Storage). Provide key support for the establishment of standards related to energy carbon peak and carbon neutrality, and expedite the formulation and revision of key standards.

Ensure energy security under open conditions. Closely monitor the follow-up impact of the Ukraine crisis, and deepen mutually beneficial cooperation with key energy resource countries. Enhance collaborative cooperation with energy-producing countries, transit countries, and consumer countries. Strengthen the capacity to ensure maritime and land energy transportation, and continuously consolidate the four major oil and gas import corridors in the northwest, northeast, southwest, and maritime regions. Actively explore new channels for energy imports, smooth the international energy industry chain and supply chain, and promote the diversification of energy imports. Intensify dynamic monitoring and risk pre-research and judgment of major overseas energy projects, strengthen the construction of overseas energy resource supply bases, and safeguard overseas interests in the energy sector.

2. Policy interpretation

The year 2023 is a key year for the implementation of the "14th Five-Year Plan", bridging the past and the future, and doing a good job in the energy sector throughout the year is of vital importance. The "2023 Energy Work Guidance" has made specific arrangements for ensuring the stable supply of energy and for continuously promoting high-quality development in the energy sector. It aims to comprehensively achieve the goal of firmly holding the energy bowl in our own hands, while at the same time deeply advancing the energy revolution, accelerating the planning and construction of a new energy system, and enhancing the level of security assurance. The "2023 Energy Work Guidance" has clearly put forward three main target guidelines, namely "Continuous Strengthening of Supply Assurance Capability," "Deepening of Structural Transformation," and "Steady Improvement of Quality and Efficiency."

Amidst the complex and evolving global geopolitical landscape, severe challenges to energy security supply, and ongoing fluctuations in international oil prices, the "Guidelines for Energy Work in 2023" prioritize energy supply and price stability.

The "Guidelines for Energy Work in 2023" first emphasize the role of "further consolidating the baseline guarantee basis of fossil fuels." In terms of natural gas development, it accelerates the construction of several key natural gas production capacity projects, promotes the application of critical technologies such as carbon capture, storage, and utilization, and strengthens oil and gas exploration, development, and storage production. The "Guidelines for Energy Work in 2023" also clearly propose to "enhance the regulatory capacity of the energy system," with underground gas storage as the mainstay and coastal LNG tanks as the auxilliary, to promote the intensive layout of gas storage facilities, and to accelerate the construction of underground gas storage in Daqing Shengping, Chongqing Tongluoxia, and Huangcaoxia.

Regarding the goal of "deepening structural transformation," the "2023 Energy Work Guidance" clearly requires: "Consolidate the development advantages of the wind power and photovoltaic industries, continuously expand the supply of clean and low-carbon energy, actively promote the low-carbon and clean use of energy in production and daily life, and consolidate and expand the strong momentum of green and low-carbon transformation through joint efforts on both the supply and demand sides." In terms of natural gas, accelerate the integrated development of oil and gas exploration and new energy, promote the intelligent and green development of the upstream oil and gas sector, and study and revise new natural gas utilization policies. Ultimately, on the basis of enhancing the supply guarantee capability of fossil fuels, gradually form a diversified supply structure driven by multiple sources, including oil, gas, coal, nuclear, new energy, and electricity.

Regarding the goal of "steadily improving quality and efficiency," the "2023 Energy Work Guidance" clearly states that a number of new energy technology innovation platforms will be established, with particular emphasis on the identification of the first batch of energy R&D innovation platforms during the "14th Five-Year Plan" period. This will accelerate the process of overcoming key technology and equipment deficiencies. Additionally, active efforts will be made to promote intelligent construction of equipment, facilities, and technological processes in areas such as power plants, power grids, oil and gas fields, oil and gas pipelines, and end-use energy.

In the "Guidelines for Energy Work in 2023", in addition to emphasizing the above three main goals, it also clearly proposes strategic measures such as "enhancing the optimal allocation of regional energy resources," "strengthening the capacity building of energy governance," and "expanding high-level open cooperation in the energy sector." These measures are aimed at providing a solid energy guarantee for the development of energy that is safer, greener, and more efficient.

(IV) "Opinions of the Ministry of Natural Resources on Several Issues Concerning Deepening the Reform of Mineral Resources Management"

1. Main policy contents

Comprehensively promote the competitive transfer of mining rights. Except for special circumstances such as agreement-based transfers, mining rights shall be publicly and competitively transferred through bidding, auction, and listing in accordance with the "Mining Rights Transfer Transaction Rules" (Natural Resources Regulations [2023] No. 1). Promote the use of letters of guarantee or deposits in mining rights transactions, and explore the establishment of relevant rules to ensure the smooth progress of mining rights transactions. For the competitive transfer of exploration rights for oil and gas (including petroleum, hydrocarbon natural gas, shale gas, coalbed methane, and natural gas hydrates), the starting price shall be determined in accordance with the guiding opinions on the starting price standards for the transfer proceeds of mining rights formulated by the Ministry of Natural Resources and the Ministry of Finance.

Implement the same-level management for the transfer and registration of exploration rights and mining rights for the same mineral species. The Ministry of Natural Resources is responsible for the transfer and registration of mining rights for petroleum, hydrocarbon natural gas, shale gas, natural gas hydrates, radioactive minerals, tungsten, rare earths, tin, antimony, molybdenum, cobalt, lithium, potash salts, and crystalline graphite.

Open up the oil and gas exploration and development market. Companies registered within the People's Republic of China, with a net asset of not less than 300 million yuan, both domestic and foreign, are eligible to obtain oil and gas mining rights in accordance with regulations. Those engaged in oil and gas exploration and development should meet the requirements and regulations for safety, environmental protection and other qualifications, and possess the corresponding technical capabilities for oil and gas exploration and development.

Implement an integrated system for oil and gas exploration and extraction. If an oil and gas mining rights holder discovers exploitable oil and gas resources and decides to continue exploitation after completing the oil (gas) testing operations, it can carry out exploitation after submitting the integrated exploration and exploitation plan (see attachment) to the natural resources department in charge with registration authority within 30 days. In areas where the geological reserves are proven during the exploration and exploitation process, a mineral resources reserve report should be compiled in a timely manner and submitted for review and filing. Within five years after reporting the integrated exploration and exploitation plan, the mining rights holder shall sign a mining rights transfer contract and handle the mining rights registration in accordance with the law. If it exceeds five years after reporting the integrated exploration and exploitation plan and continues exploitation without transferring the mining rights, it will be dealt with as illegal mining. If the mining rights holder decides not to continue exploitation after completing the oil (gas) testing operations, or if the exploitation is completed within five years or cannot be transferred to mining and exploitation is stopped, there will be no need to handle the mining rights registration.

Adjust the term of exploration rights. The registration terms for newly established, renewed, and reserved exploration rights are all five years. When applying for the renewal registration of exploration rights, 20% of the area specified in the exploration license should be deducted. The scope where resources have been submitted for non-oil and gas minerals or the scope where proven geological reserves have been submitted for oil and gas shall not be included in the deduction base. For exploration in the deep or upper parts of an existing mining right, no area will be deducted. Oil and gas exploration rights can deduct an equivalent area of other blocks of the same exploration rights holder in the same basin. However, newly transferred oil and gas exploration rights shall not be used to offset the area that should be deducted for other blocks of the same exploration rights holder within five years.

2. Policy interpretation

On July 26, the Ministry of Natural Resources issued the "Opinions of the Ministry of Natural Resources on Deepening the Reform of Mineral Resources Management," which has become an important guiding document for the management of mining rights in the next stage. The main contents of interest to natural gas exploration and extraction companies are as follows: First, comprehensively promote the competitive transfer of mining rights. Except for special circumstances such as agreement transfer, mining rights are transferred in a competitive manner through bidding, auction, and listing in accordance with the "Mining Rights Transfer Transaction Rules." The starting price for the competitive transfer of oil and gas (including hydrocarbon natural gas, shale gas, coalbed methane, natural gas hydrates) exploration rights is determined according to the "Guiding Opinions on Formulating the Starting Price Standard for Mining Rights Transfer Revenue" by the Ministry of Natural Resources and the Ministry of Finance. Second, promote the use of guarantees or deposits in mining rights transactions to guide bidders to bid rationally and reduce investment risks and operating costs. Third, on the basis of extending the renewal period of exploration rights, the retention period of exploration rights is extended from 2 to 5 years each time, further reducing the cost of handling affairs for the relevant parties. Fourth, further optimize the unified system of oil and gas exploration and extraction. The "Opinions of the Ministry of Natural Resources on Deepening the Reform of Mineral Resources Management" draws on the practices of the past three years, sorts out and refines the management requirements for each link such as self-reporting of the unified exploration and extraction plan, carrying out unified exploration and extraction work, registering mining rights, and management of non-transferred mining rights, and clarifies the content of the "Unified Exploration and Extraction Plan Form for Oil and Gas."

(V) "Notice of the Ministry of Housing and Urban–Rural Development on Issuing the Criteria for Determining Major Safety Hazards in Urban Gas Operation"

1. Main policy contents

To guide local governments in strengthening the control of safety risks and the investigation and management of hidden dangers in urban gas operations, to prevent the occurrence of major and especially serious accidents, and to effectively protect the life and property safety of the people, this standard is formulated in accordance with laws and regulations such as the "Law of the People's Republic of China on Work Safety," the "Law of the People's Republic of China on Special Equipment Safety," and the "Regulations on Urban Gas Management," as well as standards and specifications such as the "Norms for Gas Engineering Projects." The major hidden dangers referred to in this standard are those that exist in the production and operation process of gas operators, with a relatively high degree of harm and the potential to cause mass casualties or significant economic losses. Urban gas management departments of local people's governments at or above the county level may, in the course of carrying out gas safety supervision and management work, identify and determine major hidden dangers in urban gas operation safety in accordance with this standard, and urge gas operators to implement the responsibility for hidden danger rectification and eliminate hidden dangers in a timely manner in accordance with the law and regulations. This includes various situations where gas operators are judged to have major hidden dangers in safety production management, gas plant station safety management, gas pipeline and pressure regulating facility safety management, gas cylinder safety management, and safety inspections of gas users.

2. Policy interpretation

The issuance of the "Criteria for Determining Major Safety Hazards in Urban Gas Operation" is aimed at guiding local governments in strengthening the control of safety risks and the investigation and management of hidden dangers in urban gas operations, and preventing the occurrence of major and especially serious accidents. Prior to this, the gas industry had long lacked national-level criteria for determining related accident hidden dangers. The issuance of this document is of important guiding significance for gas enterprises to conduct self-inspections of safety risks and hidden dangers, and to urge the competent authorities to perform their regulatory duties. The main major hidden danger situations in the "Criteria for Determining Major Safety Hazards in Urban Gas Operation" are derived from laws, regulations, and standards such as the "Law of the People's Republic of China on Work Safety," the "Law of the People's Republic of China on Special Equipment Safety," the "Regulations on Urban Gas Management," and the "Norms for Gas Engineering Projects." These are not new gas safety production requirements set in this instance. For example, illegal acts explicitly stipulated in the "Law of the People's Republic of China on Work Safety," such as the failure to establish a safety risk grading control system, the failure to establish a system for the investigation and management of accident hidden dangers, and the failure to formulate an emergency rescue plan for production safety accidents, are all included. These are situations that would be judged as major hidden dangers in the safety production management of gas operators. Given the limited number of situations listed in the criteria, Article 10, the catch-all provision, clearly states: "Other serious violations of urban gas operation laws, regulations, and standards that pose a significant hazard, are likely to cause mass casualties, or may result in substantial economic losses are deemed major hidden dangers." This provision can provide guidance for the identification and determination of other major hidden dangers in practice.

Energy Utilization

(I) "Natural Gas Utilization Policy (Draft for Comments)"

1. Main policy contents

Based on different gas consumption characteristics, natural gas users are divided into urban gas, industrial fuel, natural gas power generation, natural gas chemical industry, and other users. Taking into account the social, environmental, and economic benefits of natural gas utilization, as well as the gas usage characteristics of different users, natural gas users are also classified into four categories: priority, permissible, restricted, and prohibited.

(1) Priority category.

Urban and rural gas: Natural gas used for cooking and domestic hot water by urban residents; natural gas used by public service facilities; natural gas used by centralized heating users; rural "coal-to-gas" clean heating projects that have been included in the national planning and programs, have secured gas sources, and have acceptable gas prices, and have strictly followed the principle of "determining the scale of conversion based on the availability of gas" to complete construction. Industrial fuel: Natural gas-fueled interruptible industrial users. Natural gas power generation: Natural gas peaking power station projects with secure gas sources and economic sustainability; natural gas combined heat and power (CHP) projects; solar thermal power generation projects with supplementary combustion. Other users: Natural gas distributed energy projects; ocean-going transport, engineering, and government vessels as well as marine engineering equipment for the development, utilization and protection of the ocean, and transport, engineering, and government vessels and equipment using liquefied natural gas (LNG) as the single fuel in inland rivers, lakes, and coastal areas; cargo trucks and intercity passenger vehicles fueled by LNG; integrated oil, gas, electricity, and hydrogen energy supply projects, and demonstration projects for terminal natural gas hydrogen blending under the premise of ensuring safety, etc., which are new forms of natural gas utilization.

(2) Permissible category.

Urban and rural gas: Households using decentralized heating in urban built-up areas without centralized heating; newly added rural "coal-to-gas" clean heating projects that have been included in the national planning and programs, have secured gas sources, and have acceptable gas prices, and have strictly followed the principle of "determining the scale of conversion based on the availability of gas" for implementation. Industrial fuel: Projects substituting natural gas for oil and liquefied petroleum gas (LPG) in industrial sectors such as building materials, machinery and electronics, light textile, petrochemical and chemical, and metallurgy. New projects using natural gas as fuel in industrial sectors such as building materials, machinery and electronics, light textile, petrochemical and chemical, and metallurgy. Projects substituting natural gas for coal with better environmental and economic benefits in industrial sectors such as building materials, machinery and electronics, light textile, petrochemical and chemical, and metallurgy. Projects for replacing fuel in industrial boilers with natural gas in the central urban areas of cities. Natural gas power generation: Natural gas power generation projects other than those in the priority and restricted categories. Natural gas chemical industry: Natural gas hydrogen production projects for supporting the hydrogenation units of refineries and chemical enterprises.

(3) Restricted category.

Urban and rural gas: Rural "coal-to-gas" clean heating projects other than those in the priority and permitted categories. Natural gas power generation: Base-load gas power generation projects in the fourteen major coal bases such as Shaanxi, Inner Mongolia, Shanxi, and Xinjiang (excluding coalbed methane (coal mine gas) power generation projects). Natural gas chemical industry: Facilities for producing methanol from natural gas and for producing downstream products from methanol, projects for replacing coal with natural gas to produce methanol; small-scale C1 chemical projects with methane as the raw material and primary products including acetylene, chloromethane, etc.; synthetic ammonia and nitrogen fertilizer projects using natural gas as the raw material, "coal-to-gas" projects for synthetic ammonia plants; natural gas hydrogen production projects other than those in the permitted category.

Safeguard measures include the following aspects:

Ensure a balance between supply and demand. Fully leverage the decisive role of the market in resource allocation while better leveraging the role of the government. Build a unified national market and accelerate the construction of the natural gas market system. The National Development and Reform Commission and the National Energy Administration will coordinate and balance the supply and demand of natural gas overall, optimize the import and sources of natural gas resources, guide and promote the efficient utilization of natural gas across the country, regulate and control the overall balance between supply and demand, and ensure the coordinated development of production, supply, storage, and marketing. The Development and Reform Commissions and Energy Bureaus of provinces (regions, municipalities) should, based on the implementation of natural gas resources and the planning and construction of regional pipeline networks and gas storage capacity, in conjunction with energy-saving and emission reduction targets, follow the priority order of natural gas utilization to earnestly carry out natural gas utilization work. They should strengthen demand-side management, optimize the gas consumption structure, develop incremental users in an orderly manner, enhance market construction, improve the efficiency of resource allocation, and ensure the balance between supply and demand.

Accelerate the construction of infrastructure. Plan comprehensively and layout natural gas infrastructure construction moderately ahead of time, enhance the interconnectivity and fair openness of infrastructure, and expedite the perfection of the natural gas "national one network" to improve the flexibility of transportation and distribution.

Ensure safe and stable supply. Local governments, natural gas resource supply companies, natural gas sales companies, natural gas infrastructure operators, and urban gas distributors must jointly ensure the safe and stable supply of natural gas, minimizing the impact of accidental supply disruptions on users. Urban gas companies and other direct supply and purchase users should sign supply contracts with upstream suppliers and use gas in accordance with the contracts.

Efficient and comprehensive utilization. On the basis of strictly adhering to the natural gas utilization policy, the application of advanced processes, technologies, and equipment is encouraged, with a focus on accelerating the phase-out of outdated natural gas utilization capacities and fostering the development of high-efficiency utilization projects. Efforts should be made to strengthen the coordinated scheduling of natural gas resources, ensuring the supply of gas for the production of potash and other fertilizers while implementing staggered production. Companies that use natural gas to produce fertilizers are encouraged to adopt coal-based technologies. The utilization of shale gas and coalbed methane (coal mine gas) on-site (for residential use, power generation, etc.) is encouraged, and connection to the pipeline network is permitted if it meets the national quality standards for commercial natural gas. In cases where connection to the pipeline network is not feasible, processing into LNG (liquefied natural gas) or CNG (compressed natural gas) for consumption and transportation is allowed. Measures should be taken to increase the commercialization rate of natural gas, boost the supply of commercial gas, and strictly prohibit wasteful venting. Natural gas with a high CO2 content (over 20%) can be comprehensively developed and utilized according to its characteristics. The utilization of cold energy from natural gas is also encouraged.

Strengthen industry regulation. Improve the industry regulatory system and enhance regulation over the entire natural gas industry chain. Upstream gas supply enterprises should implement pricing policies, taking into account the gas source structure, supply costs, market supply and demand, and other factors to reasonably determine the wholesale price level of natural gas at the gate station. Local governments should assess and supervise urban gas companies to ensure compliance with the requirements of the "Regulations on Urban Gas Management", providing gas users with a continuous, stable, and safe supply of gas that meets national quality standards. They should also standardize the entry barriers for urban gas companies, promote mergers and reorganizations, and optimize integration. Strengthen the regulation of fair and open access to infrastructure to promote orderly competition in the natural gas market and ensure the secure supply of resources. Reduce the levels of pipeline transmission and gas distribution, and encourage direct supply and direct sales of natural gas.

Improve the pricing mechanism. Implement the linkage mechanism between upstream and downstream natural gas prices effectively. Guide local governments in establishing and improving a terminal sales price formation mechanism that can rise and fall flexibly to reflect supply and demand changes, better ensuring the safe and stable supply of natural gas. Encourage regions with significant seasonal differences in natural gas consumption to actively implement differentiated pricing policies such as seasonal pricing and interruptible gas pricing, guiding rational natural gas consumption and improving the efficiency of natural gas utilization. Support exploration and innovation in natural gas trade mechanisms and contract models.

2. Policy interpretation

In 2012, the National Development and Reform Commission announced the "Natural Gas Utilization Policy (2012)". Over the past 11 years, there have been significant changes in the international and domestic energy development situation, natural gas supply and demand pattern, price level, etc. Against this background, the National Energy Administration issued the "Natural Gas Utilization Policy (Draft for Comment)" (hereinafter referred to as the "Draft for Comment"), reflecting the new direction of natural gas development in the future. Compared with before, this "Draft for Comment" has made major adjustments in the field of natural gas utilization and the order of utilization. The "Draft for Comment" still classifies natural gas into priority, permissible, restricted, and prohibited categories. However, there have been a series of changes in the content of various types of natural gas utilization within the same classification.

In the field of natural gas utilization, the "Draft for Comment" adjusts "urban gas" users to "urban and rural gas" users. Since the "Thirteenth Five-Year Plan", with the increase in China's rural "coal-to-gas" efforts, pipeline natural gas and LNG have gradually extended to counties and towns outside the city. This change represents that the policy itself no longer emphasizes the priority of large and medium-sized cities, which is more in line with the industry characteristics that urban and rural gas has entered the stage of market penetration.

In the order of natural gas utilization, the priority category has undergone the following changes: The "Draft for Comments" no longer lists "interruptible natural gas hydrogen production" as a priority category. Instead, it lists the terminal natural gas hydrogen blending demonstration project within the oil, gas, electricity, and hydrogen integrated energy supply project as a priority, giving full recognition to this new business format. Secondly, the "Draft for Comments" retains the priority category of cogeneration and adds "natural gas peaking power station projects with secured gas sources and economic sustainability" to the priority list. The role of natural gas peaking power generation is expanding within the new type of power system. In addition, the "Draft for Comments" also adjusted the content related to the development of natural gas in the water transport field. The new policy increases the use of natural gas in ocean transport and marine development.

In the permitted category, the "Draft for Comments" policy lists newly added rural "coal-to-gas" clean heating projects that are included in the national planning and programs, have secured gas sources, and have acceptable gas prices, and have strictly followed the principle of "determining the scale of conversion based on the availability of gas" for implementation, as permitted. This indicates that the mandatory requirements from the national and local governments for such projects will be reduced, and subsidies will also be correspondingly decreased. Downstream natural gas companies should abandon the mindset of "renovate first, subsidize later" and control the development pace of such users.

Regarding the restricted category, the policy of the "Draft for Comments" has changed "base-load gas-fired power generation projects in the fourteen major coal bases such as Shaanxi, Inner Mongolia, Shanxi, and Xinjiang (excluding coalbed methane (coal mine gas) power generation projects)," "new construction or expansion of facilities for producing methanol from natural gas and downstream products from methanol production," and "projects substituting natural gas for coal in methanol production" from the "prohibited category" to the "restricted category." This represents a breakthrough. Since these are now in the restricted category, it means that gas-fired power plants can be built in coal-rich areas, with greater consideration given to environmental factors and the need for emission reductions. This reflects a new development philosophy and represents progress in thinking.

Regarding the prohibited category, the "Draft for Comments" only lists the production of synthetic ammonia using natural gas atmospheric intermittent conversion process as prohibited. Other areas are not included, which is a significant reduction compared to the "Natural Gas Utilization Policy (2012)." This further relaxes the restrictions on natural gas utilization.

(II) "Guiding Opinions of the National Development and Reform Commission and Other Departments on Promoting Green Innovation and High-quality Development of the Refining Industry"

1. Main policy contents

Accelerate green and low-carbon development and guide decarbonization in the refining process. Actively guide the substitution of clean energy and green electricity, promote the substitution of existing coal-fired captive power plants (boilers) with clean energy, and encourage enterprises to actively explore and research the application of solar heating in the refining process. Encourage enterprises to adopt public electricity, centralized heating, natural gas, distributed new energy power generation, and integrated source-grid-load storage methods according to local conditions to gradually reduce the proportion of coal consumption, increase the level of renewable energy consumption, and improve the system operation efficiency and comprehensive benefits of power source development. Encourage the application of low-carbon production processes such as short process, enhanced reaction process, and catalytic cracking waste heat to generate ultra-high-pressure steam technology. Strengthen the coordinated control of methane and volatile organic compounds (VOCs).

2. Policy interpretation

At present, the Chinese economy has shifted from a stage of rapid growth to a stage of high-quality development, and the internal and external environment, supply and demand structure, and people's needs for a better life have all put forward new requirements for the refining industry. The "Guiding Opinions of the National Development and Reform Commission and Other Departments on Promoting Green Innovation and High-quality Development of the Refining Industry" revolves around the goal of green innovation and high-quality development of the refining industry. As such, it deploys four key tasks and provides a relatively comprehensive overall plan and policy guidance for the industry. In terms of industrial optimization and upgrading, it further optimizes the layout of refining capacity in relevant regions, promotes the orderly concentration of new refining projects to petrochemical industrial bases, and achieves intensive and agglomerated development. In terms of efficient use of energy resources, it guides refining enterprises to improve energy efficiency, promotes a batch of advanced energy-saving technologies in the refining industry, and realizes resource recycling through the reuse of waste plastics and waste oils. In terms of green and low-carbon development, it puts forward clear requirements for the refining industry to reduce pollution and carbon from multiple dimensions such as the overall refining process, unit process technology, CCUS demonstration application, hydrogen production and use optimization, green hydrogen refining, and carbon emission management. In terms of scientific and technological innovation, it clearly tasks the development of new refining technologies, accelerates the research and development of energy-saving and carbon-reducing refining unit technologies and equipment, and strives to deeply build a characteristic scientific and technological innovation system to achieve a high level of self-reliance. Among them, in the key tasks of accelerating the green and low-carbon development of the refining industry, it encourages enterprises to use natural gas as the energy supply in the refining process, thereby reducing the proportion of coal consumption. This is in line with China's overall goal of achieving peak carbon emissions by 2030 and helps the green development of the refining industry.

Natural Gas Prices

"Notice of the National Development and Reform Commission on the Approval of Cross-province Natural Gas Pipeline Transportation Prices"

1. Main policy contents

The National Development and Reform Commission has conducted a cost review for the pricing of cross-provincial (autonomous regions and municipalities directly under the Central Government) natural gas pipelines operated by the National Petroleum and Natural Gas Pipeline Network Group Co., Ltd., and has accordingly determined the pipeline transportation prices for the four price zones: Northwest, Northeast, Central and Eastern, and Southwest. The transportation rate for the Northwest pricing zone is set at 0.1262 yuan per thousand cubic meters per kilometer (including a 9% value-added tax, the same applies hereinafter), the Northeast pricing zone at 0.1828 yuan per thousand cubic meters per kilometer, the Central and Eastern pricing zone at 0.2783 yuan per thousand cubic meters per kilometer, and the Southwest pricing zone at 0.3411 yuan per thousand cubic meters per kilometer. The National Oil and Gas Pipeline Network Group Co., Ltd. shall calculate and determine the specific pipeline transportation price table based on the transportation distance between the natural gas entry and exit points in each pricing zone, and shall make it public through the company's official website or designated platform. The above prices will be effective from January 1, 2024.

2. Policy interpretation

The issuance of the "Notice of the National Development and Reform Commission on Determining Cross-Provincial Natural Gas Pipeline Transportation Prices" is to implement the decision-making deployment of the Party Central Committee and the State Council on deepening the reform of the petroleum and natural gas market system and accelerate the construction of a "national one-network." For the first time, the "Notice of the National Development and Reform Commission on Determining Cross-Provincial Natural Gas Pipeline Transportation Prices" has determined the cross-provincial natural gas pipeline transportation prices operated by the national pipeline company by region, and has clearly defined the cross-provincial natural gas pipeline transportation rates for the northwest, northeast, central and eastern, and southwest price zones. The national pipeline company is required to calculate the specific prices for pipeline transportation based on the transportation rates of each price zone and the transportation distance between the natural gas inlet and outlet, and make them public. This pricing is the first pricing since the reform of the natural gas pipeline operation mechanism, and it is also the first time that the state has determined the cross-provincial natural gas pipeline transportation prices according to the "one zone, one price" principle. After the price is determined, the cross-provincial natural gas pipeline transportation rates operated by the national pipeline company have been greatly reduced from 20 to 4, constructing a relatively unified transportation price structure, breaking the line division of too many transportation rates on the operation of the pipeline network, which is conducive to achieving interconnection and fair opening of pipeline network facilities, accelerating the formation of a "national one-network," promoting the free flow of natural gas resources and market competition, and helping the high-quality development of the industry.

The year 2023 was the first year after the lifting of COVID-19 pandemic control and prevention measures, and also the first year for fully implementing the guiding principles of the Party's 20th National Congress. Natural gas, as a low-carbon fossil fuel with a solid development foundation and huge development potential, occupies an important position in the national energy system. The National Development and Reform Commission and other relevant departments have cumulatively issued 12 related policies regarding green development, energy governance, energy utilization, and natural gas prices, thus planning and guiding the development of the national energy sector.

In terms of green development, a total of four policies have been issued, with the focus mainly on the green and low-carbon development during the process of natural gas exploration and utilization. On the one hand, these policies promote the integrated development of oil and gas exploration and development with new energy, emphasizing ecological priority, adaptation to local conditions, and diversified integration. They are committed to forming a new development pattern in the upstream oil and gas sector that features the integration with new energy and new industries and the complementary use of multiple energy sources. Natural gas has become the "best partner" that coexists and prospers with renewable energy sources such as wind and solar energy. On the other hand, the policies strengthen the updating, upgrading, and application of energy-saving standards in the energy field, thereby promoting the standardization work on energy conservation and efficiency in the energy field and effectively enhancing the application and implementation of energy-saving standards. Energy conservation, carbon reduction, and efficiency improvement are the basic principles, important supports, and key means for achieving the "dual carbon" goals.

In terms of energy governance, policies mainly focus on two aspects: ensuring supply and stabilizing prices, as well as enabling digital and intelligent empowerment. Firstly, amid the profound changes unseen in a century, it is urgent to enhance the resilience and security level of China's energy industry chain and supply chain. In 2023, both energy supervision and guidance work gave top priority to the stable and secure supply of energy. On the one hand, they consolidated the role of fossil energy as the foundation for a bottom-line guarantee, accelerated the construction of multiple important natural gas production capacity projects, and strengthened oil and gas exploration, development, as well as the increase of reserves and production. On the other hand, they strengthened the monitoring, analysis, and early warning of the supply and demand situation of energy sources such as thermal coal, electricity, and natural gas, so as to firmly hold the "energy rice bowl" in our own hands in all aspects. Secondly, the policies promoted the in-depth integration of digital technologies with the development of the energy industry, strengthened the construction of new infrastructure that combines traditional energy with digital and intelligent technologies, unleashed the potential value of energy data elements, and drove the formation of new quality productivity in the energy sector.

In terms of energy utilization, the "Natural Gas Utilization Policy (Draft for Comments)" released after an 11-year interval provides a new basis for encouraging, guiding, and regulating the downstream use of natural gas. Regarding natural gas pricing, the implementation of a "one-region-one-price" system for determining the transportation prices of cross-provincial natural gas pipelines helps to break the excessive segmentation of pipeline operations caused by numerous transportation rates. This measure is conducive to achieving interconnectivity and fair access to pipeline facilities and accelerates the formation of a "national one-network."

Local Natural Gas Policy

In 2023, local governments issued a total of 37 policies related to the natural gas industry, mainly involving natural gas prices, power grid connection, and pipeline transportation. The local natural gas-related policies in 2023 are shown in Table 2.

A thorough examination and control of policy content were conducted, and an analysis of local natural gas policies in 2023 was performed. With the help of Nvivo12 software, a word cloud is formed to visually show the policy focus, as shown in Fig. 2.

Table 2 Local natural gas policies in 2023

Province/ City	Document name	Release date
Beijing	"Beijing's 14th Five-Year Plan for Gas Development"	2023–01-01
	"Notice on Adjusting the Sale Price of Natural Gas for Non-residential Use in Our City"	2023–11-28
Shanghai	"Notice on Adjusting the Benchmark Price for Non-residential Natural Gas Sales in Our City"	2023–03-31
	"Shanghai Natural Gas Pipeline Distribution Price Management Measures"	2023–10-10
Tianjin	"Notice on Further Strengthening the Price Supervision of Gasification Services at Imported Liquefied Natural Gas Receiving Stations"	2023–01-10
	"Notice on Announcing the Non-residential Natural Gas Sales Price of the City Gas Network in the Non-heating Season of 2023"	2023–05-30
	"Notice on Announcing the Gasification Service Price of Our City's Imported Liquefied Natural Gas Receiving Station"	2023–09-28
	"Notice on Announcing the Non-residential Natural Gas Sales Price of the City Gas Network in the Heating Season of 2023–2024"	2023–12-04
Guizhou	Notice on Issuing the "Implementation Plan for Carbon Peak in the Energy Field of Guizhou Province"	2023–01-10
	"Notice on Reducing the Price of Non-residential Gas and Related Matters"	2023–05-20
	Notice on Issuing the "Management Measures for the Distribution Price of Urban Pipeline Gas in Guizhou Province"	2023–07-06
Hunan	"Notice on Improving the Price Linkage Mechanism of Natural Gas Upstream and Downstream in Hunan Province"	2023–11-16
Jiangsu	"Implementation Opinions on Establishing and Improving the Price Linkage Mechanism of Natural Gas Upstream and Downstream"	2023–09-28
Shandong	Notice on Issuing the "Three-Year Action Plan for Green, Low-Carbon, High-Quality Energy Development in Shandong Province (2023–2025)" and the "Key Work Tasks for Green, Low-Carbon, High-Quality Energy Development in Shandong Province in 2023"	2023–03-08
Shaanxi	"Notice on Issuing the '14th Five-Year' Comprehensive Energy Conservation and Emission Reduction Implementation Plan"	2023–02-08
	Baoji City "Notice on Linkage Adjustment of Non-residential Pipeline Natural Gas Terminal Sales Price in Our City"	2023–08-08
	Notice on Issuing the "Interim Measures for the Construction and Operation Management of Long-distance Oil and Gas Pipelines in Shaanxi Province"	2023–10-26
Zhejiang	"Notice on Matters Related to the Provincial Gate Station Price of Natural Gas"	2023–01-06

(continued)

Table 2 (continued)

Province/City	Document name	Release date
Anhui	"Notice on Reducing the Short-distance Pipeline Transportation Price of Natural Gas in Some Areas of the Province and Other Related Matters"	2023–12-27
Fujian	Notice on Issuing the "Fuzhou City Pipeline Natural Gas Price Linkage Mechanism"	2023–05-13
Shanxi	"Notice of the People's Government of Shanxi Province on Issuing the Implementation Plan for Carbon Peak in Shanxi Province"	2023–01-09
	"Notice on the Issuance of Shanxi Province's Long-Distance Oil and Natural Gas Pipeline Construction and Protection Emergency Plan"	2023–04-07
Sichuan	"Notice on the Issuance of the Implementation Plan for Carbon Peak in the Energy Sector of Sichuan Province"	2023–01-11
Hubei	"Notice on the Establishment and Improvement of the Natural Gas Upstream and Downstream Price Linkage Mechanism"	2023–06-21
Hainan	"Notice on the Issuance of Hainan Province's 14th Five-Year Comprehensive Energy Conservation and Emission Reduction Work Plan"	2023–01-20
Chongqing	"Notice on the Issuance of Chongqing's Natural Gas Operation Dispatch Management Measures"	2023–09-04
	"Notice on the Natural Gas Sales Price in the Central Urban Area during the 2023 Heating Season"	2023–10-30
Inner Mongolia	"Notice on Adjusting the Sales Price of Pipeline Natural Gas for Residents and Non-residents"	2023–04-04
Xinjiang	"Notice on the Issuance of the Autonomous Region's Natural Gas (Coalbed Methane) Exploration and Development Incentive Policy"	2023–07-04
Shanghai	"Notice on Adjusting the Base Price of Non-residential Natural Gas Sales in the City"	2023–03-31
Henan	"Notice on the Issuance of Henan Province's New Energy and Renewable Energy Development '14th Five-Year' Plan"	2023–04-17
Guangxi	"Notice on the Issuance of Guangxi Zhuang Autonomous Region's Carbon Peak Implementation Plan"	2023–01-30
Ningxia	"Notice on the Issuance of the Interim Regulations on the Management of Bottled Liquefied Petroleum Gas Distribution in Ningxia Hui Autonomous Region"	2023–08-31
Yunnan	"Notice on the Issuance of Yunnan Province's Green Energy Development '14th Five-Year' Plan"	2023–01-04
Hebei	"Notice on the Issuance of Hebei Province's 2023–2024 Heating Season Natural Gas Supply Work Plan"	2023–04-18

(continued)

Table 2 (continued)

Province/City	Document name	Release date
	"Notice on Adjusting the Sales Price of Pipeline Natural Gas for Residents in the Main Urban Area"	2023–06-14
Jilin	"Notice on the Issuance of the Method for Linking Upstream and Downstream Prices of Urban Pipeline Natural Gas in Jilin City"	2023–03-29

Data source Provincial and municipal governments websites, Development and Reform Commission websites.

Fig. 2 Word cloud of local natural gas policies

From the word cloud in Fig. 2, it can be seen that "natural gas prices" and "natural gas pipelines" are the focus of policy attention. Therefore, local natural gas policies are divided into two major categories: natural gas pricing policies and natural gas pipeline policies. The terms "short-haul," "linkage," and "adjustment" reflect the policy direction in 2023, and the newly added keyword "gasification" reflects the new policy on gasification service prices in 2023. To further interpret the policies, the natural gas-related policies of various provinces and cities are classified based on the word cloud, resulting in Table 3.

Table 3 Classification of local natural gas policies in 2023

Policy theme	Policy category	Document name	Release time
Natural gas prices	Gasification service prices	Tianjin: "Notice on Further Strengthening the Price Supervision of Imported Liquefied Natural Gas Receiving Station Gasification Services"	2023–01-10
		Tianjin: "Notice on Announcing the Gasification Service Price of Our City's Imported Liquefied Natural Gas Receiving Station"	2023–09-28
		Zhejiang: "Notice on Natural Gas Provincial Gate Station Price and Other Related Matters"	2023–01-06
		Shanghai: "Shanghai City Pipeline Natural Gas Distribution Price Management Measures"	2023–10-10
		Ningxia: Notice on Issuing "Interim Regulations on the Distribution Management of Bottled Liquefied Petroleum Gas in Ningxia Hui Autonomous Region"	2023–08-31
Natural gas prices	Non-residential gas Prices	Tianjin: "Notice on Announcing the Non-residential Natural Gas Sales Price of the City Gas Network for the 2023–2024 Heating Season"	2023–12-04
		Beijing: "Notice on Adjusting the Non-residential Natural Gas Sales Price in Our City"	2023–11-28
		Guizhou: "Notice on Reducing Non-residential Gas Prices and Related Matters"	2023–05-20
		Shanghai: "Notice on Adjusting the Non-residential Natural Gas Sales Base Price in Our City"	2023–03-31

(continued)

Table 3 (continued)

Policy theme	Policy category	Document name	Release time
		Tianjin: "Notice on Announcing the Non-residential Natural Gas Sales Price of the City Gas Network for the 2023 Non-heating Season"	2023–05-30
		Hubei: "Notice on Establishing and Improving the Natural Gas Upstream and Downstream Price Linkage Mechanism"	2023–06-21
		Hunan: "Notice on Improving the Natural Gas Upstream and Downstream Price Linkage Mechanism in Hunan Province"	2023–11-16
		Jiangsu: "Implementation Opinions on Establishing and Improving the Natural Gas Upstream and Downstream Price Linkage Mechanism"	2023–09-28
		Shaanxi: Baoji City "Notice on Linkage Adjustment of Non-residential Pipeline Natural Gas Terminal Sales Price in Our City"	2023–08-08
		Fujian: Notice on Issuing "Fuzhou City Pipeline Natural Gas Price Linkage Mechanism"	2023–05-13
		Inner Mongolia: "Notice on Adjusting the Sales Price of Pipeline Natural Gas for Residents and Non-residents"	2023–04-04
	Residential gas prices	Chongqing: "Notice on the Natural Gas Sales Price in the Central Urban Area for the 2023 Heating Season"	2023–10-30
		Hebei: "Notice on Issuing the Hebei Province 2023–2024 Heating Season Natural Gas Supply Work Plan"	2023–04-18

(continued)

Table 3 (continued)

Policy theme	Policy category	Document name	Release time
		Hebei: "Notice on Adjusting the Sales Price of Residential Pipeline Natural Gas in the Main Urban Area"	2023–06-14
		Inner Mongolia: "Notice on Adjusting the Sales Price of Pipeline Natural Gas for Residents and Non-residents"	2023–04-04
		Jilin: Notice on Issuing "Jilin City Urban Pipeline Natural Gas Upstream and Downstream Price Linkage Measures"	2023–03-29
Natural gas Pipeline transportation	Pipeline transportation prices	Anhui: "Notice on Lowering the Price of Short-Distance Natural Gas Pipeline Transportation in Some Areas of the Province"	2023–12-27
		Guizhou: Notice on Issuing "Guizhou Province Urban Pipeline Gas Distribution Price Management Measures"	2023–07-06
	Pipeline construction	Shaanxi: Notice on Issuing "Shaanxi Province Petroleum and Natural Gas Long-Distance Pipeline Construction and Operation Management Measures (Interim)"	2023–10-26
	Pipeline protection	Shanxi: "Notice on Issuing Shanxi Province Petroleum and Natural Gas Long-Distance Pipeline Construction and Protection Emergency Plan"	2023–04-07
Others		Beijing: "Beijing's 14th Five-Year Plan for Gas Development"	2023–01-01
		Guizhou: Notice on Issuing "Guizhou Province Energy Sector Carbon Peak Implementation Plan"	2023–01-10

(continued)

Table 3 (continued)

Policy theme	Policy category	Document name	Release time
		Shandong: Notice on Issuing "Shandong Province Energy Green Low-Carbon High-Quality Development Three-Year Action Plan (2023–2025)" and "Shandong Province Energy Green Low-Carbon High-Quality Development 2023 Key Work Tasks"	2023–03-08
		Shaanxi: "Notice on Issuing the '14th Five-Year' Energy Conservation and Emission Reduction Comprehensive Work Implementation Plan"	2023–02-08
		Shanxi: "Notice of Shanxi Provincial People's Government on Issuing Shanxi Province Carbon Peak Implementation Plan"	2023–01-09
		Sichuan: Notice on Issuing "Sichuan Province Energy Sector Carbon Peak Implementation Plan"	2023–01-11
		Hainan: "Notice on Issuing Hainan Province '14th Five-Year' Energy Conservation and Emission Reduction Comprehensive Work Plan"	2023–01-20
		Xinjiang: Notice on Issuing "Autonomous Region Natural Gas (Coalbed Methane) Exploration and Development Incentive Policy"	2023–07-04
		Henan: Notice on Issuing "Henan Province New Energy and Renewable Energy Development '14th Five-Year' Plan"	2023–04-17

(continued)

Table 3 (continued)

Policy theme	Policy category	Document name	Release time
		Guangxi: "Notice on Issuing Guangxi Zhuang Autonomous Region Carbon Peak Implementation Plan"	2023–01-30
		Yunnan: "Notice on Issuing Yunnan Province Green Energy Development '14th Five-Year' Plan"	2023–01-04

Local Natural Gas Prices

Some provinces and cities in China have adjusted natural gas prices. The pricing policies introduced are mainly divided into gasification service pricing policies, non-residential gas pricing policies, and residential gas pricing policies.

(I) Policy contents

1. Gasification service pricing policy

The gasification service price refers to the fees charged by the receiving station for gasifying liquefied natural gas (LNG) and providing related necessary services to users. This includes costs related to the unloading, temporary storage, and gasification of LNG. The temporary storage period generally does not exceed 45 days. To implement the national requirements for improving the pricing mechanism of gasification services at imported LNG receiving stations, Tianjin has issued a notice on the management of gasification service prices at natural gas receiving stations.

According to the notification titled "Notice on the Announcement of the Gasification Service Prices at the Import Liquefied Natural Gas (LNG) Receiving Stations in Our City" issued by Tianjin Municipality, the gasification service prices at the import LNG receiving stations are as follows: The gasification service price at the Sinopec Tianjin LNG Co., Ltd. receiving station is set at 0.20 yuan per cubic meter. The price at the National Oil & Gas Pipeline Network Group Tianjin LNG Co., Ltd. receiving station is 0.30 yuan per cubic meter. The Beijing Gas Group (Tianjin) LNG Co., Ltd. receiving station charges 0.32 yuan per cubic meter.

The "Measures of Shanghai Municipality on the Administration of Gas Distribution Price of Pipeline Natural Gas" stipulate that the annual distribution volume shall be determined based on the annual sales volume of gas companies and the load rate of the pipeline network. If the pipeline network load rate reaches or exceeds 50%, the distribution volume shall be verified according to the actual distribution volume; if the pipeline network load rate is below 50%, the distribution volume shall be calculated and determined based on a 50% load rate. The gas distribution price shall, in principle, be reviewed and adjusted every three years. However, if there are significant changes in pipeline facility investment, distribution volume, costs, etc., an early review and adjustment may be made. These measures came into effect on December 1, 2023, and will remain valid until November 30, 2028.

2. Non-residential natural gas pricing policy

Beijing, Shanghai, Chongqing, Guizhou, Jiangsu and other provinces and cities have adjusted the prices for non-residential gas use, and a total of 9 documents have been issued. The natural gas prices in many places are showing an upward trend due to upstream price fluctuations or the arrival of the heating season, while the natural gas prices for non-residential users in Shanghai are showing a downward trend.

The Beijing "Notice on Adjusting the Sales Price of Non-Residential Natural Gas in Our City" stipulates: The sales price of non-residential pipeline natural gas will increase by 0.43 yuan/m^3 from November 19, 2023 to March 15, 2024. During the heating season, the adjusted price for gas used for power generation is 2.70 yuan/m^3, the price at the compressed natural gas mother station is 2.76 yuan/m^3, and the price for gas used for heating and cooling, and commercial gas in the six urban districts is 2.88 yuan/m^3 and 3.30 yuan/m^3 respectively. In other areas, it is 2.64 yuan/m^3 and 3.06 yuan/m^3. During the non-heating season, the adjusted price for gas used for power generation is 2.27 yuan/m^3, the price at the compressed natural gas mother station is 2.33 yuan/m^3, and the price for gas used for heating and cooling, and commercial gas in the six urban districts is 2.45 yuan/m^3 and 2.87 yuan/m^3 respectively. In other areas, it is 2.21 yuan/m^3 and 2.63 yuan/m^3. The price increase will be canceled on March 16, 2024.

The Shanghai "Notice on Improving the Upstream and Downstream Price Linkage Mechanism for Non-Residential Natural Gas Users in Our City" stipulates that the base price for non-residential natural gas sales Shanghai is uniformly reduced by 0.30 yuan/m^3. The adjusted base price for natural gas for Caojing Thermal Power in the city's pipeline company's direct supply users is 3.44 yuan/m^3, the adjusted base price for natural gas power generation is 3.44 yuan/m^3, the adjusted base price for the chemical industry area is 3.69 yuan/m^3, and the adjusted base price for others is 3.69 yuan/m^3. For general non-residential users who use more than 5 million cubic meters of natural gas, the price of natural gas is 4.23 yuan/m^3; from 1.2 million to 5 million cubic meters, the price of natural gas is 4.71 yuan/m^3; below 1.2 million cubic meters, the price of natural gas is 4.88 yuan/m^3. The above prices will be implemented from April 1, 2023. The actual sales price is based on the benchmark price set by the government, and is determined by the supply company within a fluctuation range of 5%.

Tianjin City stipulates in the "Notice on Announcing the Non-Heating Season Urban Gas Pipeline Non-Residential Natural Gas Sales Price for 2023" that the non-heating season urban gas pipeline non-residential natural gas sales price for 2023 is: 3.55 yuan/m^3 for general industrial and commercial and other gas uses. Each natural gas operating company can independently set specific sales prices on the premise that they do not exceed the above prices. This regulation will be implemented from April 1, 2023.

Tianjin City stipulates in the "Notice on Announcing the Heating Season Urban Gas Pipeline Non-Residential Natural Gas Sales Price for 2023–2024" that the heating season urban gas pipeline non-residential natural gas sales price for 2023–2024 is: 4.11 yuan/m^3 for general industrial and commercial and other gas uses, and 3.82 yuan/m^3 for centralized heating gas. Each natural gas operating company can independently set specific sales prices on the premise that they do not exceed the above prices, taking into account factors such as the company's gas purchase cost and distribution price. The above prices will be implemented from November 1, 2023.

Guizhou Province stipulates in the "Notice on Reducing Non-Residential Gas Prices and Related Matters" that non-residential gas prices will be reduced. The non-residential gas price in Guiyang city area and the area sharing the same gas distribution network, which is priced at the provincial level, will be reduced from 3.96 yuan/m^3 to 3.72 yuan/m^3. For those priced at the city and county level, each should ensure that low-priced gas sources are prioritized to guarantee residential living gas on the basis of urgently calculating the non-residential gas source purchase cost (the amount that should be reduced in April should be considered in the calculation), and according to the local natural gas upstream and downstream price linkage mechanism, organize the implementation of non-residential gas price reduction work as soon as possible. The implementation period for this linkage cycle is from May 1, 2023 to October 31, 2023.

In the "Notice on Improving the Upstream and Downstream Natural Gas Price Linkage Mechanism in Hunan Province," Hunan Province stipulates the scope of linkage. The upstream and downstream natural gas price linkage (hereinafter referred to as gas price linkage) means that the terminal sales price is linked to the gas source procurement price (including transportation costs) of gas companies. The terminal sales price in the pricing area using piped natural gas consists of the average gas source procurement price, the pipeline transportation price, and the distribution price. The terminal sales price in the pricing area using only liquefied natural gas (LNG) or compressed natural gas (CNG) is composed of the distribution price and the average gas source procurement price. The average procurement price of piped natural gas refers to the weighted average price of all gas sources (including tax) procured by urban gas companies. The average procurement price of LNG and CNG refers to the weighted average price of all LNG and CNG procurement prices (including tax) and transportation costs procured by urban gas companies. In the same city (district, county) where there are multiple gas companies with unified pricing, the procurement price is determined by the weighted average of the procurement prices of all gas companies.

Taking the pricing area specified in the "Hunan Province Pricing Catalogue" as a unit, the gas price linkage mechanism will be activated when the fluctuation of the average gas procurement price reaches or exceeds 5% of the benchmark gate price. When the conditions for gas price linkage are met, the terminal sales price for residential gas is, in principle, adjusted once a year, while the terminal sales price for non-residential gas is adjusted twice a year. If the fluctuation of the average gas procurement price does not meet the activation criteria, the terminal sales price of natural gas will not be adjusted, and the increase or decrease will be carried over to the next adjustment cycle for increase or decrease calculation. When the benchmark gate price, pipeline transportation price, or distribution price is adjusted, the terminal sales price of natural gas will be adjusted synchronously, in the same amount and direction, without being restricted by the frequency of linkage. The terminal sales price is adjusted up or down according to the changes in the procurement price. The increase in the residential gas price linkage is subject to an upper limit management. In principle, the increase in the residential gas price linkage in this period shall not exceed 10% of the terminal sales price in the previous period. Any excess will be considered comprehensively, while there is no limit on the decrease. The non-residential gas price linkage is, in principle, adjusted up or down according to the changes in the average gas procurement price. When the upstream price rises too high, take social tolerance into consideration, and in accordance with the principle of balancing the interests of urban gas companies and consumers and maintaining stable economic operation, the adjustment amplitude of the residential and non-residential gas price linkage can be moderately controlled. When the market price continues to rise significantly, which may have a serious adverse impact on the normal life of residents and the stable operation of the economy, the linkage can be temporarily suspended. The amount that should have been adjusted but was not adjusted will be considered comprehensively in the next adjustment cycle.

The specific formula for natural gas upstream and downstream price linkage is as follows: The terminal sales price after the gas price linkage adjustment = the terminal sales price of the previous period + the price linkage adjustment amount.

The price linkage adjustment amount = (the average purchase price of the gas source in this period—the average purchase price of the gas source in the previous period) ÷ (1—the supply and sales difference rate).

The supply and sales difference rate (loss rate) is calculated according to the annual supply and the sales difference rate of the distribution price in the pricing area, and in principle, it shall not exceed 4%.

When the linkage mechanism is first established, the terminal sales price formula: terminal sales price = average purchase price of gas source (including transportation costs) + distribution price.

Among them, the gas price linkage is implemented by the provincial and municipal (prefectural) price authorities in accordance with the price management authority specified in the pricing catalogue.

The "Notice on Establishing and Improving the Natural Gas Upstream and Downstream Price Linkage Mechanism" in Hubei Province indicates that the terminal sales price in various places is linked to the procurement price (including transportation costs) of gas companies. The procurement price does not distinguish the form of gas source prices, and is determined in principle according to the weighted average price of all gas sources purchased by gas companies in the same region, including pipeline natural gas, liquefied natural gas (LNG), compressed natural gas (CNG), etc. When the procurement price of a gas company is significantly higher than the procurement price in the surrounding area or the average price of the main local gas source, it may not be linked or the linkage standard may be reduced. When the procurement price of gas sources outside the contract has a significant impact on the terminal sales price in this region, the principle of voluntary commission by users may be followed to implement a consignment purchase and sales price policy for the gas volume purchased outside the contract. The difference between the purchase and sales prices shall not exceed the gas distribution price in this region.

The terminal sales price for non-residential gas use is linked in principle on a quarterly or monthly basis. Those that were linked for half a year or more should gradually transition to quarterly linkage, and those with the capacity to do so can be linked on monthly basis. The linkage cycle of the terminal sales price for residential gas use should not exceed one year in principle, and those with a large price difference between the peak and off-peak seasons can be linked semi-annually or differentiated by peak and off-peak seasons. The linkage formula is as follows:

When the linkage mechanism is first established, the terminal sales price is determined by the following formula: Terminal sales price = weighted average procurement price + gas distribution price.

After the linkage mechanism is established, the adjustment of the terminal sales price is determined by the following formula: Terminal sales price = last period terminal sales price + price linkage adjustment amount.

Price linkage adjustment amount = (current period weighted average procurement price - last period weighted average procurement price) / (1 - supply and sales difference rate) ± last period's adjustment amount and deviation amount.

The supply and marketing difference rate is, in principle, determined at 4.5% for newly constructed pipelines and 3.5% for pipelines that have been in operation for three years (inclusive) or more.

Jiangsu Province pointed out in the "Implementation Opinions on Establishing and Improving the Natural Gas Upstream and Downstream Price Linkage Mechanism" that the scope of natural gas upstream and downstream price linkage is the terminal sales price and the procurement price (including transportation costs through pipelines, vehicles, ships, etc.) of gas companies being linked. The procurement price is determined in principle according to the weighted average price of all gas sources purchased by gas companies. When the procurement of market-oriented gas sources has a significant impact on the terminal sales price of non-residential areas in this region, the price difference model can be implemented for the terminal sales price of non-residential areas in accordance with the principle of voluntary entrustment by users, and the price difference between buying and selling should not be higher than the local gas distribution price.

The terminal sales price for non-residential gas use is linked in principle on a quarterly or monthly basis. The current linkage cycle is half a year or more, and it will gradually transition to quarterly linkage by 2025. Areas with the capacity to do so can be linked on a monthly basis. The terminal sales price for residential gas use can be linked semi-annually or annually. The terminal sales price is adjusted according to changes in procurement prices, and gas sources with lower prices are given priority to ensure residential gas use. Local governments can combine the actual situation and clearly link the terminal sales price with the actual procurement price of the previous period or the predicted procurement price of the current period.

For those linked to the actual procurement price, the gas company's gas purchase contract and invoice should be strictly audited, and the procurement price should be calculated based on the actual procurement cost incurred by the enterprise. For those linked to the predicted procurement price, a reasonable prediction of the procurement price should be made based on the future natural gas procurement quantity, price, and gas demand in the same period of previous years specified in the contract signed by the gas company. At the same time, a deviation verification mechanism should be established, and the difference between the predicted price and the actual procurement price should be taken into consideration in the later linkage.

The linkage formula is: When the linkage mechanism is first established, the terminal sales price = weighted average procurement price + gas distribution price.

After the linkage mechanism is established, for those linked to the actual procurement price, the terminal sales price = last period terminal sales price + (current period weighted average procurement price - last period weighted average procurement price) / (1 - supply and sales difference rate) ± last period's adjustment amount / current period's gas supply.

For those linked to the predicted procurement price, the terminal sales price = last period terminal sales price + (current period weighted average procurement price - last period weighted average procurement price) / (1 - supply and sales difference rate) ± last period's adjustment amount / current period's gas supply ± deviation amount / current period's gas supply.

The supply and sales difference rate is determined according to the actual supply and sales difference of the gas company in the previous year, and the maximum should not exceed 4%.

The adjustment of the terminal sales price for residential gas use should adhere to the principle of stability and prudence, with limits set on the extent of price increases to avoid unduly increasing the burden on residential gas users. Local governments may take into account the local residents' ability to bear costs and refer to the price adjustment situations of previous years to set upper limits on the amount or extent of price increases. In principle, the price of the first tier of residential gas should not be increased by more than 0.50 yuan in a single adjustment, while there is no limit on the extent of price decreases. When the market price of natural gas continues to rise significantly, which may have a seriously adverse impact on the normal life of residents and the stable operation of the economy, the linkage should be temporarily suspended. The amount that should have been adjusted but was not in this period may be included in the later linkage for comprehensive consideration or implemented in phases gradually. When the distribution price is adjusted, the terminal sales price should be adjusted accordingly, without being restricted by the linkage mechanism.

The notice issued by Fujian Province on the "Fuzhou City Pipeline Natural Gas Price Linkage Mechanism" points out that the sales price of pipeline natural gas for non-residential users is linked to the procurement price of natural gas. The adjusted sales price = base period sales price +linkage adjustment amount, where the linkage adjustment amount = average unit gas procurement price during the calculation period—average unit gas procurement price during the base period. The sales price for non-residential gas use is in principle adjusted every three months in a linkage adjustment cycle. When the composite gas purchase price changes by 5% within the cycle, the terminal sales price of natural gas for non-residential users is in principle linked to the gas procurement price.

In the "Notice on Adjusting the Sales Price of Pipeline Natural Gas for Residential and Non-residential Use," Inner Mongolia proposed: From April 1, 2023 to October 31, 2023, according to the price adjustment plan of PetroChina Natural Gas Sales Inner Mongolia Branch, the terminal sales price of non-residential natural gas supplied through short-distance pipelines in the region will be uniformly adjusted. The non-residential sales price in Hohhot City has been adjusted to 3.593 yuan/m^3, in Baotou City and Guyang County it is 2.975 yuan/m^3 and 3.275 yuan/m^3 respectively, in Ulanqab City it is 3.58 yuan/m^3, and in Ordos City there are different regulations for different areas. The prices in Dongsheng District, Kangbashi District, and Yijinhuoluo Banner are 2.87 yuan/m^3, in Hantai Town (Dongsheng District) it is 3.06 yuan/m^3, in Dalad Banner it is 2.88 yuan/m^3, in Hangjin Banner it is 2.87 yuan/m^3, in Uxin Banner it is 2.57 yuan/m^3, in Etoke Front Banner it is 2.76 yuan/m^3, in Ulan Town it is 2.79 yuan/m^3, in Chessboard Well it is 2.81 yuan/m^3, and in Mongolian West it is 2.95 yuan/m^3. The price in Wuhai City has been adjusted to 3.025 yuan/m^3, in Bayannur City it is 3.225 yuan/m^3, and in Alxa League it is 3.17 yuan/m^3.

3. Residential natural gas pricing policy

Residential gas use includes gas for household living, gas for school teaching and student life, gas for elderly care and welfare institutions, etc. There are relatively few policies for adjusting the price of residential gas. In most provinces and cities, the price of residential gas remains unchanged. However, the price of residential gas in Hebei Province has increased significantly.

The "Notice on the Natural Gas Sales Price in the Central Urban Area during the Heating Season in 2023" in Chongqing City indicates that the maximum sales price for residential gas will not be adjusted. The maximum sales prices for the first, second, and third tiers of residential gas will still be implemented at 2.196 yuan/m^3, 2.366 yuan/m^3, and 2.716 yuan/m^3, respectively. The maximum sales prices for gas used by low-income users, extremely poor people, schools, elderly and welfare institutions, and military canteens that implement residential gas prices will not be adjusted. The maximum sales price for gas used by low-income users and extremely poor people will still be implemented at 1.96 yuan/m^3, and the maximum sales price for gas used by schools, elderly and welfare institutions, and military canteens that implement residential gas prices will still be implemented at 2.246 yuan/m^3.

Hebei Province proposed in the "Notice on Adjusting the Sales Price of Pipeline Natural Gas for Residential Use in the Main Urban Area" that the first tier price for residential gas will be adjusted from 2.78 yuan/m^3 to 3.15 yuan/m^3, an increase of 0.37 yuan/m^3. The second and third tier prices will be adjusted to 3.62 yuan/m^3 and 4.25 yuan/m^3, respectively, which are 1.15 times and 1.35 times the first tier price.

Inner Mongolia proposed in the "Notice on Adjusting the Sales Price of Pipeline Natural Gas for Residential and Non-residential Use": From April 1, 2023 to March 31, 2024, according to the price adjustment plan of PetroChina Natural Gas Sales Inner Mongolia Branch, the terminal sales price of residential natural gas supplied through short-distance pipelines in the region will be uniformly adjusted. The residential sales price in Hohhot City has been adjusted to 2.252 yuan/m^3, in Baotou City and Guyang County it is 2.252 yuan/m^3 and 2.502 yuan/m^3, respectively. In

Ulanqab City it is 2.782 yuan/m^3, and in Ordos City (Dongsheng District, Kangbashi District, Yijinhuoluo Banner), Hantai Town (Dongsheng District), Dalad Banner, Hangjin Banner, Uxin Banner, Etoke Front Banner, Ulan Town, Chessboard Well, and Mongolian West it is 2.192 yuan/m^3, 2.192 yuan/m^3, 2.222 yuan/m^3, 2.292 yuan/m^3, 2.072 yuan/m^3, 2.082 yuan/m^3, 2.122 yuan/m^3, 2.172 yuan/m^3, 2.422 yuan/m^3, respectively. The price in Wuhai City has been adjusted to 2.422 yuan/m^3. It is 2.522 yuan/m^3 in Bayannur City, and 2.382 yuan/m^3 in Alxa League.

The "Jilin City Urban Pipeline Natural Gas Upstream and Downstream Price Linkage Method" by Jilin Province stipulates that the sales price of pipeline natural gas consists of the comprehensive procurement price and the urban pipeline gas distribution price. The comprehensive procurement price refers to the weighted average price of various gas sources purchased by gas companies to meet the urban gas demand. The urban pipeline gas distribution price (hereinafter referred to as the distribution price) refers to the price of the gas distribution link provided by the gas company to users through the public gas pipeline network system within its business scope. The distribution price is set by the government, formulated and adjusted by the city and county governments in accordance with the law, and specifically implemented by the price authorities of the same administrative rank. The distribution price should be checked regularly, and the price adjustment cycle should not exceed 3 years in principle. If there are significant changes in pipeline investment, gas distribution volume, and costs, it can be checked and adjusted in advance. For new companies that implement the initial distribution price, after the design production period, the price authorities should take the initiative to conduct a price check and adjust the distribution price in a timely manner.

(II) Policy interpretation

By combing through policies on natural gas prices and considering the domestic and international situations, three characteristics of natural gas prices in 2023 are summarized as follows:

1. Natural gas industry development and price adjustment

Facing the extremely complex international situation, China's natural gas industry has actively responded by vigorously enhancing exploration and development efforts, insisting on both supply and demand sides, and ensuring stable supply and prices. The import of pipeline gas has seen steady growth, infrastructure construction has been continuously advanced, a "national one-network" has been preliminarily formed, and gas storage capacity has doubled. These efforts have made a positive contribution to stabilizing the global natural gas industry supply chain. The diversification of upstream gas sources, market entities, and market-oriented pricing in China have essentially taken shape. The external dependence on natural gas exceeds 40%, with nearly 60% of the natural gas share procured from the market. In the remaining 40% of the volume subject to government price regulation, a government-guided pricing method that can be appropriately adjusted according to market demand has been adopted. An elastic pricing mechanism that can rise and fall in response to market conditions has been established. Prices must rise when the market rises and fall when the market falls, ensuring both protection and stability. It is essential to both alleviate the burden on enterprise users and relieve the pressure on gas supply companies. Several provinces in China have introduced gas price adjustment plans, combining upstream and downstream linkage mechanisms to adjust terminal natural gas prices. In 2023, natural gas prices generally showed an upward trend. To cope with the rise in natural gas prices, Beijing proposed measures to continuously improve a multi-source, multi-directional gas supply system and completed the construction of the Tianjin Nangang LNG emergency reserve project. Shaanxi Province has been advancing the "three-retrofits" for existing coal-fired power unit, which include energy-saving and consumption reduction retrofitting, heating retrofitting, and flexibility retrofitting, and has implemented clean electricity and natural gas substitution.

2. Establish and improve the natural gas upstream and downstream price linkage mechanism

The National Development and Reform Commission issued guidance on the natural gas upstream and downstream price linkage mechanism, clearly demanding the acceleration of the promotion and implementation of the price linkage mechanism, and the deepening of the reform of the oil and natural gas market system. In August 2023, the State Council Safety Production Committee issued the "National Urban Gas Safety Special Rectification Work Plan," and the National Development and Reform Commission issued opinions on reforming and improving the price formation mechanism of important resources. Both institutions demand the establishment and improvement of the natural gas upstream and downstream price linkage mechanism and the rational guidance of terminal sales prices.

At present, all regions have basically established a natural gas upstream and downstream price linkage mechanism, and in order to adapt to the new situation of frequent and violent fluctuations in the natural gas source market price, many local governments have actively supplemented and improved the urban gas price linkage mechanism. The main practices include: First, flexibly shortening the cycle for price linkage adjustments. The notice issued by Fujian Province on the "Fuzhou City Pipeline Natural Gas Price Linkage Mechanism" points out that the sales price of non-residential gas is, in principle, linked and adjusted on a 3-month cycle. When the change in the comprehensive gas purchase price reaches 5% within the cycle, the terminal sales price of natural gas for non-residential users is, in principle, linked and adjusted with the gas procurement price. The notice issued by Hubei Province on "Establishing and Improving the Natural Gas Upstream and Downstream Price Linkage Mechanism" indicates that the terminal sales price of non-residential gas is, in principle, linked on a quarterly or monthly basis. For those with a previous linkage cycle of half a year or more, a gradual transition to quarterly linkage should be made, and areas with the conditions to do so may link on a monthly basis. Second, implementing different price linkage methods for off-peak and peak seasons. In response to the significant price difference between off-peak and peak seasons in the gas market, Beijing, based on the seasonal supply and demand structure of the natural gas market, has divided the year into off-peak and peak seasons, with the peak season price increase system being canceled during the off-peak season.

3. Receiving station gasification service pricing policy

The Tianjin City policy document "Notice on Announcing the Gasification Service Prices of Our City's Imported Liquefied Natural Gas (LNG) Receiving Stations" further stipulates that operating enterprises of imported LNG should publicly disclose specific gasification service prices through portal websites, fair and open information platforms or other relevant platforms. This will strengthen price transparency, facilitates user inquiries, usage, and supervision, and ensures openness and transparency. Enterprises should strictly implement pricing policies, strengthen policy interpretation and user service work, improve service and management quality, and promote orderly competition and fair openness. The policy clarifies the definition and connotation of gasification service prices and shifts them from government-set prices to government-guided prices, implementing a maximum price cap management. This helps to foster an orderly competitive market environment and supports the fair opening of LNG receiving stations. A more flexible and elastic pricing mechanism can stimulate the enthusiasm of receiving stations and promote the healthy development of the natural gas industry. The policy provides guidance for local governments to set and adjust gasification service prices, which helps to enhance the scientific and standardized nature of pricing and ensures national energy security. A rational gasification service pricing policy helps to ensure the stable supply of natural gas and meet the energy demands of socio-economic development.

Natural Gas Pipeline Construction

Local natural gas pipeline construction plays an important role in ensuring energy security, promoting economic development, improving environmental quality, improving the living standards of residents, and optimizing the energy structure.

(I) Policy contents

1. Natural gas pipeline transportation price

In the "Notice on Reducing the Short-Haul Pipeline Transportation Prices for Some Natural Gas within the Province and Other Related Matters," Anhui Province stipulates the following adjustments to the short-haul transportation prices for natural gas: The short-haul transportation prices for 10 branches, including the Feixi Lu'an branch line, Liying branch line, Lihao branch line, Wutong branch line, Jiangnan liaison line, Lihuai branch line, Jiangbei Concentration Zone line, and Jiangbei liaison line owned by Anhui Natural Gas Development Co., Ltd., as well as the Lujiang to Wuhu line owned by Hefei Huanran China Resources Gas Co., Ltd. and Anhui United Natural Gas Co., Ltd., and the Hefei to Jinzhai line owned by Hefei Huanran China Resources Gas Co., Ltd. and Huaihe Energy Gas Group Co., Ltd., are reduced from the current

0.21 yuan/m^3 to 0.20 yuan/m^3. The short-haul transportation prices for 8 branches, including the Guangde branch line, Héchāo branch line, Lifu branch line, Jiangnan Concentration Zone (Chitong) branch line owned by Anhui Natural Gas Development Co., Ltd., the Lifu pipeline owned by Lixin County Guozhen Gas Co., Ltd., the Taihe Shui Zhen branch line owned by Anhui PetroChina Kunlun Huishang Gas Co., Ltd., the Dingmingfeng pipeline owned by Chuzhou PetroChina Kunlun Gas Co., Ltd., and the Bozhou Qiaocheng to Fuyang Taihe line owned by Huaihe Energy Gas Group Co., Ltd., are reduced from the current 0.20 yuan/m^3 to 0.18 yuan/m^3. The short-haul transportation price for the Xuanning Huang branch line owned by Anhui Waneng Portua Natural Gas Co., Ltd. is set at 0.20 yuan/m^3 according to the third price zone. The short-haul transportation prices for other branches remain at the current prices. New pipelines will be subject to the prices specified for their respective price zones. These adjustments will be effective from January 1, 2024.

In the notice on the issuance of the "Measures for the Administration of Urban Pipeline Gas Distribution and Sales Prices in Guizhou Province," Guizhou Province stipulates that the distribution price of pipeline gas is the price at which urban gas companies provide gas distribution services to users through the gas pipeline network. The sales price refers to the price at which gas companies sell gas to users through the gas pipeline network. Where gas companies operate distribution and sales as an integrated business, the gas procurement price should be "bought in and sold out at the same price," and they shall not profit from the difference between the gas purchase and sales prices. These measures will be implemented from August 1, 2023.

2. Natural gas pipeline construction

In the notice issued by Shaanxi Province on the "Interim Measures for the Administration of the Construction and Operation of Long-Distance Oil and Gas Pipelines in Shaanxi Province," it is stated that for oil and gas long-distance pipeline projects that need to be included in the planning, the lower-level oil and gas authorities should submit an application to the higher-level oil and gas authorities. The project should clarify the current situation of resource supply and market consumption, as well as demand forecasts, land use requirements, investment scale, and safeguard measures. New loading and unloading branch pipelines added to existing mainline long-distance pipelines should comply with the provincial and municipal development plans for oil and gas long-distance pipelines. It is encouraged that the corresponding mainline long-distance pipeline entities invest in and construct these pipelines to reduce the levels of pipeline transportation and lower gas usage costs. In areas where pipeline gas is already available, new planned pipelines must ensure the utilization efficiency of existing pipelines, and duplicate construction is strictly prohibited to avoid waste of resources. State-owned large oil and gas enterprises and various types of social capital are allowed to invest in and construct branch pipeline facilities in accordance with the unified provincial planning. The openings of branch pipelines must comply with the provincial plan. These interim measures will be implemented from October 20, 2023, with a validity period of 2 years.

3. Natural gas pipeline protection

The "Notice on the Issuance of Shanxi Province's Emergency Plan for the Construction and Protection of Long-Distance Oil and Natural Gas Pipelines" applies to the administrative region of Shanxi Province. It addresses serious accidents affecting the safety of oil and natural gas transportation caused by third-party human activities. This includes the destruction of oil and gas pipelines and their ancillary facilities by third parties using mechanical tools, as well as accidents affecting the safe operation of oil and gas pipelines due to activities such as sand excavation and stone quarrying by third parties. In the event that natural disasters (such as earthquakes and geological disasters), industrial accidents (such as hazardous chemical accidents and special equipment accidents), social security incidents (such as terrorist attacks and criminal cases), or other factors affect or may affect the normal operation of oil and gas pipelines in Shanxi Province, the relevant special emergency plans of Shanxi Province shall be implemented, with this plan ensuring proper coordination. The accidents related to the construction and protection of oil and gas pipelines are categorized into four levels based on the degree of harm, scope of impact, and extent of destruction: extremely serious, serious, significant, and general. The plan requires pipeline transportation companies to enhance their awareness of the primary responsibility for pipeline protection, establish and improve risk early warning mechanisms, and improve their capabilities in safety risk prevention and emergency management. They should establish and perfect daily safety and risk management systems to promptly identify and address various risks and hidden dangers in oil and gas pipelines and implement risk control measures. They must also identify key pipeline protection areas, such as high-consequence areas, key flood prevention sections, and important crossing sections, establish records, and conduct focused inspections and patrols. Additionally, they should identify external risks and hidden dangers for key sections and implement corresponding prevention and control measures. Pipeline transportation companies are also required to strengthen communication with local governments and relevant departments to effectively implement pipeline protection measures. They must also ensure proper emergency duty, promptly deal with emergencies, and report them in the first instance to ensure the safe and stable operation of oil and gas pipelines.

(2) Policy interpretation

In 2023, local natural gas pipeline construction policies mainly focus on ensuring the supply and stable pricing of natural gas, reforming the oil and gas pipeline system, building gas storage capacity, and improving pipeline protection management. The reduction in natural gas pipeline transportation prices in Anhui Province lightens the burden on enterprises and residents. The reduction in natural gas pipeline prices means that enterprises and residents need to pay less when using natural gas, which helps to reduce their production and living costs, improving their economic output and quality of life. Reducing natural gas prices will stimulate more enterprises and residents to use natural gas, thereby increasing its consumption. This helps to expand the scale of the natural gas market and promote the development of the natural gas industry. As a clean and efficient energy source, the price reduction of natural gas helps to encourage more enterprises and residents to use it as an alternative to other energy sources, thereby optimizing the energy structure and reducing environmental pollution. The reduction in natural gas pipeline prices may lower the costs of related industries, such as natural gas power generation and chemical industries. This helps to enhance the competitiveness of these industries and promote their development and growth.

In 2023, local governments spared no effort in ensuring the supply and stabilizing the prices of natural gas, securing the demand for natural gas in people's livelihoods, and guaranteeing the stable supply and rational pricing of natural gas. They deepened the reform of the oil and gas pipeline network system, resolutely implemented the central government's reform decisions and plans, and promoted the reform of the oil and gas pipeline network system to serve the high-quality development of the industry. They paid attention to the construction of gas storage capacity, strengthened market regulation and risk response, and enhanced gas storage capacity to cope with potential natural gas supply shortages.

The main focus is on increasing the efforts to renovate and upgrade aging urban gas pipelines, with the aim of enhancing urban safety resilience and improving the quality of life for residents. First, with facility safety as the goal, a comprehensive investigation is carried out to get a clear understanding of the situation. Safety inspections and assessments are conducted, and the organic renewal of urban gas pipelines is systematically promoted. The responsibility for controlling safety risks and eliminating hidden dangers of urban gas pipelines is strictly implemented, and immediate renovations are carried out on pipelines with safety hazards. Second, all necessary renovations should be made. Based on the safety inspection and assessment report of urban gas pipelines, an update and renovation plan is reasonably formulated according to the principle of making all necessary renovations. Management systems are standardized, and management throughout the process is strengthened to ensure project quality and safety. Third, coordinated promotion is emphasized. Each town and street is responsible for coordinating the update and renovation of urban gas pipelines, which is integrated with the renovation of old communities, municipal construction, urban development, and other work. Fourth, intelligent management is implemented. Big data, cloud computing, and other technologies are used to establish and improve the comprehensive management information platform for urban underground municipal infrastructure, thereby enhancing the level of intelligent operation and maintenance management.

Accompanying the global energy structure transition and the increasing demand for clean energy in China, natural gas, as a clean and efficient energy source, is playing an increasingly important role in optimizing local energy consumption structures. In 2023, local governments introduced a series of policies around natural gas supply security, price mechanisms, infrastructure construction, and energy conservation and environmental protection, providing strong support for the stable development of the natural gas market.

In terms of supply assurance and price mechanism optimization, local governments actively cooperated with natural gas suppliers to strengthen supply-demand docking and ensure a stable supply of natural gas. At the same time, by improving the pricing mechanism, the linkage between upstream and downstream natural gas prices and the marketization process were steadily promoted, making the prices more reasonably reflect market supply and demand changes. Some regions have also implemented price control measures such as seasonal differences to balance supply and demand contradictions at different times.

In terms of energy conservation, emission reduction, and the promotion of clean utilization, local governments actively responded to national policies on energy conservation and emission reduction, and promoted the widespread application of natural gas in areas such as residential life and industrial production. By implementing clean energy substitution programs, enterprises were encouraged to use natural gas to replace high-pollution energy sources such as coal, thereby reducing pollutant emissions. At the same time, supervision over the efficiency of natural gas use was strengthened to promote clean utilization and sustainable development.

In terms of infrastructure construction and interconnectivity, local governments have increased investment in the construction of natural gas infrastructure and accelerated the construction of pipeline and gas storage facilities. At the same time, they strengthened the interconnectivity of natural gas pipelines between regions to achieve resource sharing and optimal allocation. These measures effectively enhanced the capacity for natural gas transportation and storage, providing strong support for the stable development of the market.

In terms of market regulation and safety management reinforcement, the government intensified its supervision of the natural gas market, standardized market order, and combated illegal and criminal activities. By establishing and improving market regulation mechanisms, fair competition and consumer rights were safeguarded. At the same time, safety management was strengthened, the safety production responsibility system was implemented, and safety inspections and hidden danger investigations were enhanced to ensure the safety of natural gas supply and use.

In 2023, local natural gas policies achieved positive results in multiple aspects. First, the capacity for natural gas supply assurance was enhanced, effectively alleviating supply and demand contradictions. Second, the optimization of the pricing mechanism made natural gas prices more rational, promoting the healthy development of the market. Additionally, the promotion of clean energy and the strengthening of infrastructure construction also made positive contributions to energy conservation, emission reduction, and sustainable development.

The Natural Gas Industry in China

In 2023, as the effect of the global "COVID-19 pandemic" continued to weaken, the domestic economy began to recover. The production, consumption, and import and export of natural gas in China all showed positive trends, and the development momentum of the domestic natural gas industry was robust.

In 2022, domestic oil and gas companies saw a year-on-year increased of 3.1% in exploration investment, with a slower growth rate. The remaining technically recoverable reserves of natural gas, coalbed methane, and shale gas increased slightly compared to 2021. In 2023, the national natural gas production grew by 5.58% year-on-year, with continued increases in coalbed methane and shale gas production. The output of liquefied natural gas has risen for seven consecutive years.

By the end of 2023, China had built a cumulative total of 43 underground gas storage facilities (including groups), with 11 more planned and under construction. A total of 32 LNG receiving stations had been constructed, with 15 more planned and under construction, and 4 LNG receiving stations were expanded. The mileage of natural gas pipelines reached 116,300 km, an increase of 2.8% year-on-year.

In 2023, the average import price of LNG was 628.36 US dollars per ton, a decrease of 22.52% compared to the previous year; the average import price of PNG was 400.95 US dollars per ton, an increase of 3.24% year-on-year. The total import volume of natural gas was 119,971,200 tons, an increase of 9.9% year-on-year, of which LNG imports amounted to 71,316,600 tons, an increase of 12.55% year-on-year; PNG imports were 48,654,500 tons, an increase of 6.22% year-on-year.

In 2023, China's apparent consumption of natural gas was 394.5 billion cubic meters, an increase of 7.6% year-on-year. Urban gas, industrial fuel, power generation, and chemical industries increased by 7.86%, 6.29%, 7.00%, and 2.73% year-on-year respectively, with urban gas and industrial fuel being the main consumers, accounting for a combined total of 74.99%.

Driven by the "dual carbon" goals, China's energy structure is continuously being adjusted. As a clean energy source, natural gas will play an increasingly important role. In 2024, China needs to continue to enhance the supply level of natural gas to

© Petroleum Industry Press 2025

L. Zhang et al., *Annual Report on Operation of China's Natural Gas Industry: 2023*,
https://doi.org/10.1007/978-981-95-2371-9_3

ensure that energy supply maintains a reasonable elasticity margin and to safeguard China's energy security.

China's Natural Gas Exploration

In 2022, the situation of natural gas exploration in China continued to improve. Despite a significant reduction in drilling work, the total investment in oil and gas exploration still increased slightly due to a substantial increase in 2D and 3D seismic work. The remaining technically recoverable reserves of natural gas and shale gas increased, but the increase is significantly lower than that of 2021. The remaining technically recoverable reserves of natural gas in 10 provincial regions increased compared to last year, with Sichuan, Ningxia, Xinjiang, and Inner Mongolia showing higher increases. Among the three provincial regions with remaining technically recoverable reserves of shale gas, Guizhou Province has a significant increase in reserves. The increase in the remaining technically recoverable reserves of coalbed methane mainly relied on Shanxi Province.

The Overall Situation of Oil and Gas Exploration

(I) Investment in Oil and Gas Exploratio

The situation of China's oil and natural gas exploration investment from 2017 to 2022 is shown in Table 1.

Table 2 shows that in 2022, the investment in oil and gas exploration continued the upward trend of the previous two years, reaching 82.387 billion yuan, an increase of 3.1% over the previous year. In terms of oil and gas exploration work, the number of exploration wells was 2,510, a decrease of 277 from the previous year, with a growth rate of −9.94%; the 2D seismic exploration workload was 19,500 km, an increase of 6,600 km compared to the previous year, with a growth rate of 51.16%; the 3D seismic exploration workload was 47,000 square kilometers, an increase of 9,300 square kilometers from the previous year, with a growth rate of 24.67%. Although the decrease in exploration well work led to a reduction in oil and gas exploration investment, the significant increase in 2D and 3D seismic work ultimately led to a net increase in overall oil and gas exploration investment.

The trend of changes in oil and gas exploration investment from 2017 to 2022 is shown in Fig. 1.

(II) Overall reserves situation

The overall situation of natural gas reserves from 2017 to 2022 is shown in Table 2.

Table 2 shows that in 2022, the remaining technically recoverable reserves of natural gas continued to increase, reaching 6,569.012 billion cubic meters, an increase

Table 1 Investment in oil and gas exploration in China from 2017 to 2022

Project		2017	2018	2019	2020	2021	2022
Exploration investment	Billion yuan	58.449	63.658	82.129	71.024	79.906	82.387
Growth rate	%	10.80	8.91	29.02	−13.52	12.51	3.10
Well exploration	Well	2727	2955	2919	2956	2787	2510
Well exploration changes	Well	12	228	−36	37	−169	−277
Well exploration change rate	%	0.44	8.36	−1.22	1.27	−5.72	−9.94
2D Seismic	Kilometers	38,000.00	44,000.00	51,400.00	30,000.00	12,900.00	19,500.00
2D Seismic changes	Kilometers	−14,869.8	6,000	7,400	−21,400	−17,100	6,600
2D Seismic change rate	%	−28.13	15.79	16.82	−41.63	−57.00	51.16
3D Seismic	Square kilometers	33,000.00	34,000.00	47,100.00	42,700.00	37,700.00	47,000.00
3D Seismic changes	Square kilometers	6,547.91	1,000	13,100	−4,400	−5,000	9,300
3D Seismic change rate	%	24.75	3.03	38.53	−9.34	−11.71	24.67

Data source China Mineral Resources Report (2018–2023)

of 229.745 billion cubic meters from 2021, with a growth rate of 3.62%; in 2022, the remaining technically recoverable reserves of coalbed methane increased slightly, reaching 365.969 billion cubic meters; in 2022, the remaining technically recoverable reserves of shale gas reached 560.559 billion cubic meters, an increase of 16.497 billion cubic meters from 2021, with an increase of 3.03%.

The trend of changes in the remaining technically recoverable reserves of natural gas, coalbed methane, and shale gas from 2017 to 2022 is shown in Fig. 2.

According to Fig. 2, from 2017 to 2022, the remaining technically recoverable reserves of natural gas, coalbed methane, and shale gas generally showed a growth trend. The change rate of the growth of remaining technically recoverable reserves was relatively stable for natural gas, slightly fluctuating for coalbed methane, and quite volatile for shale gas. For example, it decreased from a growth rate of 61.98% in 2017 to 8.94% in 2018, and then increased again to 77.84% in 2019. The reason for the large fluctuation in the growth rate of the remaining technically recoverable reserves of shale gas is, on the one hand, due to the improvement in the level of shale gas exploration and development technology, which has led to a significant increase in the scale of proven reserves of shale gas. On the other hand, Sichuan Province

Table 2 Overall situation of natural gas reserves (2017–2022)

Project	2017	2018	2019	2020	2021	2022
Remaining technically recoverable reserves of natural gas	55,220.96	57,936.08	59,665.80	62,665.78	63,392.67	65,690.12
Remaining technically recoverable reserves of coalbed methane	3025.36	3046.30	3040.70	3315.54	3659.68	3659.69
Remaining technically recoverable reserves of shale gas	1982.88	2160.20	3841.80	4026.17	5440.62	5605.59
Increase in remaining technically recoverable reserves of natural gas	5553.79	8311.57	15,800.00	12,600.00	16,284.00	11,323.00
Increase in remaining technically recoverable reserves of conventional gas	–	–	8090.90	10,357.00	8051.00	–
Increase in remaining technically recoverable reserves of coalbed methane	–	147.08	64.90	673.13	779.00	–
Increase in remaining technically recoverable reserves of shale gas	–	1246.78	7644.20	1918.27	7454.00	–

Unit: hundred million cubic meters
Data source China Mineral Resources Report (2018–2023) and China Natural Gas Development Report (2018–2023), China Economic Data

and Chongqing City are the main areas for natural gas exploration and development in our country, and also the main areas for shale gas exploration and development. When calculating the remaining technically recoverable reserves of natural gas, statistical discrepancies may lead to inaccuracies in the remaining technically recoverable reserves of shale gas, thereby causing more pronounced fluctuations in the growth rate.

Natural Gas Reserves in Provinces, Autonomous Regions, and Municipalities

(I) Natural gas

The statistics of the remaining technically recoverable reserves of natural gas in various provinces, autonomous regions, and municipalities in mainland China from 2020 to 2022 are shown in Table 3.

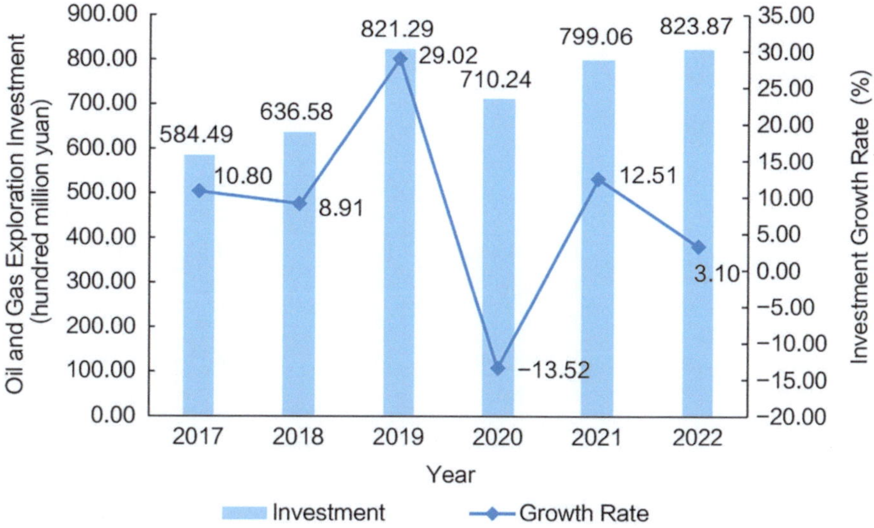

Fig. 1 Trends in oil and gas exploration investment (2017–2022). *Data source* 2018–2023 "China Mineral Resources Report"

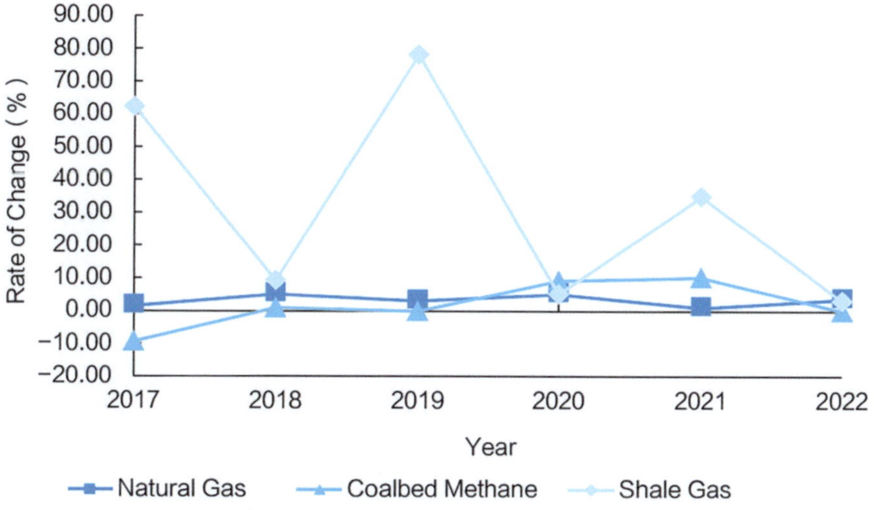

Fig. 2 Trends in the remaining technically recoverable reserves of natural gas, coalbed methane, and shale gas (2017 to 2022). *Data source* China Mineral Resources Report (2018–2023) and China Natural Gas Development Report (2018–2023), China Economic Data

Table 3 Statistics of the remaining technically recoverable reserves of natural gas in 31 provincial regions in China (2020 to 2022)

Region	2020	2021	2022
Nationwide	62,665.78	63,392.67	65,690.12
Beijing	–	0.04	0.04
Tianjin	293.09	293.05	294.50
Hebei	372.26	361.95	336.49
Shanxi	1402.04	1247.02	1210.59
Inner Mongolia	10,123.53	9887.52	10,115.95
Liaoning	164.53	157.32	152.52
Jilin	767.11	826.16	804.53
Heilongjiang	1494.64	1388.11	1343.88
Shanghai	–	–	–
Jiangsu	21.39	21.05	21.90
Zhejiang	–	–	–
Anhui	0.24	0.24	0.24
Fujian	–	–	–
Jiangxi	–	–	–
Shandong	343.52	339.64	347.25
Henan	62.82	63.49	61.39
Hubei	44.59	43.96	44.18
Hunan	–	–	–
Guangdong	0.97	0.97	0.97
Guangxi	1.38	1.37	1.36
Hainan	21.47	21.29	19.35
Chongqing	2500.73	2639.42	2562.83
Sichuan	15,274.98	15,556.37	16,546.46
Guizhou	6.10	6.10	6.10
Yunnan	0.47	0.47	0.47
Tibet	–	–	–
Shaanxi	11,096.45	11,630.03	11,770.37
Gansu	588.00	643.81	729.76
Qinghai	1055.32	1093.51	1034.42
Ningxia	280.67	334.60	931.94
Xinjiang	11,237.85	11,175.29	11,482.11

Unit: hundred million cubic meters

Data source National Mineral Resources Reserve Statistics (2020–2022), Ministry of Natural Resources

Natural gas resources in China are widely distributed. As of 2022, 25 provinces, autonomous regions, and municipalities directly under the Central Government have proven natural gas reserves. The number of provinces, autonomous regions, and municipalities with remaining technically recoverable reserves exceeding 10 billion cubic meters, 20 billion cubic meters, 30 billion cubic meters, 40 billion cubic meters, 50 billion cubic meters, 100 billion cubic meters, 200 billion cubic meters, 500 billion cubic meters, and 1 trillion cubic meters are 15, 14, 13, 11, 11, 8, 5, 4, and 4, respectively. The regions with reserves exceeding 1 trillion cubic meters are Sichuan (1.6546 trillion cubic meters), Shaanxi (1.1770 trillion cubic meters), Xinjiang (1.1482 trillion cubic meters), and Inner Mongolia (1.0116 trillion cubic meters). These four provinces account for 76% of China's total remaining technically recoverable natural gas reserves.

(II) Shale gas

The statistics of the remaining technically recoverable reserves of shale gas in 31 provinces, autonomous regions, and municipalities in China from 2020 to 2022 are shown in Table 4.

The distribution of exploitable shale gas resources in China is relatively concentrated. As of 2022, shale gas has been discovered in three provinces and municipalities, all concentrated in the southwestern region of China, namely Sichuan (accounting for 67.39%), Chongqing (accounting for 31.31%), and Guizhou (accounting for 1.30%). With the advancement of shale gas exploration technology, shale gas resources may be discovered in more regions.

(III) Coalbed methane

The remaining technically recoverable reserves of coalbed methane in China's 31 provinces, autonomous regions, and municipalities from 2020 to 2022 are shown in Table 5.

In 2022, the recoverable resources of coalbed methane in China were mainly distributed across six provinces. These provinces and their respective shares are as follows: Shanxi (332.643 billion cubic meters, accounting for 90.89%), Shaanxi (22.230 billion cubic meters, accounting for 6.07%), Sichuan (3.818 billion cubic meters, accounting for 1.04%), Guizhou (3.204 billion cubic meters, accounting for 0.88%), Liaoning (2.501 billion cubic meters, accounting for 0.68%), and Anhui (1.573 billion cubic meters, accounting for 0.43%). Compared to 2021, there was a slight increase in the national remaining technically recoverable reserves of coalbed methane, with an increment of 34.415 billion cubic meters. This growth was primarily attributed to the contribution from Shanxi, which amounted to 39.076 billion cubic meters.

Table 4 Statistics of the remaining technically recoverable reserves of shale gas in 31 provincial regions in China (2020–2022)

Region	2020	2021	2022
Nationwide	4026.17	5440.62	5605.59
Beijing	–	–	–
Tianjin	–	–	–
Hebei	–	–	–
Shanxi	–	–	–
Inner Mongolia	–	–	–
Liaoning	–	–	–
Jilin	–	–	–
Heilongjiang	–	–	–
Shanghai	–	–	–
Jiangsu	–	–	–
Zhejiang	–	–	–
Anhui	–	–	–
Fujian	–	–	–
Jiangxi	–	–	–
Shandong	–	–	–
Henan	–	–	–
Hubei	–	–	–
Hunan	–	–	–
Guangdong	–	–	–
Guangxi	–	–	–
Hainan	–	–	–
Chongqing	1498.65	1620.54	1754.96
Sichuan	2527.52	3803.03	3777.69
Guizhou	–	17.05	72.94
Yunnan	–	–	–
Tibet	–	–	–
Shaanxi	–	–	–
Gansu	–	–	–
Qinghai	–	–	–
Ningxia	–	–	–
Xinjiang	–	–	–

Unit: hundred million cubic meters

Data source National Mineral Resources Reserve Statistics (2020–2022), Ministry of Natural Resources

Table 5 Statistics of remaining technically recoverable reserves of coalbed methane in China's provincial regions (2020–2022)

Region	2020	2021	2022
Nationwide	3315.54	3315.54	3659.69
Beijing	–	–	–
Tianjin	–	–	–
Hebei	–	–	–
Shanxi	2935.67	2935.67	3326.43
Inner Mongolia	–	–	–
Liaoning	25.47	25.47	25.01
Jilin	–	–	–
Heilongjiang	–	–	–
Shanghai	–	–	–
Jiangsu	–	–	–
Zhejiang	–	–	–
Anhui	15.74	15.74	15.73
Fujian	–	–	–
Jiangxi	–	–	–
Shandong	–	–	–
Henan	–	–	–
Hubei	–	–	–
Hunan	–	–	–
Guangdong	–	–	–
Guangxi	–	–	–
Hainan	–	–	–
Chongqing	–	–	–
Sichuan	40.64	40.64	38.18
Guizhou	32.04	32.04	32.04
Yunnan	–	–	–
Tibet	–	–	–
Shaanxi	265.98	265.98	222.30
Gansu	–	–	–
Qinghai	–	–	–
Ningxia	–	–	–
Xinjiang	–	–	–

Unit: hundred million cubic meters

Data source National Mineral Resources Reserve Statistics (2020–2022), Ministry of Natural Resources

Natural Gas Development in China

In 2023, China's natural gas development continued to show a promising situation. The national natural gas output reached 232.4 billion cubic meters, an increase of 12.29 billion cubic meters compared to 2022. Following this trend, the natural gas production plan for the "14th Five-Year Plan" period is expected to be achieved ahead of schedule. In 18 provincial regions, natural gas production increased compared to 2022, with Sichuan, Xinjiang, and Shaanxi being the regions with the largest production share and growth rate. The Changqing Oilfield, Southwest Oil and Gas Field, and Tarim Oilfield under China National Petroleum Corporation (CNPC) still ranked as the top three in natural gas production, with their combined output accounting for more than half of China's total.

The Overall Situation of Natural Gas Development

The statistics of China's natural gas output from 2017 to 2023 are shown in Table 6.

Data from the "Monthly Aggregate" and "Annual Output" categories in Table 6 were taken separately to plot the trends in natural gas output and growth rate changes from 2017 to 2023, as shown in Figs. 3 and 4.

Whether based on the monthly total or the annual production statistics, the trend in natural gas is basically the same. Since 2017, annual natural gas output has been on the rise. Taking 2020 as the turning point, the growth rate first increased year by year and then declined rapidly year by year. In 2023, the monthly total production reached 230.682 billion cubic meters, exceeding the predicted value of 222.725 billion cubic meters (see the "China Natural Gas Industry Annual Operation Report Blue Book (2022)"). The monthly total annual growth rate was 5.81%, while the predicted annual growth rate was 4.31%. According to this trend, natural gas production is expected to reach the planned target of 240 billion cubic meters ahead of schedule by the end of the "14th Five-Year Plan" period (2025).

Natural Gas Output of Provinces, Autonomous Regions, and Municipalities

The natural gas output of mainland China's provinces, autonomous regions, and municipalities from 2017 to 2023 is shown in Table 7.

Based on the data in Table 7, the distribution of natural gas production in mainland China's provincial-level regions in 2023 is shown in Table 8.

According to the statistical results in Table 8, the six provincial regions in mainland China with natural gas output greater than 10 billion cubic meters are, in descending order of production volume: Sichuan, Xinjiang, Shaanxi, Inner Mongolia, Shanxi,

Table 6 Statistics of China's natural gas output (2017–2023)

Month	2017	2018	2019	2020	2021	2022	2023
1–2	251.50	261.90	286.80	314.00	348.30	372.00	397.99
3	135.80	135.20	151.00	169.00	184.70	197.00	204.64
4	122.00	128.90	141.00	161.00	168.60	177.00	189.40
5	119.90	126.20	144.00	159.00	168.80	177.00	189.70
6	115.50	121.80	139.00	152.00	172.20	173.00	182.72
7	117.40	129.60	139.00	142.00	157.60	171.00	183.80
8	119.50	129.00	138.00	142.00	158.70	170.00	180.97
9	111.50	121.80	135.00	146.00	156.80	164.00	179.40
10	124.10	134.20	146.00	163.00	164.50	185.00	190.20
11	126.30	142.70	151.00	169.00	177.30	189.00	199.40
12	136.10	152.50	160.00	187.00	191.90	204.00	208.60
Monthly total	1479.60	1583.80	1730.80	1904.00	2049.40	2179.00	2306.82
Monthly total difference	113.23	104.2	147	173.2	145.4	129.6	127.82
Monthly total growth rate	8.27%	7.04%	9.18%	9.88%	7.55%	6.24%	5.81%
Annual	1480.35	1601.59	1753.62	1924.95	2075.84	2201.10	2324.00
Annual difference	111.70	121.24	152.03	171.33	150.89	125.26	122.90
Annual growth rate	8.16%	8.19%	9.49%	9.77%	7.84%	6.03%	5.58%

Unit: hundred million cubic meters
Data source China Economic Data, National Bureau of Statistics
Note The monthly total refers to the annual natural gas output calculated based on the monthly data released by the National Bureau of Statistics and China Economic Data. The annual figure refers to the natural gas output released by China Economic Data on an annual basis. Due to differences in statistical methods, there are certain discrepancies.

and Guangdong. These regions account for a cumulative production share of 82.53%. Among them, Sichuan, Xinjiang, and Shaanxi have an annual natural gas production exceeding 30 billion cubic meters, while Inner Mongolia's annual production is close to 30 billion cubic meters (it exceeded 30 billion cubic meters in 2022). The combined production of these four provincial regions reaches 162.687 billion cubic meters, accounting for 70.81% of the total. They are expected to be the main contributors to China achieving its natural gas production targets of 240 billion cubic meters by the end of the "14th Five-Year Plan" period (2025) and 330 billion cubic meters by the end of the "16th Five-Year Plan" period (2035).

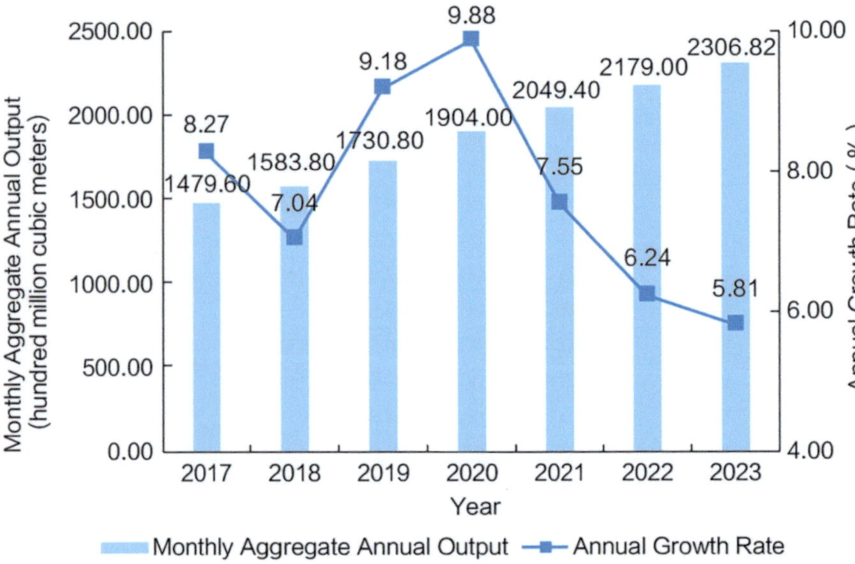

Fig. 3 Trends in annual output based on monthly aggregates (2017–2023). *Data source* China Economic Data, National Bureau of Statistics

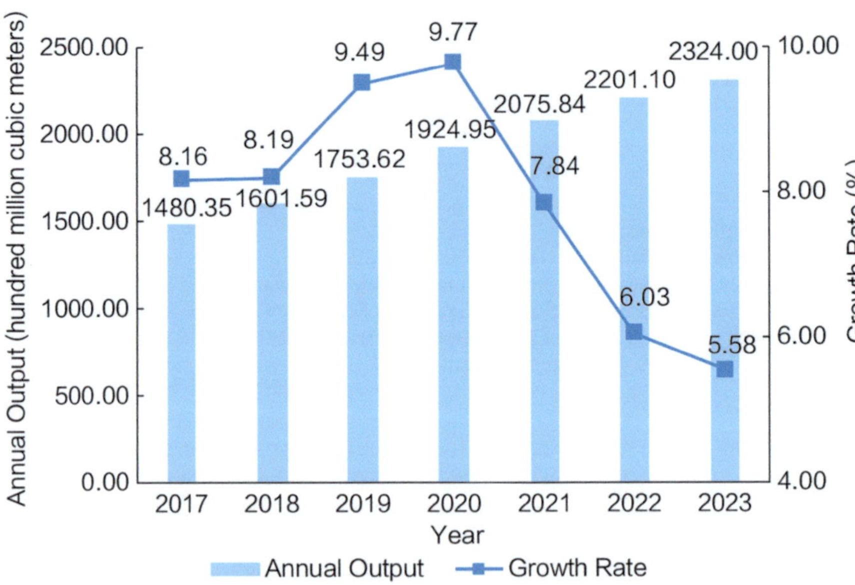

Fig. 4 Trends in annual output (2017–2023). *Data source* China Economic Data, National Bureau of Statistics

Table 7 Natural gas output of mainland China's provinces, autonomous regions, and municipalities (2017–2023)

Region	2017	2018	2019	2020	2021	2022	2023
Beijing	15.41	17.28	29.73	19.98	4.28	8.00	1.20
Tianjin	21.50	33.90	34.90	36.30	39.02	41.17	41.67
Hebei	7.39	6.15	5.84	5.61	5.28	5.66	5.90
Shanxi	46.76	52.42	64.62	85.93	123.35	132.10	145.50
Inner Mongolia	0.19	16.07	22.07	25.55	289.75	307.22	286.30
Liaoning	5.11	5.87	6.18	7.37	7.91	8.74	8.10
Jilin	18.58	18.43	10.33	19.81	21.38	20.50	19.30
Heilongjiang	40.54	43.54	45.66	46.78	50.50	55.75	59.40
Shanghai	1.71	14.54	12.47	15.11	17.24	19.47	36.10
Jiangsu	2.94	9.97	12.19	4.24	0.94	0.85	0.70
Zhejiang	6.14	0.00	0.01	0.00	0.00	0.00	0.00
Anhui	2.60	2.25	2.12	2.24	2.32	2.46	2.60
Fujian	–	–	–	–	–	–	–
Jiangxi	0.21	0.20	0.03	0.00	0.00	0.00	0.10
Shandong	4.15	4.80	5.10	5.77	6.22	7.97	8.50
Henan	2.98	2.90	2.96	2.90	2.93	3.88	4.40
Hubei	1.27	5.13	4.89	1.01	1.29	1.38	1.70
Hunan	0.00	0.00	0.00	0.03	0.03	0.01	0.00
Guangdong	89.23	102.50	112.08	131.59	132.48	124.39	123.70
Guangxi	0.21	0.19	0.25	0.22	0.22	0.23	0.40
Hainan	1.10	1.06	1.03	1.04	8.03	31.54	35.00
Chongqing	60.70	61.17	65.10	79.96	87.11	88.42	95.60
Sichuan	356.39	369.85	441.35	463.34	522.21	554.06	594.80
Guizhou	4.15	2.98	3.23	5.04	5.22	7.92	10.50
Yunnan	0.04	0.00	0.00	0.00	0.00	0.00	0.20
Tibet	–	–	–	–	–	–	–
Shaanxi	419.40	444.48	473.42	527.38	294.13	307.11	328.50
Gansu	0.60	1.03	0.12	3.90	4.18	5.44	7.50
Qinghai	64.01	64.05	64.00	64.01	62.00	60.00	60.00
Ningxia	0.00	0.00	0.00	0.00	0.21	0.16	2.20
Xinjiang	307.04	321.85	342.03	369.83	387.59	406.68	417.27

Unit: hundred million cubic meters
Data source China Economic Data

Table 8 Distribution of natural gas in China's 31 provincial regions in 2023

Category	Number of regions (individuals)	Natural gas production (hundred million cubic meters)	Production share (%)	Provincial regions
Undeveloped	2	0.00	0.00	Fujian, Tibet
No production	2	0.00	0.00	Zhejiang, Hunan
Production greater than 0 less than 5	9	13.50	0.59	Beijing, Jiangsu, Anhui, Jiangxi, Henan, Hubei, Guangxi, Yunnan, Ningxia
Production greater than 5 less than 10	4	30.00	1.31	Hebei, Liaoning, Shandong, Gansu
Production greater than 10 less than 30	2	29.80	1.30	Jilin, Guizhou
Production greater than 30 less than 50	3	112.77	4.91	Tianjin, Shanghai, Hainan
Production greater than 50 less than 100	3	215.00	9.36	Heilongjiang, Chongqing, Qinghai
Production greater than 100 less than 200	2	269.20	11.72	Shanxi, Guangdong
Production greater than 200 less than 300	1	286.30	12.46	Inner Mongolia
Production greater than 300 less than 400	1	328.50	14.30	Shaanxi
Production greater than 400 less than 500	1	417.27	18.16	Xinjiang
Production greater than 500	1	594.80	25.89	Sichuan
Total quantity	31	2297.14	100.00	

The Development of Major Oil and Gas Fields

The natural gas production of the main gas-producing oil and gas fields from 2017 to 2023 is shown in Table 9.

Table 9 Natural gas production from major gas fields (2017–2023)

Oilfield company	Oil and gas field enterprises	2017	2018	2019	2020	2021	2022	2023
China petroleum	Southwest Oil and Gas Field Branch	210.20	226.30	268.60	318.20	354.20	383.40	450.00
	Changqing Oilfield Branch	369.40	387.50	412.50	448.50	465.40	493.40	506.50
	Tarim Oilfield Branch	253.30	266.20	285.50	311.00	319.30	323.00	326.00
	Daqing Oilfield Limited Liability Company	40.10	43.40	45.50	46.60	50.20	55.40	58.00
	Xinjiang Oilfield Branch	28.40	29.20	29.30	30.00	34.90	38.50	40.00
	Qinghai Oilfield Branch	64.00	64.10	64.00	64.00	62.00	60.00	60.00
China petrochemical	Southwest Oil & Gas Branch	60.70	61.50	66.20	67.10	80.00	84.00	92.02
	Zhongyuan Oilfield Branch	59.80	65.60	70.70	64.00	69.40	66.20	88.21
	Jianghan Oilfield Branch	61.30	61.40	64.40	68.10	73.00	73.40	73.30
	North China Oil & Gas Branch	37.30	40.60	44.30	47.50	50.90	50.00	50.58
China offshore oil	Bohai	54.50	60.20	57.80	54.30	61.50	65.60	69.40
	Eastern South China Sea	86.90	126.10	142.70	146.70	137.40	147.10	156.40
	Western South China Sea	99.80	96.80	116.10	161.70	181.90	234.90	246.20

(continued)

Table 9 (continued)

Oilfield company	Oil and gas field enterprises	2017	2018	2019	2020	2021	2022	2023
Shaanxi Yanchang Petroleum (Group) Co., Ltd		25.80	33.70	48.10	57.10	71.40	75.80	80.00
National natural gas production		1480.35	1601.59	1753.62	1924.95	2075.84	2201.10	2324.00

Unit: hundred million cubic meters
CNOOC unit: billion cubic feet
Data sources (1) Xiao Lu. China's Natural Gas Production from 2017 to 2022. International Petroleum Economics, 2023.31(04), p.103; (2) Pipeline networks of PetroChina Southwest Oil & Gas Field Branch, PetroChina Changqing Oilfield Branch, and PetroChina Tarim Oilfield Branch; (3) Pipeline networks of Sinopec Southwest Oil & Gas Branch, Sinopec Zhongyuan Oilfield Branch, and Sinopec Jianghan Oilfield Branch; (4) Annual performance reports of China National Offshore Oil Corporation from 2017 to 2023; (5) Pipeline network of Shaanxi Yanchang Petroleum (Group) Co., Ltd.
Note 1 cubic foot = 0.0283168 cubic meters.

According to Table 9, the annual natural gas production of the major oil and gas fields in 2023 has increased compared to 2022. The comparison of natural gas production of the main oil and gas fields in 2023 and 2022 is shown in Table 10.

The annual natural gas production structure of oil and gas fields in 2023 was not much different from that in 2022. The total annual natural gas production of the Southwest Oil and Gas Field, Changqing Oil Field, and Tarim Oil Field, which belong to China National Petroleum Corporation (CNPC) accounted for more than 55% in both years. These oil and gas fields are the mainstays for achieving the near-term planning target of 240 billion cubic meters by the end of the "14th Five-Year Plan" in 2025 and the medium and long-term planning target of 330 billion cubic meters by the end of the "16th Five-Year Plan" in 2035.

The Development Situation by Product Type

(I) Product attributes

Natural gas is divided into conventional natural gas and unconventional natural gas according to its attributes. Unconventional natural gas includes coalbed methane and shale gas. This section mainly reports and analyzes the development of unconventional natural gas.

1. Coalbed methane

The production of coalbed methane from 2017 to 2023 is shown in Table 11.

Table 10 Comparison of natural gas production of major oil and gas fields in 2022 and 2023

Oilfield company	Oil and gas field enterprises	2022 Production	2023 Production	2022 Proportion	2023 Proportion	Annual production Change number	Annual production Change rate (%)
PetroChina	Southwest Oil and Gas Field Branch	383.40	420.00	17.42%	19.36%	66.60	17.37
	Changqing Oil Field Branch	506.50	458.18	22.42%	21.79%	13.10	2.66
	Tarim Oil Field Branch	323.00	326.00	14.67%	14.03%	3.00	0.93
	Daqing Oilfield Co., Ltd	55.40	58.00	2.52%	2.50%	2.60	4.69
	Xinjiang Oilfield Branch	38.40	40.00	1.75%	1.72%	1.50	3.90
	Qinghai Oilfield Branch	60.00	60.00	2.73%	2.58%	0.00	0.00
Sinopec	Southwest Oil and Gas Branch	84.01	92.02	3.82%	3.96%	8.02	9.55
	Zhongyuan Oilfield Branch	84.59	88.21	3.01%	3.80%	22.01	33.25
	Jianghan Oilfield Branch	71.96	73.3	3.33%	3.15%	−0.10	−0.14
	North China Oil and Gas Branch	50.01	50.59	2.27%	2.18%	0.58	1.16
CNOOC	Bohai	65.60	69.40	–	–	3.80	5.79
	Eastern South China Sea	147.10	156.40	–	–	9.30	6.32
	Western South China Sea	234.90	246.20	–	–	11.30	4.81

(continued)

Table 10 (continued)

Oilfield company	Oil and gas field enterprises	2022 Production	2023 Production	2022 Proportion	2023 Proportion	Annual production Change number	Annual production Change rate (%)
Shaanxi Yanchang Petroleum (Group) Co., Ltd		75.60	80.00	3.44%	3.44%	4.20	5.54
National Natural Gas Production		2201.10	2324.00	100.00%	100.00%	122.90	5.58

Unit: hundred million cubic meters
CNOOC unit: billion cubic feet
Data source (1) Xiao, Lu. Natural Gas Production in China from 2017 to 2022. "International Petroleum Economics," 2023.31 (04), page 103; (2) Official websites of PetroChina Southwest Oil and Gas Field Branch, PetroChina Changqing Oil Field Branch, PetroChina Tarim Oil Field Branch; (3) Official websites of Sinopec Southwest Oil and Gas Branch, Sinopec Zhongyuan Oil Field Branch, Sinopec Jianghan Oil Field Branch; (4) CNOOC Annual Performance Report from 2017 to 2023; (5) Official website of Shaanxi Yanchang Petroleum (Group) Co., Ltd
Note 1 cubic foot = 0.0283168 cubic meters

Table 11 Coalbed methane output and growth rate (2017–2023)

Month	2017	2018	2019	2020	2021	2022	2023
1–2	12.50	11.80	13.40	14.80	16.50	18.40	20.70
3	6.50	5.70	6.70	7.90	8.80	9.30	10.50
4	6.40	6.00	6.90	7.90	8.30	9.30	10.60
5	6.20	5.80	6.90	8.10	8.56	9.50	10.90
6	5.70	6.00	7.10	7.40	8.18	9.10	10.80
7	5.40	5.80	7.20	7.90	8.20	9.50	10.80
8	5.90	6.00	7.00	8.10	8.50	9.50	11.00
9	5.70	6.00	7.10	8.40	8.50	9.80	11.60
10	5.90	5.80	6.90	9.10	8.70	9.50	12.00
11	6.40	6.30	7.50	9.00	9.10	10.20	12.70
12	6.40	6.40	7.70	11.50	9.70	11.10	13.80
Total	73.00	71.60	84.40	100.10	103.04	115.20	135.40
Growth rate		−1.92%	17.88%	18.60%	2.94%	11.80%	17.53%

Unit: hundred million cubic meters
Data source China Economic Data

The changes in coalbed methane production and growth rate from 2017 to 2023 are shown in Fig. 5.

After a slight decline in 2017–2018, the production of coalbed methane has been growing. From 2019 to 2023, except for 2021 when the production growth rate was lower than the average growth rate of 9.46%, the other years maintained a high

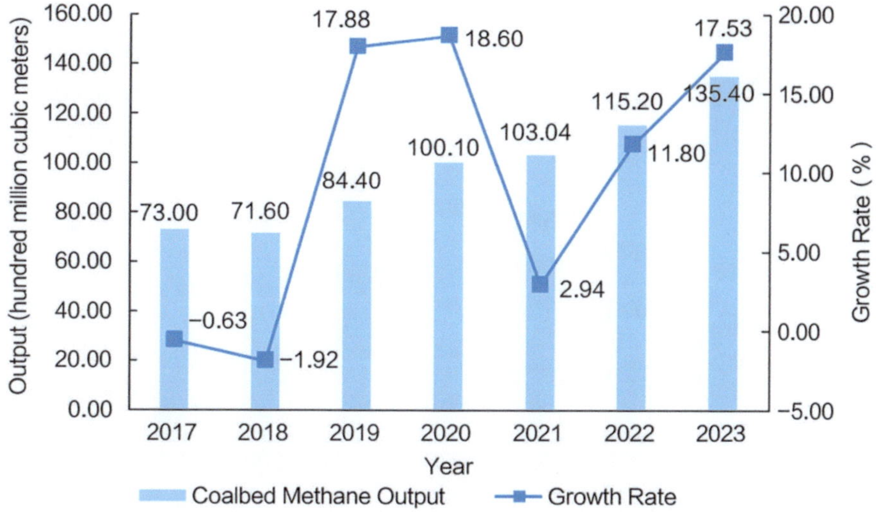

Fig. 5 Annual coalbed methane output and growth rate (2017–2023). *Data source* China Economic Data

level of growth, indicating that the growth momentum of annual coalbed methane production is robust.

2. Shale gas

The production and growth rate of shale gas from 2017 to 2023 are shown in Table 12 and Fig. 6.

The production of shale gas has been continuously increasing from 2017 to 2023. The annual production of shale gas showed an accelerating growth trend from 2017 to 2019. After 2019, the growth rate of shale gas production has been declining year by year, and it has remained stable in the last two years.

3. Development outlook for unconventional natural gas

The proportions of coalbed methane and shale gas—these two types of unconventional natural gas—in the total natural gas production from 2017 to 2023 are shown in Table 13.

The proportions of unconventional natural gas from 2017 to 2023 is shown in Fig. 7.

Table 12 Annual shale gas output and growth rate (2017–2023)

Project	2017	2018	2019	2020	2021	2022	2023
Output	92.00	109.00	154.00	200.40	230.00	240.00	250.00
Growth rate	16.46%	18.48%	41.28%	30.13%	14.77%	4.35%	4.17%

Unit: hundred million cubic meters
Data source National Energy Administration

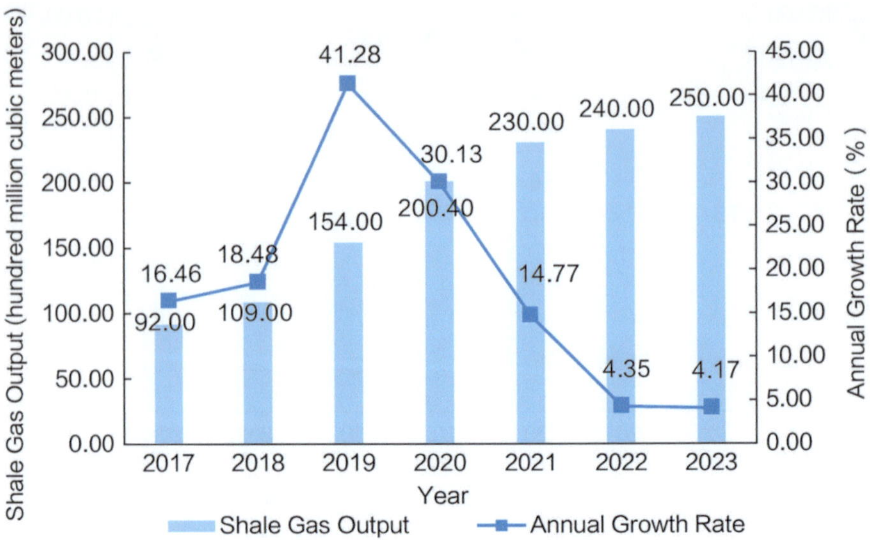

Fig. 6 Annual shale gas production and growth rate (2017–2023). *Data source* National Energy Administration

Table 13 Changes in the proportion of unconventional natural gas (2017–2023)

Project	2017	2018	2019	2020	2021	2022	2023
Natural gas production	1480.35	1601.59	1753.62	1924.95	2075.84	2201.10	2324.00
Coalbed methane output	73.00	71.60	84.40	100.10	103.04	115.20	135.40
Shale gas output	92.00	109.00	154.00	200.40	230.00	240.00	250.00
Unconventional natural gas	165.00	180.60	238.40	300.50	333.04	355.20	385.40
Unconventional proportion (%)	11.15	11.28	13.59	15.61	16.04	16.14	16.58
Coalbed methane proportion (%)	4.93	4.47	4.81	5.20	4.96	5.23	5.83
Shale gas proportion (%)	6.21	6.81	8.78	10.41	11.08	10.90	10.76

Unit: hundred million cubic meters
Data source China Economic Data

Over time, the production of unconventional natural gas and its proportion in total natural gas production show an increasing trend, and its supplementary role in China's natural gas resource development is becoming more and more prominent. Among them, the proportion of shale gas production in the total unconventional natural gas production first rises rapidly and then remains stable, while the proportion of coalbed methane production shows a slow and gentle upward trend.

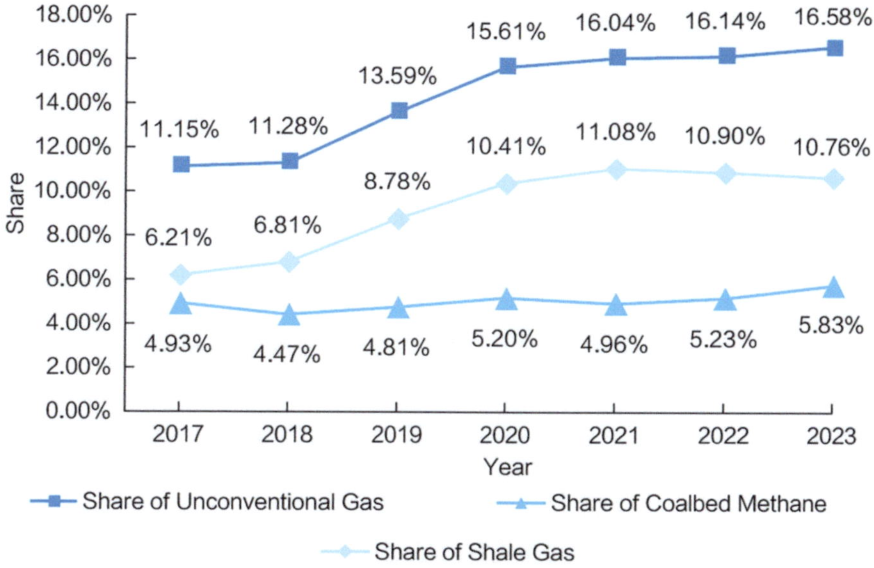

Fig. 7 Trends in the proportion of unconventional natural gas (2017–2023). *Data source* China Economic Data

(II) Product state

Natural gas can be divided into four types according to the state of the product: ordinary natural gas (NG), compressed natural gas (CNG), liquefied natural gas (LNG), and natural gas hydrate (NGH). The main commodity forms in the market are ordinary natural gas (NG) and liquefied natural gas (LNG). The previous section mainly focused on the development of NG, and this section mainly reports and analyzes the development of LNG.

The annual production of LNG from 2017 to 2023 is shown in Table 14.

According to Table 14, the annual production and growth rate of LNG from 2017 to 2023 are shown in Fig. 8.

As can be seen from Fig. 8, the annual production scale of LNG from 2017 to 2023 has shown a year-on-year increasing trend. The annual production growth rate has remained at a relatively high level, reaching an annual growth rate of 29.42% in 2019. In the past three years, the production has shown a steady growth trend.

Table 14 Annual LNG output and growth rate statistics (2017–2023)

Project	2017	2018	2019	2020	2021	2022	2023
Annual output	829.00	900.20	1165.00	1332.90	1545.10	1742.70	2037.10
Growth rate (%)	19.23	8.59	29.42	14.41	15.92	12.79	16.89

Unit: 10 kilotons
Data source China Economic Data

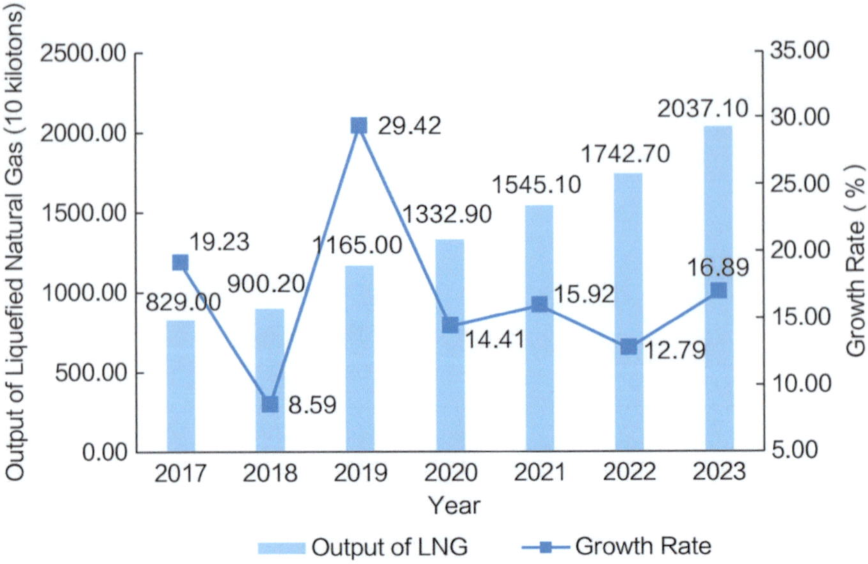

Fig. 8 Trends in LNG output and growth rate (2017–2023). *Data source* China Economic Data

Natural Gas Imports and Exports in China

In 2023, as international natural gas prices gradually returned to normal, China's natural gas consumption steadily increased, and the import volume resumed its growth trend. The total import volume reached 119,971,200 tons, an increase of 10.8 million tons compared to 2022, representing a 9.9% increase. With the domestic economic recovery and the decline in international gas prices, it is expected that the natural gas import volume will continue to grow in the coming year to meet the constantly increasing natural gas consumption demand.

Natural Gas Imports

The "Action Plan for Carbon Peak by 2030," released by the State Council in October 2021, pointed out the need to gradually increase the proportion of natural gas consumption, promote the joint development of natural gas with multiple energy sources, and reasonably guide the use of natural gas in industry and chemical industries, thereby changing the way enterprises use energy. In 2023, the proportion of natural gas in China's primary energy consumption was about 8.5%, which increased by 0.05 percentage points compared to 2022. According to the target set forth in the Energy Production and Consumption Revolution Strategy (2016–2030), by 2030, the proportion of natural gas consumption in China's primary energy is expected to be raised to around 15%. Judging from this data, there is still a huge potential for the increase in China's natural gas consumption. In 2023, China's natural gas production volume was 232,430 million cubic meters, approximately 16,716 thousand tons, while the consumption volume was 28,375 thousand tons. The import volume of natural gas was 11,997.12 thousand tons, with an external dependence rate of about 42.3%. This indicates that China's current annual natural gas production is far below the domestic consumption volume. The extraction volume is still unable to meet the consumption volume, and the production growth rate cannot keep pace with the consumption growth rate. Therefore, China's dependence on natural gas imports will continue to exist in the long term.

(I) Natural gas import situation

As of December 2023, China's natural gas import source countries reached 26. The sources of import are characterized by diversification across multiple regions and countries, and a synchronized model of trade in both pipelined natural gas (PNG) and liquefied natural gas (LNG) is gradually taking shape. In terms of the total volume of natural gas imports in China, compared with 109.17 million tons in 2022, it increased by 9.9% year-on-year. Looking at the entire international natural gas market, since 2023, the demand for natural gas in European countries has not been high, the demand growth in the Asia–Pacific region was not significant, resulting in weak international natural gas market demand. The supply side, mainly composed of the United States and Middle Eastern countries, still maintains a high growth rate of natural gas production. The international natural gas market has an adequate supply, and overall, the supply and demand of the international natural gas market are relatively relaxed. Although in the second half of the year, several unexpected accidents occurred in natural gas supply countries such as Australia and Norway, increasing the risk of supply interruption and causing a short-term increase in international natural gas prices, the natural gas prices showed a downward trend for the whole year and are gradually returning to normal. With the decline of international natural gas prices and the continuous increase of domestic gas demand, China's natural gas imports rose significantly in 2023.

1. Sources of natural gas imports

Currently, the natural gas imported by China is mainly divided into two categories based on transportation methods: PNG (Pipeline Natural Gas) and LNG (Liquefied Natural Gas). Compared to the 28 natural gas import source countries in 2022, the number decreased by 2 in 2023. China's PNG imports are mainly transported via pipelines. Thanks to its geographical advantages, China's pipeline natural gas can be imported more reliably and safely through overland natural gas pipelines. The western route can reach the main natural gas producing countries in Central Asia via Khorgos, the eastern route can access Russia's Chayandinskoye gas field via Heihe, and the southern route can reach Myanmar's Kyaukphyu via Kunming. In recent years, with the completion of the main natural gas pipelines in the eastern and western routes, China's PNG imports have seen further growth. LNG is mainly imported through maritime transportation, with specialized LNG fleets transporting ultra-low-temperature liquefied natural gas. Amid geopolitical impacts such as the Russia-Ukraine conflict and the Israel-Palestine conflict, maritime freight prices have repeatedly hit new highs, and the capacity of LNG fleets is relatively insufficient. In this international context, as the world's largest LNG importer, China's stability and security of liquefied natural gas supply have also been somewhat affected.

It should be noted that since 2022, the General Administration of Customs has made statistical adjustments to PNG, and no longer publishes multi-dimensional statistical group data other than the total import volume. Therefore, it is impossible to obtain accurate data on the import quantities of PNG from different source countries. The existing import quantities from various source countries are calculated based on the average import price of PNG. The calculation method mainly involves dividing the total import price of different importing countries in each month by the average import price of PNG in that month to obtain the monthly import volume. Then, the annual import volume is obtained by summing up the import volumes of each source country from January to December. The calculation results are for reference only. As shown in Fig. 9, the main import source countries of China's PNG in 2023 remained the same as in 2022, namely Kazakhstan, Turkmenistan, Uzbekistan, Russia, and Myanmar. Among them, the import volume from Turkmenistan still ranked first, with an import volume of about 24.0153 million tons, accounting for 49.36%, a year-on-year decrease of 14.41% in the proportion. The import volume of PNG from Russia was about 16.0741 million tons, with a year-on-year increase of 48.55% in the proportion, reaching 33.04% in 2023 and ranking second. With the continuation of the Russia-Ukraine conflict, Russia has basically stopped supplying PNG to Europe, while its supply of PNG to China has been increasing continuously. Currently, Russia is actively preparing for the construction of the new pipeline "Power of Siberia 2" and the addition of the "Far East route" for PNG transportation to China. It can be imagined that the proportion of PNG imports from Russia will continue to rise in the future. The proportion of imports from Kazakhstan also shows an upward trend, with a year-on-year increase of 14.51%. In contrast, the proportion of imports from Uzbekistan has dropped to 3.02%, about half of that in 2022. The import proportions from the remaining countries remained basically the same as in 2022.

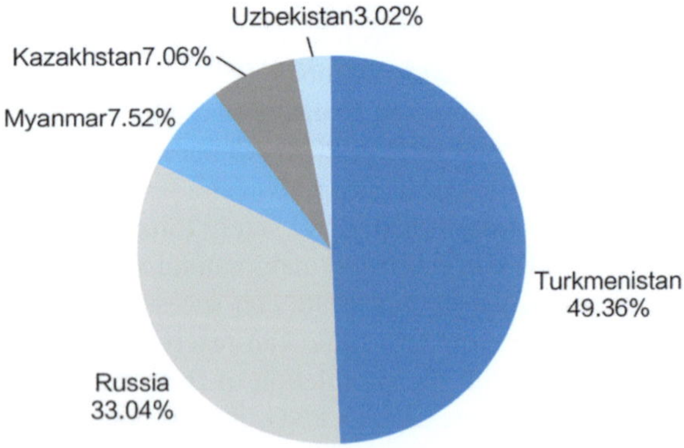

Fig. 9 Ranking of major import source countries of China's PNG in 2023. *Data source* General Administration of Customs

China imported LNG from 21 countries in 2023, 3 fewer than in 2022. As shown in Fig. 10, the main LNG import source countries for China are Australia, Qatar, Russia, and Malaysia. The total import LNG quantity of the four countries has dropped from 81.06% in 2022 to 78.48%. Including these 4 countries, there are 7 countries in total with an import volume of more than 2 million tons, including Indonesia, the United States, and Papua New Guinea, with their import volumes accounting for 5.59%, 4.42%, and 3.50% respectively. The country with the largest import volume in 2023 is still Australia, with a quantity of 24.161 million tons, accounting for 33.88%. The import proportion of the United States has increased, with an import volume accounting for 4.42%, a year-on-year increase of 34.23%. Compared with 2022, Russia's import proportion continues to grow, with its LNG import volume accounting for 11.29% in 2023, up from 10.25% in 2022, an increase of 10.17%, ranking third. It can be seen that Russia's import proportions in both PNG and LNG have significantly increased, which is related to the further strengthening of the energy cooperation between China and Russia in 2023. The two sides signed a new agreement to increase Russia's natural gas supply to China. Apart from these changes, the LNG import proportions of other source countries are basically the same as in 2022. Comparing the natural gas import and export data in recent years, it can be found that the proportion of LNG imported from Australia by China has declined significantly, having dropped from 43.91% in 2020 to 33.88% in 2023. At the same time, the import proportions from the 4 main importing countries have also been adjusted compared to 2022, and the total proportion is also declining. This shows that the situation where Australia's LNG imports once dominated is being reversed, and China is striving to reduce its dependence on the main importing countries for natural gas, and the import sources for LNG is becoming gradually diversified. With the continuous optimization of the import structure, China will have more initiative in international LNG import price negotiations.

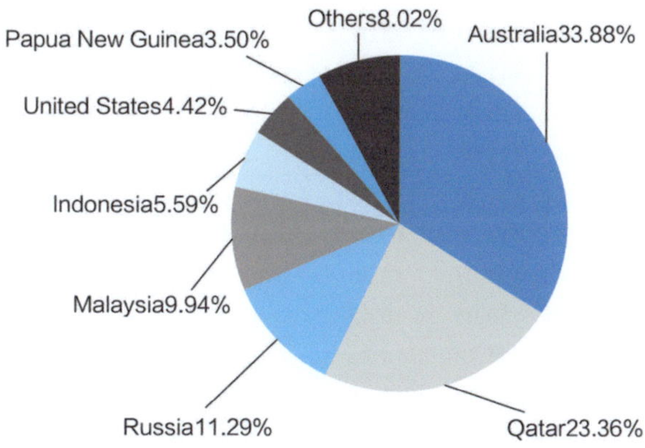

Fig. 10 Ranking of major import source countries of China's LNG in 2023. *Data source* General Administration of Customs

Power.

2. Natural gas import volume

The statistical table of China's Natural Gas Import Volume and the chart of China's Natural Gas Import Volume and Year-on-Year Growth Rate from 2022 to 2023 are shown in Table 15 and Fig. 11 respectively. The data shows that China's total annual natural gas imports reached 119.97 million tons, an increase of 10.8 million tons compared to 2022, a year-on-year increase of 9.90%. Among them, the import volume of LNG was 71.3166 million tons, a year-on-year increase of 12.55%, with the largest monthly import volume occurring in December, at 8.4012 million tons. The import volume of PNG was 48.65 million tons, a year-on-year increase of 6.22%, with the largest monthly import volume appearing in August, at 4.56 million tons. It can be seen that both PNG and LNG imports increased compared with 2022, with a more significant growth in LNG. As China's epidemic prevention and control policies entered a new phase, various industries gradually recovered, leading to a continuous increase in natural gas demand, coupled with the continuous decline in international natural gas prices, China's natural gas import volume reversed the decline in 2022 and returned to a growth trend in 2023.

In addition, it can be seen that in the first half of 2023, China's natural gas import volume was 56.54 million tons, a year-on-year increase of 5.7%, while in the second half of the year, the natural gas import volume was 63.44 million tons, a year-on-year increase of 14.0%. This is in line with China's natural gas import pattern: the

Table 15 Statistics of China's natural gas imports in 2023

Month	PNG		LNG		Natural gas	
	Import volume (10 kilotons)	Year-on-Year growth (%)	Import volume (10 kilotons)	Year-on-Year growth (%)	Import volume (10 kilotons)	Year-on-Year growth (%)
1	336.33	−3.01	589.83	−24.35	926.15	−17.78
2	344.56	−7.20	516.71	7.30	861.28	0.99
3	350.53	4.43	531.56	15.88	882.09	11.04
4	421.08	12.64	473.12	9.44	894.20	10.93
5	421.70	1.73	635.19	30.23	1056.89	17.14
6	443.08	13.75	589.81	23.16	1032.89	18.94
7	444.87	12.36	583.70	23.84	1028.57	18.60
8	456.04	10.45	622.21	32.49	1078.25	22.18
9	445.78	4.86	562.00	−3.98	1007.78	−0.26
10	361.76	1.18	512.76	28.46	874.52	15.57
11	414.93	6.59	674.66	5.79	1089.59	6.09
12	424.79	15.40	840.12	28.41	1264.91	23.72
Total	4865.45	6.22	7131.66	12.55	11,997.12	9.90

Data source General Administration of Customs

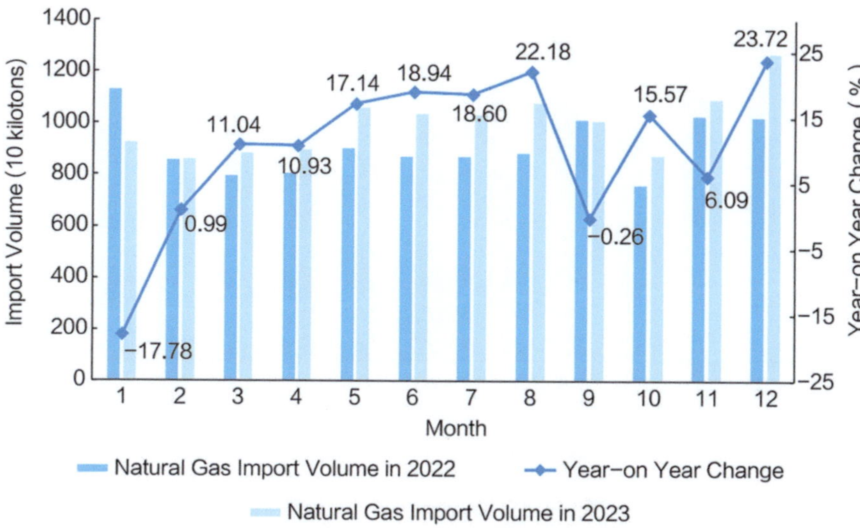

Fig. 11 China's natural gas import volume and year-on-year growth rate (2022–2023). *Data source* General Administration of Customs

natural gas import cycle is generally divided into off-peak and peak seasons, with less imports in the first half of the year and more imports in the second half of the year due to the combined factors of supply assurance and increased electricity demand. Moreover, after experiencing a negative growth in import volume in 2022 compared with 2021, China's import volume returned to a growth trend in 2023. Firstly, except for January and September, the monthly natural gas import volumes in 2023 were all higher than those in 2022. Secondly, in 2023, there were eight months with a year-on-year growth rate exceeding 10%, indicating a relatively significant growth trend.

Compared to 2022, the growth trend of natural gas import volume in 2023 was more evident, with the largest increases in August and December. The average growth rate from January to December was 10.60%. Throughout the year, seven months—May to September, November, and December—saw import volumes exceeding 10 million tons. The highest import volume was 12.65 million tons in December, a year-on-year increase of 23.72%. The lowest import volume was in February, at 8.61 million tons, with a year-on-year increase of 0.99%. The reasons for this trend are twofold. On one hand, cold waves and the need for winter heating led to increased electricity use in households and gas consumption in industries, driving up the demand for natural gas and consequently the import volume. On the other hand, with the comprehensive lifting of COVID-19 restrictions in 2023, the gradual resumption of normal production and living activities, and the increased gas demand from tourism, transportation, and public service sectors all contributed to the rise in natural gas imports.

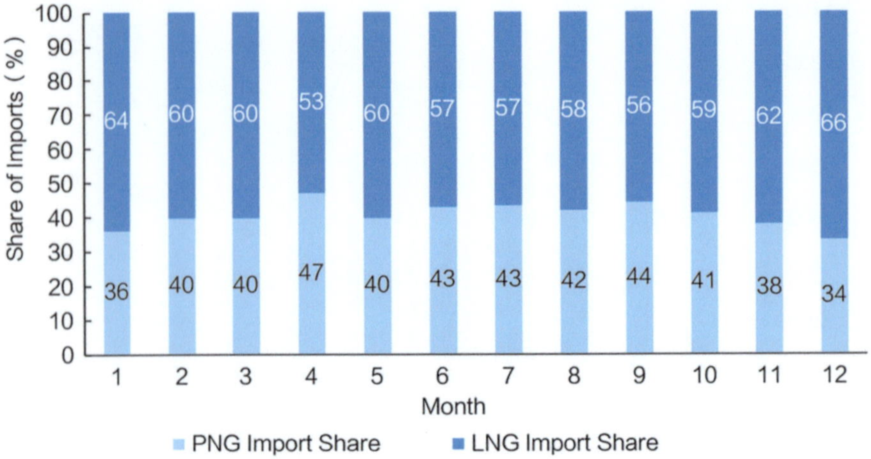

Fig. 12 Structure of China's natural gas imports in 2023. *Data source* General Administration of Customs

The structure of China's natural gas imports in 2023 is shown in Fig. 12. Compared to 2022, there was no significant change in the structure of natural gas imports in 2023. The average proportion of PNG decreased from 42.42% in 2022 to 40.73%, a decrease of 3.98%. The average proportion of LNG increased from 57.58% to 59.27%. LNG still occupied a dominant position.

In recent years, China has accelerated the process of building a diversified natural gas import structure, which is currently in its initial stage. The import ratio of major importing countries such as Australia and Turkmenistan is around 40%. If international disputes occur, there is a risk of having insufficient gas sources to cover shortages. Because LNG has the advantages of not being constrained by pipeline infrastructure and flexible allocation, and China's LNG receiving station facilities are relatively complete, in the foreseeable future, LNG imports will continue to hold a dominant position. However, with the increase in uncertainty in the international natural gas market, the proportion of LNG imports may gradually decrease. In terms of PNG imports, because most of China's import agreements are "long-term agreements," the gas price is linked to international crude oil prices. As a result, the space for price fluctuation is smaller than in the international spot natural gas market, making the gas price relatively stable. Accordingly, the proportion of PNG imports should increase.

3. Natural gas import prices

When talking about natural gas import prices, it is first necessary to distinguish the structure of natural gas imports, as the prices of different types of natural gas and their underlying factors vary. China's imported natural gas is divided into PNG and LNG, and LNG can be further divided into long-term contracts, medium and short-term contracts, and spot LNG. At present, most of China's PNG and LNG are signed as "long-term agreements," where the buyers and sellers enter into supply and demand agreements lasting around 20 years based on the "take-or-pay" principle. The import prices of long-term and medium-to-short-term LNG contracts are linked with international oil prices, primarily using the natural gas pricing mechanism of the Asia–Pacific region led by Japan, with the benchmark price referring to the Japan Crude Cocktail (JCC) index, and adjusted according to factors such as inflation and periodic revision clauses. In contrast, the import price of PNG is partly determined through government-negotiated prices under a bilateral monopoly and partly linked to oil prices. This can be simply summarized as "$P = \alpha \cdot x + \beta$," where P represents the import price of natural gas, α denotes the variables affecting the price, such as negotiable factors like policy, production, and demand, x stands for the oil price index, and β is the adjustment constant.

Unlike long-term contracts, the price of spot LNG is mainly affected by supply and demand in the international market, and the changes are more flexible. According to the literature, there is no correlation between the spot LNG price and the spot crude oil price in China, and the two spot markets cannot form a price transmission relationship. However, the volatility of international natural gas price indices has both direct and indirect transmission effects on China's spot LNG prices, with the potential to influence these prices by up to 40%. Therefore, China's spot LNG imports primarily reference international natural gas price indices.

Due to the regional nature of natural gas trade, there is currently no unified standard for natural gas pricing internationally. The spot price of LNG mainly follows the market trends and refers to the three major global natural gas trading prices. The three main pricing methods are as follows: (1) The North American natural gas pricing mechanism, led by the United States, which is characterized by typical market competition. The volume of domestic natural gas production and the trading price at the Henry Hub (HH) directly affect the quantity and price of imported natural gas. (2) The European natural gas pricing mechanism, led by the United Kingdom, which primarily uses a pricing method linked to competing energy sources such as refined oil products, supplemented by reference to trading center prices like NBP (National Balancing Point) and TTF (Title Transfer Facility). (3) The Platts JKM (Japan-Korea Marker) price index, led by Japan and South Korea in the East Asian region. The Platts JKM pricing system, after years of development, has become a widely accepted price index that reflects the level of LNG spot trading in Asia.

Referring to the price trends of HH, NBP, TTF, and JKM, the international gas price overall showed a downward trend in 2023. The relaxed supply and demand situation in the international liquefied natural gas market, along with the weak global macroeconomic environment, contributed to the decline in global natural gas prices. From January to December 2023, the average spot price of Europe's TTF was $13.06 per million British thermal units (Btu), a decrease of 65.62% compared with the same period in 2022. The average spot price of Northeast Asia LNG spot delivered was $15.75 per million Btu, a year-on-year decrease of 54.26%. The average spot price of the US HH fell from $6.5 per million Btu in 2022 to $2.61 per million Btu, a year-on-year decrease of 60.43%.

In the first half of 2023, from the demand side, due to higher temperatures during the heating season and the global economic recession, European countries' inventory drawdown was slower than in previous years, with stock levels reaching above the five-year average. Meanwhile, demand growth in the Asia–Pacific region was not significant, resulting in weak international natural gas market demand. On the supply side, LNG production in the Middle East steadily increased. Driven by the Russia-Ukraine conflict and high natural gas prices, the United States vigorously developed LNG extraction in 2023, leading to a substantial increase in production. Consequently, the international natural gas market supply was abundant. As a result, the market supply and demand relationship was relaxed in the first half of the year, and international gas prices continued to fall. In the second half of 2023, supply-side incidents such as abnormal maintenance of Norwegian gas fields, strikes declared at Australian LNG export plants, and leaks in the Finland-Estonia natural gas pipeline, as well as geopolitical impacts from the Israel-Palestine conflict, caused energy supply panic and led to an increase in international LNG prices. However, with the onset of winter and warmer-than-usual weather, coupled with continued weak demand, although international natural gas prices rose, the extent of the increase was not significant.

Looking at the current trend of natural gas prices, it is expected that international natural gas prices will continue to decline. The average monthly price of Northeast Asian spot LNG in 2023 is shown in Table 16. It can be observed that the average price of Northeast Asia LNG spot from January to December 2023 was \$15.75 per million British thermal units (Btu). The lowest average price occurred in July at \$9.49 per million Btu, and the highest year-on-year decrease was in October, reaching as high as 76.16%. Figure 13 presents a comparison of the monthly average delivered prices of Northeast Asia LNG spot between 2022 and 2023. As can be clearly seen from Fig. 13, the monthly average spot price in Northeast Asia in 2023 has decreased compared to 2022, with an average decline of 51.87%. It can be predicted that the international natural gas market will gradually return to the state before the Russia-Ukraine conflict, and the average spot price of Northeast Asia LNG will gradually return to the level before the surge.

In order to facilitate comparison with domestic prices, the unit "US dollars / million British thermal units" has been converted to the unit "US dollars / ton" based on the relevant formula. This conversion is based on the equivalent heat released by the complete combustion of LNG, and the calculated results are for reference only. As shown in Fig. 14, the comparison of Northeast Asia LNG spot prices and China's LNG import prices in 2023 reveals that the monthly average spot price of Northeast Asia LNG first dropped sharply and then rebounded. In contrast, the average import price of China's LNG was relatively stable with a smaller fluctuation range. In the first quarter, the average import price of China's LNG was significantly lower than the monthly average spot price of Northeast Asia LNG. In the second and third quarters, as the monthly average spot price of Northeast Asia LNG continued to decline and the average import price of China remained relatively stable, the two gradually converged. In the fourth quarter, both prices increased, but the rise in the monthly average spot price of Northeast Asia LNG was more pronounced.

Table 16 Year-on-year comparison of monthly average spot prices of LNG in Northeast Asia in 2023

Month	Average Spot Price of LNG in Northeast Asia in 2022 (USD/MMBtu)	Average Spot Price of LNG in Northeast Asia in 2023 (USD/MMBtu)	Year-on-Year change (%)	Month-on-Month change (%)
1	36.04	31.30	−13.15	12.19
2	34.50	25.30	−26.67	−19.17
3	24.90	18.40	−26.10	−27.27
4	37.30	14.30	−61.66	−22.28
5	32.50	12.70	−60.92	−11.19
6	23.90	11.03	−53.85	−13.15
7	22.90	9.49	−58.56	−13.96
8	39.40	11.61	−70.53	22.34
9	44.50	10.96	−75.37	−5.60
10	54.10	12.90	−76.16	17.70
11	35.3	14.20	−59.77	10.08
12	27.9	16.82	−39.71	18.45

Data source "International Natural Gas Market Monthly Review"

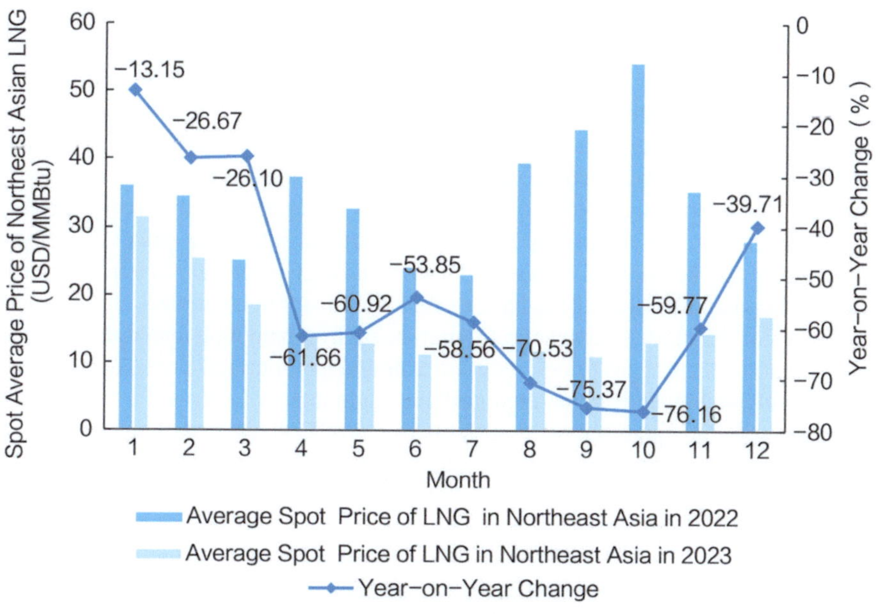

Fig. 13 Comparison of monthly average spot prices of LNG in Northeast Asia in 2022 and 2023. *Data source* "International Natural Gas Market Monthly Review"

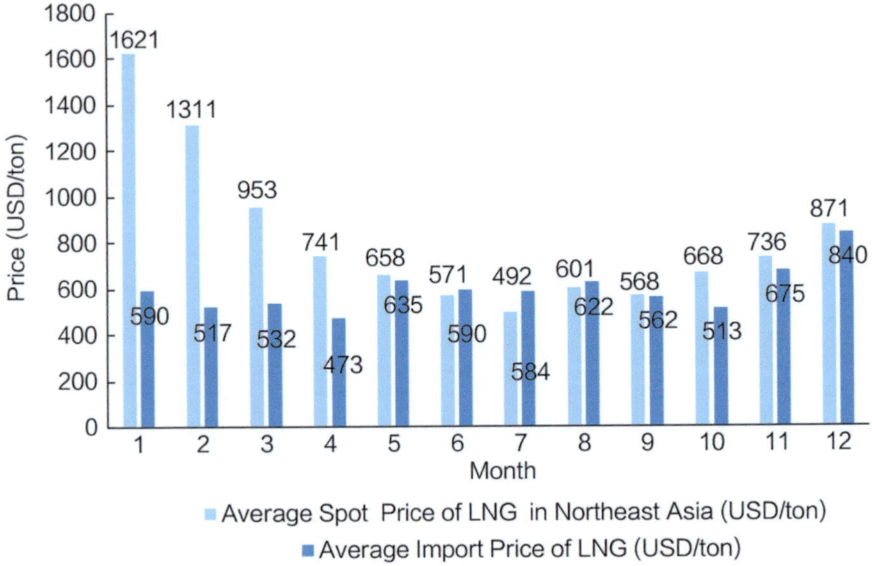

Fig. 14 Comparison of spot prices of LNG in Northeast Asia in 2023 with China's LNG import prices. *Data source* "International Natural Gas Market Monthly Review"

From the above fluctuations in international LNG spot prices, although international natural gas prices are highly volatile, the impact on China's LNG import prices is generally controllable. According to relevant data, most of China's LNG import prices are based on long-term agreements, with LNG spot purchases accounting for only about 8% of the total natural gas consumption. While this supply method mitigates the pressure of rising international natural gas prices, it also carries certain risks. For example, from January to July 2023, if international natural gas prices continued to decline as they had been, China would have faced higher natural gas import prices. Since 2023, geopolitical conflicts have caused significant fluctuations in the international natural gas market. In the current market environment with uncertain changes in international natural gas prices, long-term natural gas contracts linked to oil prices are relatively stable, and "contracted gas" has undoubtedly become the "stabilizer" for ensuring the adequate supply and stable prices of China's natural gas market.

The monthly average import prices of LNG in China and their changes in 2022 and 2023 are shown in Table 17 and Fig. 15. It can be seen that, compared with 2022, the domestic LNG import prices in 2023 generally showed a downward trend, with the exception of a price increase in February. The prices of LNG decreased to varying degrees in other months. From January to December 2023, the cumulative average import price of LNG was 628.36 US dollars per ton, compared with the average import price of 811.04 US dollars per ton in 2022. The price per ton decreased by 182.68 US dollars, with a year—on—year decrease of 22.52%. The average price in the first half of the year was 664.89 US dollars per ton, with an average decrease of 10.89%, while the average price in the second half of the year was 591.83 US dollars per ton, with an average decrease of 32.43%. Overall, the prices in the first half of the year were higher, with the average decrease far lower than that in the second half of the year. The import prices of natural gas generally showed a downward trend. From January to March, the import prices were relatively high and showed a downward trend. China maintained a wait-and-see attitude and did not purchase a large amount of natural gas, which also corresponded to the fact that the import volume in the first three months was not large as mentioned earlier. The largest decrease appeared in October, reaching 41.5%, but the import price was not very low, at 567.84 US dollars per ton. This was related to the fact that the import price in the same period of 2022 was the highest of the year. The lowest price appeared in July, at 541.35 US dollars per ton.

The monthly import prices of PNG (Pipeline Natural Gas) in China and their changes in 2022 and 2023 in China are shown in Table 18 and Fig. 16. It can be seen that the import prices of PNG in 2023 showed a downward trend throughout the year. Compared with the same period in 2022, the prices increased year-on-year in the first half of the year, while they decreased year-on-year in the second half. From January to December, the average import price of PNG was 400.95 US dollars per ton. Compared with 2022, the price per ton increased by 12.60 US dollars, with a year-on-year increase of 3.24%. It can be found from the bar chart that the monthly import prices of PNG in 2023 showed a downward trend. However, the monthly average import prices of PNG in 2023 were higher than those in 2022. This was because the high gas prices at the end of 2022 continued into the beginning of 2023. Although the prices continued to decline after entering 2023, the increase in domestic gas demand in the second half of the year led to an insignificant decrease in import

Table 17 Comparison of monthly LNG import prices in China in 2022 and 2023

Month	Import price of LNG in 2022 (USD/ton)	Import price of LNG in 2023 (USD/ton)	Year-on-Year change (%)	Month-on-Month change (%)
1	969.55	820.62	−15.36	−9.12
2	679.72	786.24	15.67	−4.19
3	662.38	619.90	−6.41	−21.16
4	694.41	612.15	−11.85	−1.25
5	757.33	590.83	−21.99	−3.48
6	713.68	559.58	−21.59	−5.29
7	705.21	541.35	−23.24	−3.26
8	887.13	563.43	−36.49	4.08
9	927.07	561.18	−39.47	−0.40
10	970.65	567.84	−41.50	1.19
11	862.39	620.71	−28.03	9.31
12	902.97	696.48	−22.87	12.21

Data source General Administration of Customs

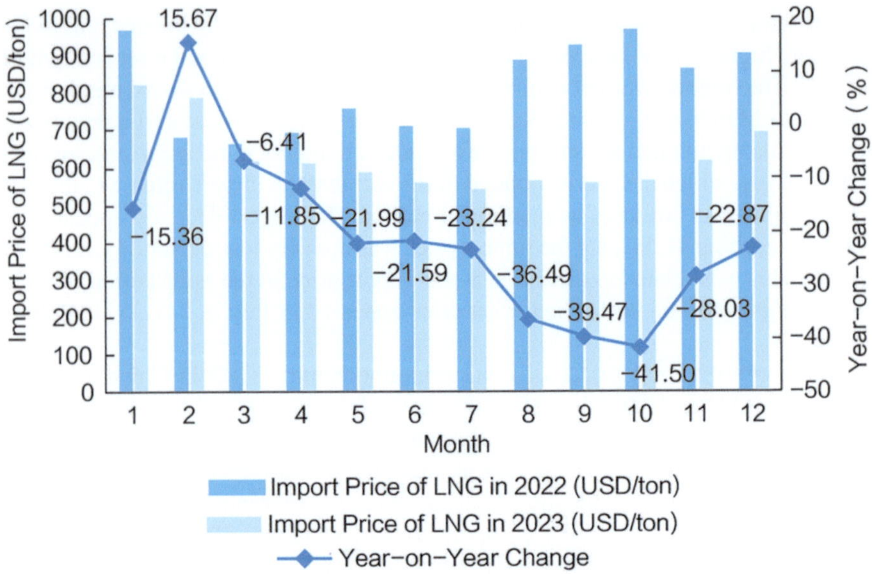

Fig. 15 Monthly LNG import prices in China in 2022 and 2023. *Data source* General Administration of Customs

prices, with the year-on-year increase being greater than the year-on-year decrease. The highest import price of PNG appeared in January, at 458.35 US dollars per ton, while the lowest import price appeared in September, at 366.31 US dollars per ton.

(I) External dependence rate of natural gas in China

The natural gas external dependence rate refers to the proportion of natural gas imports in a given period to the total natural gas consumption in the same period. It is used to measure the degree of domestic natural gas consumption's reliance on the international market. It is the percentage of a country's net import volume of natural gas to the country's total domestic natural gas consumption. The level of this rate not only reflects the degree of integration between the country and the international natural gas market but also reflects the degree of dependence on foreign natural gas resources and the level of security in supply.

Since the beginning of the twenty-first century, with the rapid growth of natural gas consumption in China and the gradual improvement of people's living standards, and driven by the urgent need for atmospheric environmental governance, the domestic market's demand for natural gas resources has been continuously increasing, resulting in a persistently high external dependence rate. Since 2018, when China first became a net importer of natural gas, the external dependence rate for natural gas has continued to rise. In recent years, natural gas has increasingly become a strategically important part of China's energy consumption structure. The increasing gap between domestic natural gas production and market demand undoubtedly poses a more severe challenge to China's energy security. According to

Table 18 Comparison of monthly PNG import prices in China in 2022 and 2023

Month	Import Price of PNG in 2022 (USD/ton)	Import Price of PNG in 2023 (USD/ton)	Year-on-Year Change (%)	Month-on-Month Change (%)
1	318.50	458.35	43.91	0.25
2	334.03	443.24	32.69	−3.30
3	341.70	442.94	29.63	−0.07
4	334.73	441.52	31.90	−0.32
5	373.70	387.88	3.79	−12.15
6	370.61	388.31	4.78	0.11
7	371.07	390.24	5.17	0.50
8	432.69	366.72	−15.25	−6.03
9	431.61	366.31	−15.13	−0.11
10	434.89	370.17	−14.88	1.05
11	459.44	377.27	−17.89	1.92
12	457.18	378.41	−17.23	0.30

Data source General Administration of Customs

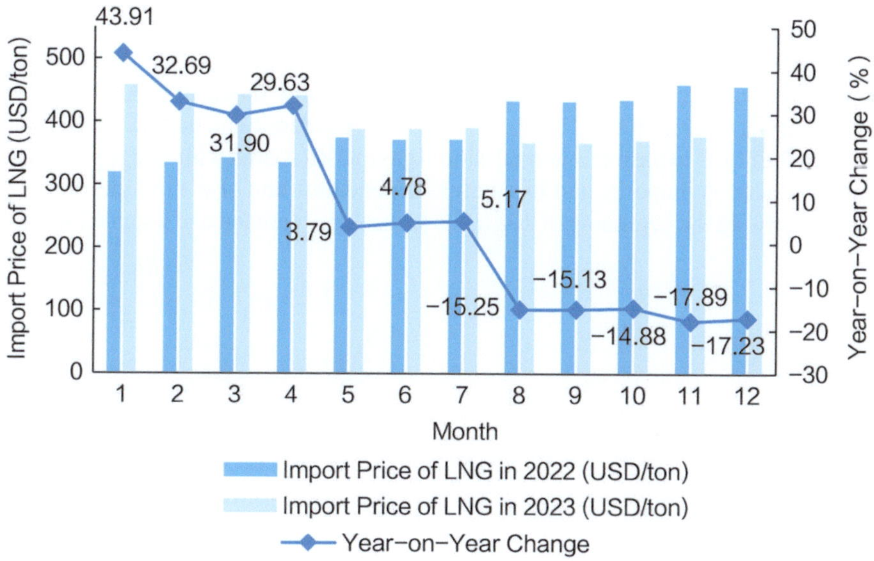

Fig. 16 Monthly PNG import prices in China in 2022 and 2023. *Data source* General Administration of Customs

Table 19 Natural gas external dependence in China (2013–2023)

Year	External dependence (%)
2013	31.6
2014	31.7
2015	31.8
2016	35.0
2017	40.0
2018	45.3
2019	43.1
2020	43.0
2021	44.9
2022	40.9
2023	42.3

Data source General Administration of Customs, National Development and Reform Commission

the data published on the official website of the National Development and Reform Commission, China's external dependence rate for natural gas from 2013 to 2023 is shown in Table 19 and Fig. 17.

It can be seen that, after a sharp increase from 2013 to 2018, China's external dependence rate of natural gas has remained basically stable at around 40% in recent years. In 2022, affected by the "COVID-19" pandemic and the Russia-Ukraine conflict, the prices in the international natural gas market soared with a large premium. China reduced its imports, and its external dependence rate of natural gas dropped significantly. At the beginning of 2023, due to weak gas demand and a stable increase

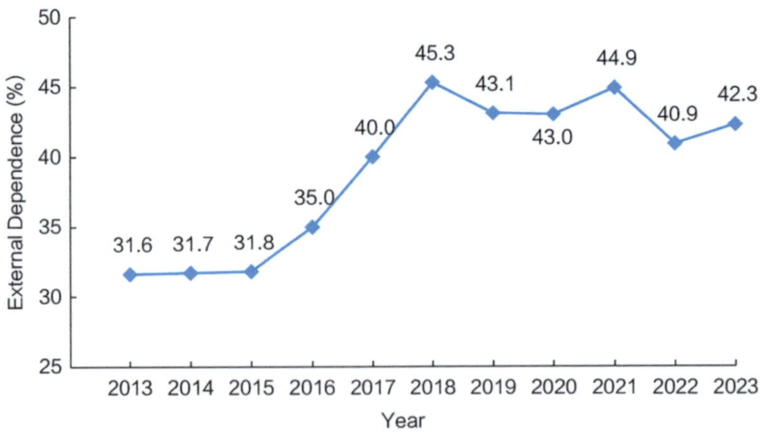

Fig. 17 Natural gas external dependence in China (2013–2023). *Data source* General Administration of Customs, National Development and Reform Commission

in market supply, international natural gas prices continued to fall. In the second half of the year, accidents in several natural gas-supplying countries caused international prices to rise and fall several times, making the market fluctuate unstably. Therefore, China cautiously increased its natural gas import activities. The external dependence rate of natural gas rose from 40.9% in 2022 to 42.3% in 2023, an increase of about 1.4%, which is still lower than the level before the Ukraine crisis. It is believed that, in the future, with the gradual recovery of China's economic activities and the fall of gas prices, the demand for natural gas will grow steadily, and the external dependence rate of natural gas is very likely to continue to rise.

Based on the monthly natural gas import data published by the General Administration of Customs and the monthly apparent consumption data of natural gas released by the National Development and Reform Commission, the monthly external dependence rate of natural gas can be calculated. Since the apparent consumption data published by the National Development and Reform Commission for January and February 2023 were aggregate figures, the same treatment was applied. The specific monthly data for China's external dependence rate of natural gas in 2023 is shown in Table 20 and Fig. 18.

As can be seen, the China's external dependence rate of natural gas in 2023 shows experienced an upward trend, with relatively stable fluctuations from May to September, and larger fluctuations at the beginning and the end of the year. The minimum external dependence rate is 36.19%, which occurred in March. At that time, the international natural gas price had been continuously declining for several months and showed a continuing downward trend, and China increased its import volume. The maximum external dependence rate was 47.32%, which occurred in September. In October, the external dependence rate plummeted to 38.09%, a decrease of about

Table 20 Monthly natural gas external dependence in China in 2023

Month	Import volume (10 kilotons)	Consumption volume (10 kilotons)	External dependence (%)
1	1787.43	4573.39	39.08
2			
3	882.09	2437.37	36.19
4	894.20	2286.34	39.11
5	1056.89	2405.00	43.95
6	1032.89	2338.84	44.16
7	1028.57	2336.68	44.02
8	1078.25	2350.35	45.88
9	1007.78	2129.55	47.32
10	874.52	2295.69	38.09
11	1089.59	2515.76	43.31
12	1264.91	2707.79	46.71

Data source General Administration of Customs, National Development and Reform Commission

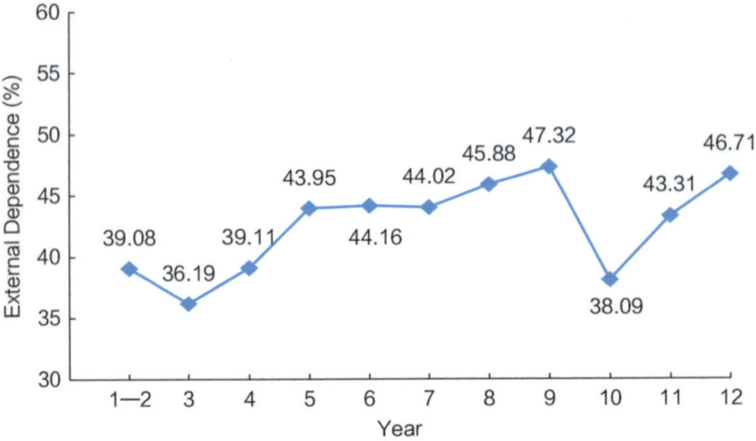

Fig. 18 Monthly natural gas external dependence in China in 2023. *Data source* General Administration of Customs, National Development and Reform Commission

19.5%. This was due to factors such as the strike at Australian LNG plants and the Israel-Palestine conflict, which affected the international natural gas market supply, leading China to reduce its import volume and mainly rely on domestic gas for consumption. In terms of trend, the average external dependence rate for natural gas in the first half of the year was 40.5%, which was lower than the 44.22% in the second half of the year. The main reason is that gas use in the first half of the year was mainly for industrial and commercial purposes, while in the second half of the year, the demand for natural gas increased slightly due to the addition of winter heating gas consumption. Overall, China's external dependence rate is relatively high and has certain seasonal fluctuations. It is also greatly affected by international natural gas prices, which poses certain challenges to China's gas storage and peak-shaving capabilities.

(III) Reflections on natural gas imports

1. Existing Problems

(1) The import channels for natural gas are relatively concentrated, and the external dependence on natural gas is relatively high. China is currently building a diversified resource pool to reduce the proportion of imports from single-source countries. As of December 2023, the number of countries from which China imports natural gas has reached 26. However, the main sources of import are still Australia, Turkmenistan, Qatar, and Russia, accounting for nearly 70% of China's total natural gas imports. This indicates that while the number of countries from which China imports natural gas has diversified, the import channels remain relatively concentrated. The primary reason for this is the geographical distribution of natural gas resources worldwide and the significant extraction

costs. These four countries have vast territories with sparse populations, abundant natural resources, and convenient natural gas storage, which gives them a natural advantage in terms of low costs. Additionally, their geographical proximity to China makes them the optimal choice for natural gas imports. Therefore, the situation of concentrated natural gas import channels in China is likely to persist in the long term.

Moreover, China's external dependence on natural gas is relatively high. In 2023, it increased from 40.9% in 2022 to 42.3%. On one hand, China's total natural gas resources are relatively limited and unevenly distributed. The country's relatively limited exploration and development technologies restrict the output and speed of natural gas production. On the other hand, as China's economic structure continues to adjust and optimize, its energy consumption structure is also changing. Natural gas, as a clean energy source, is expected to gradually increase its share in China's energy consumption structure, with demand rising along with economic development. Therefore, at this stage, China's production growth rate is temporarily unable to keep pace with the increase in consumption demand, and the external dependence on natural gas may continue to rise in the future. Faced with the complex international situation, China's concentrated import channels and high external dependence pose challenges to the country's energy security. If the main source countries experience political turmoil, economic crises, or natural disasters, it could lead to disruptions in China's natural gas supply. High external dependence means that China's natural gas supply is relatively passive. In the event of supply shortages or interruptions, China's energy security could be threatened.

(2) The natural gas import pricing mechanism is unreasonable, and China has weak bargaining power in the international natural gas market.

As an important clean energy source, natural gas is not only an essential energy source for China to achieve its carbon peak and carbon neutrality goals but also in short supply in developed countries such as those in Europe. In the context of a closely interconnected global bulk energy market, the price transmission of natural gas in the international market is strong, and significant market fluctuations are influenced by a variety of factors, including capital, market expectations, market structure, geopolitics, international situation, and unexpected events. These factors are beyond the control of any single country or enterprise. Currently, China is in a weak position in international energy pricing, with a "premium" on imported natural gas and a lack of "bargaining power." In addition to the objective reasons that China's large natural gas demand and domestic production capacity cannot keep up with consumption, leading to a seller's market for natural gas imports, long-term energy price controls and an underdeveloped energy financial system have also indirectly led to the current situation.

In recent years, China's natural gas industry, among other energy sectors, has been gradually transitioning from a government-controlled stage to a market-oriented stage. There are currently two types of natural gas pricing mechanisms in place: government-guided pricing and market-regulated pricing. The implementation of government-regulated pricing sometimes fails to quickly and accurately reflect the domestic natural gas supply and demand relationship, resulting in a price inversion at the domestic natural gas consumption end. This weakens the enthusiasm for domestic gas production and further exacerbates the dependence on imported gas. Secondly, due to historical and developmental reasons, there is a lack of a well-developed natural gas spot market. The essence of natural gas market pricing power is to establish a benchmark price, and the key to forming a benchmark price is free trade. The formation of a benchmark price is not the result of careful design but rather a natural evolution of the development of the spot market. To form a benchmark price recognized by the market, a futures exchange alone is far from sufficient. More importantly, efforts should be made to improve the basic spot market. A well-developed spot market should be one where buying and selling are free, prices are fair, and logistics are smooth. At this stage, China's natural gas import prices are mainly linked to foreign price benchmarks for pricing, and domestic sales are mainly priced according to the "city gate price." Abolishing the bundled city gate pricing is the future development direction of China's natural gas pricing management system.

(3) The pricing standards for natural gas are not unified.

Unlike the majority of countries worldwide that use energy content for natural gas measurement and pricing, China currently mainly employs gas volume for both measurement and pricing. However, this pricing method has caused some inconveniences as China's import volume of natural gas increases, such as the lack of uniformity in measurement units and price conversion issues in natural gas imports. Moreover, pricing natural gas by volume itself has certain drawbacks. According to relevant research, volume measurement only reflects the volume of natural gas but does not reflect the amount of usable heat content in the gas. It cannot fully reflect the market commodity trading principle of pricing natural gas according to its quality and offering better prices for higher quality. This is not conducive to the scientific and rational use of limited natural gas resources and is unfair to both the supply and demand sides. Furthermore, with the increasing demand for clean energy in China, the current natural gas supply structure is diversified, with PNG, LNG, coalbed methane, and shale gas all occupying a certain proportion in the total natural gas supply. The heat content per unit volume of natural gas from different sources varies greatly, making it impossible to accurately reflect the economic value of natural gas. Additionally, to meet the needs of natural gas market reform and promote the development of natural gas trading centers, it is necessary to establish benchmark prices that are in line with international standards. Internationally, natural gas is measured and priced based on its energy content and heat value. Clearly, pricing by volume is not suitable for international transactions. Moreover, since pricing by volume cannot accurately reflect the actual economic value of natural gas, it affects the fair competition of natural gas of different qualities on trading platforms and hinders the internationalization process of China's natural gas trading center construction.

2. Suggestions

(1) Strengthen international trade and cooperation in natural gas and improve domestic natural gas extraction capabilities.

In the face of the current situation where natural gas import channels are concentrated and the dependence on foreign natural gas is high, China should base its actions on national conditions and ensure the safety of natural gas imports in these two aspects. First, in combination with the characteristics of international natural gas trade and taking the "Belt and Road" initiative as an opportunity, China should enhance investment in the natural gas sector of Central Asian and African countries, which are rich in natural gas resources. By doing so, China can increase its overseas equity gas volume, diversify procurement risks, and build a resource pool with diversified structure in terms of import sources, channels, trade forms, and price benchmarks. This will help mitigate the the impact of international political and economic factors on China's natural gas supply. Secondly, China needs to boost domestic natural gas development. This involves increasing investment to accelerate the exploration and development of domestic upstream natural gas resources. It also includes strengthening the capacity for gas storage and peak shaving, as well as the construction of a responsible system. The orderly implementation of the coal-to-gas conversion should be continued. Moreover, enterprises should collaborate with national scientific research institutes and universities to form a research and innovation mechanism that is enterprise-led and features complementary roles among industry, academia, and research. Efforts should be made to achieve theoretical breakthroughs and technological innovations in the exploration of unconventional natural gas resources such as coalbed methane and shale gas, thereby enhancing extraction capabilities. Furthermore, China should coordinate the development of natural gas exploitation with ecological and environmental protection. This includes strengthening the research and development capabilities for related facilities and equipment. To ensure the dominant position of natural gas in the primary energy mix, China should actively participate in international natural gas trade, reduce the cost of natural gas resources, and diversify import channels. This will decrease dependence on the international market. The role of the national pipeline company should be fully utilized to enhance gas storage capacity and promote the development of the domestic natural gas market. This will increase domestic natural gas production capabilities and safeguard China's natural gas energy security.

(2) Deepen the market-oriented reform of natural gas and establish and improve the domestic futures and spot markets.

Looking at the historical experience of natural gas marketization in countries such as Europe, America, and Japan, the reform of natural gas prices is not accomplished overnight and requires a sustained effort over a long period of time. At present, China's natural gas marketization reform has entered deep waters, and it is necessary to further deepen and reform the natural gas pricing mechanism, develop China's natural gas financial market, and accelerate the improvement of the functions of the Shanghai Petroleum and Natural Gas Trading Center and the Chongqing Petroleum and Natural Gas Trading Center. The marketization reform should be promoted from the following aspects: First, China should adopt a "pilot-first and phased liberalization" approach to gradually eliminate the price control at the city gate level. Regions where gas supply competition is already robust should take the lead in removing price controls at the city gate, integrating price reform with the development of competitive markets. Secondly, it is necessary to rationally determine the pricing of natural gas infrastructure. China should study the applicability and feasibility of implementing a two-part tariff system domestically, learning from developed countries in Europe and America in pricing natural gas infrastructure. This involves dividing service fees into capacity charges and usage charges. The advantage of this pricing method is that it can address the issue of reasonable charges caused by uneven gas usage under a single-part tariff, and it also promotes investments in gas storage facilities, thereby improving the overall utilization efficiency of pipelines. Thirdly, efforts should be made to advance the construction of gas storage facilities to improve peak-shaving capabilities during seasonal transitions. Increasing investment in infrastructure to boost storage and transportation capacity is essential. This will fundamentally enhance China's ability to import and regulate natural gas, and flexibly apply strategies such as "storing more during off-peak seasons and extracting more during peak seasons" to mitigate price fluctuations between peaks and off-peaks. Then, strengthening market regulation is crucial. A comprehensive regulatory system for the natural gas market should be established to enhance supervision and restraint over market entities. Relevant laws and regulations should be refined to standardize the behavior of market participants and prevent illegal activities such as market manipulation and fraud. At the same time, supervision over natural gas quality and safety should be intensified to protect consumer rights and ensure fair market competition. Lastly, the construction of natural gas trading hubs is essential to form benchmark prices. A pricing benchmark that reflects China's natural gas market demand should be established, shifting from the original pricing method linked to oil and gas to a market-based pricing method driven by "gas-to-gas competition." This will enable nationwide linkage of natural gas prices and gradually replace government pricing. The ultimate goal of China's natural gas price reform is to rationalize the value distribution along the natural gas industry chain, enhance the level of resource market allocation, reduce societal gas usage costs, and boost the substitution capacity of natural gas for coal and oil. Achieving domestic natural gas market reform will increase market transparency, improve the price formation mechanism, and strengthen the competitiveness and innovation capabilities of market entities, thereby promoting the healthy development of the natural gas market.

(3) Accelerate the reform of energy-based measurement and pricing for natural gas in China.

As a commonly used fuel for combustion, the value of natural gas lies in the heat it provides. Adopting energy-based pricing can better reflect its commodity attributes and value, facilitate accurate measurement, ensure fairness, and reduce billing disputes. The National Development and Reform Commission (NDRC) has also explicitly proposed the implementation of energy-based measurement and pricing for natural gas in the "Regulatory Measures for the Fair Openness of Oil and Gas Pipeline Facilities." Therefore, a gradual and phased approach can be adopted to advance this reform on the premise of market stability. Consider promoting the implementation of energy pricing from the aspects of pricing schemes, reform plans, and implementation plans. First, establish a unified standard for heat content measurement methods and pricing units. Accurately grasp the heat value levels of natural gas at the main nodes of the natural gas industry chain. Use a scientific price conversion method to transform the pricing from yuan/m ^{3} to yuan/joule (megajoule). At this stage, how to calculate the average heat content under the conditions of annual market fluctuations of various types of gas sources and determine the benchmark heat content is a key issue that needs to be considered. Secondly, it is necessary to formulate and release a reform plan for energy-based measurement and pricing of natural gas as soon as possible. Although energy-based measurement and pricing of natural gas have been put on the reform agenda, the formal plan has not yet been introduced. Under the current natural gas market situation, the demand for reform is very strong. Experts, enterprises, and research institutions in the relevant fields need to combine the actual conditions of China's natural gas market and make decisions on energy measurement and pricing methods, pricing units, benchmark heat content of natural gas, natural gas prices, and pipeline transportation prices as soon as possible, formulating and releasing a reform plan. Finally, in terms of implementation strategies, first consider pilot work in developed areas, such as in coastal areas with a high degree of internationalization of natural gas, like the Yangtze River Delta and the Pearl River Delta. The reform pilot work should be carried out in these regions. The promotion of the reform of natural gas measurement and pricing should be carried out in the order of industrial users, non-residential users, and residential users. At the same time, it is necessary to establish and improve the regulatory system for the natural gas market, strengthen supervision and restraint over pricing and measurement, enhance information disclosure, increase market transparency, and prevent the occurrence of information asymmetry and market manipulation.

Natural Gas Exports

At this stage, China's natural gas supply is still insufficient, and there is basically no surplus for export to foreign countries. In terms of China's natural gas exports, it mainly supplies a small amount to Hong Kong and Macau. When the inventory is sufficient and international prices fluctuate, it also exports a small amount to other countries to earn a price difference.

(I) Natural gas exports

The export volume of natural gas in China in 2023 slightly increased compared to 2022, with a cumulative export of 4.666 million tons from January to December, mainly in the form of PNG exports. PNG exports totaled 3.513 million tons, accounting for 75.3% of the total, with the main export destinations being Hong Kong and Macau in China. LNG exports increased from 580,000 tons in 2022 to 1.153

million tons, accounting for 24.7% of the total, with the main export destinations being South Korea, Japan, and Thailand (Table 21, Fig. 19).

Table 21 China's natural gas exports in 2022 and 2023

Month	Natural gas export volume in 2022 (10 kilotons)	Natural gas export volume in 2023 (10 kilotons)	Year-on-Year change (%)
1	34	26	−22.97
2	33	34	5.26
3	20	47	135.91
4	23	27	14.91
5	30	31	2.81
6	33	36	10.70
7	41	42	2.40
8	44	42	−3.46
9	53	42	−21.55
10	47	37	−21.38
11	38	47	25.33
12	28	55	100.48

Data source General Administration of Customs

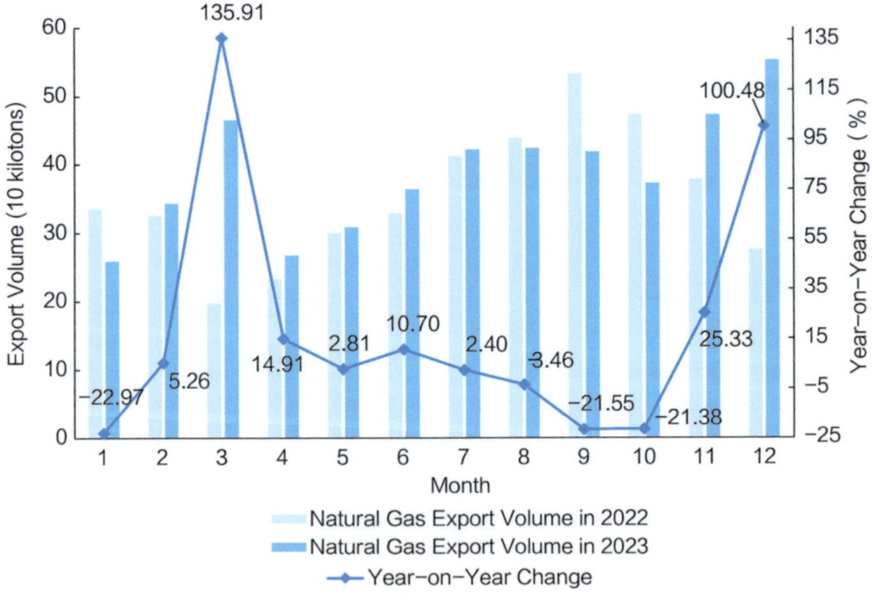

Fig. 19 China's natural gas exports and year-on-year growth rate (2022–2023). *Data source* General Administration of Customs

From Table 21 and Fig. 19, it can be seen that the export volume in 2023 increased by 10.34% compared to 2022, with the increase mainly concentrated at the beginning and end of the year. The highest monthly export volume occurred in December, at 550,000 tons, a year-on-year increase of 100.48%. The lowest monthly export volume was 260,000 tons, which occurred in January, a decrease of 22.97% year-on-year. The month with the largest increase was March, with a year-on-year increase of 135.91% and an export volume of 470 kilotons.

The main destinations for China's current natural gas exports are Hong Kong and Macau in China, with the main export customs being Guangzhou Customs, Shenzhen Customs, Haikou Customs, and Gongbei Customs. The main exporting companies are CNOOC and PetroChina, with CNOOC and PetroChina responsible for supplying gas to Hong Kong, and CNOOC responsible for supplying gas to Macau. Looking at the overall trend of LNG spot prices in the international market in 2023, the prices were high at the beginning of the year due to the continuation of the sharp rise in 2022, and at the end of the year due to panic caused by frequent accidents on the supply side. Under these circumstances, China is reselling LNG to earn a price difference while ensuring a stable supply of natural gas domestically, which also explains the reason for the larger year-on-year increase in LNG export volume in 2023.

(II) Natural gas export prices

From January to December 2023, the average monthly export price of natural gas in China is shown in Table 22 and Fig. 20.

As can be seen, the overall export price of natural gas in 2023 was lower than that in 2022, with an average export price of 655.4 USD/ton, a year-on-year decrease of

Table 22 China's natural gas export prices (2022–2023)

Month	Natural gas export prices in 2022 (USD/ton)	Natural gas export prices in (USD/ton)	Year-on-Year change (%)
1	777.4	656.2	−15.59
2	727.1	945.8	30.08
3	493.1	913.9	85.35
4	539.8	611.3	13.23
5	505.2	615.5	21.83
6	583.2	620.2	6.35
7	737.6	582.6	−21.01
8	698.0	554.9	−20.50
9	876.8	572.6	−34.69
10	1188.8	568.9	−52.14
11	878.1	586.6	−33.20
12	621.2	635.8	2.36

Data source General Administration of Customs

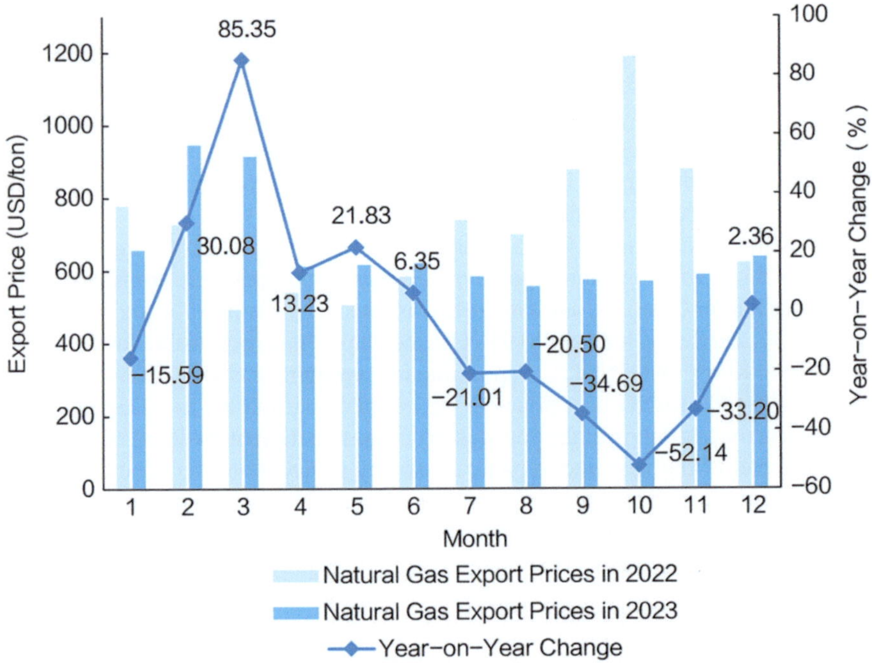

Fig. 20 China's natural gas export prices (2022–2023). *Data source* General Administration of Customs

8.83%. The highest export price was 945.8 USD/ton, which occurred in February, while the lowest was 554.9 USD/ton, occuring in August. The largest increase was in March, reaching 85.35%. Overall, the growth of China's export volume in 2023 was not significant, the export unit price decreased, but the overall price increase was similar to the trend of international natural gas prices, indicating that China's natural gas export prices have a significant positive correlation with international natural gas prices.

Analysis of Natural Gas Import and Export Trends

In recent years, China's demand for natural gas has been continuously strong. The domestic production of natural gas cannot meet the consumption demand, resulting in a large gap. Therefore, China has always been cautious in terms of natural gas exports, with a relatively small export volume. This trend will continue in the foreseeable future. Hence, the analysis of the imports and exports trends of natural gas in China will mainly focus on import trend. China's dependence on foreign natural gas has remained above 40% in the past five years. In the import structure, the proportion of LNG, including long-term agreements and spot purchases, accounts for 59.27%, of which LNG spot purchases account for about 20% of China's natural gas supply. Currently, the import price of LNG spot in the international natural gas market is largely constrained by the international supply and demand situation of natural gas. Therefore, in addition to considering the domestic natural gas demand, it is also necessary to pay attention to the international market supply of natural gas. Based on the supply and demand situation of the international natural gas market, the fluctuation of international natural gas prices, and the domestic natural gas production situation, the following analysis of China's natural gas import trend is presented.

Over the past two years, the international natural gas market has been full of uncertainties. In 2022, Russia was originally the main source of natural gas for EU countries. After Russia announced its withdrawal from the European market due to the conflict between Russia and Ukraine, European countries began to hoard large amounts of natural gas to ensure energy security. This led to a rapid increase in demand in the natural gas market and a surge in natural gas prices. Entering 2023, affected by a mild winter and economic recession, the demand for natural gas in Europe and the Asia–Pacific region was weak. European countries had higher natural gas inventories due to previous hoarding and lower consumption, reaching a historically high level. Compared with the weak demand side, the global natural gas market supply rose steadily. The natural gas output in major exporting regions such as the Middle East and the United States increased steadily. Although supply-side incidents such as the second-half-year strike at Australian LNG plants and geopolitical events like the Israel-Palestine conflict occurred, they brought several uncertain price increases to the natural gas market, but the degree and duration of these increases were limited and did not reach the level of the 2022 price surge. Overall, the international natural gas market in 2023 was characterized by loose supply and demand, with international natural gas prices fluctuating downward.

From the current situation, the impact of the Russia-Ukraine situation on the supply and demand pattern of the international natural gas market is gradually weakening. European countries are gradually turning to regions other than Russia to import natural gas, and international natural gas prices are returning to normal. In recent years, the uncertainty in the international natural gas market has increased, the Russia-Ukraine conflict has not ended, and the Israel-Palestine conflict has erupted again. In addition to the occurrence of geopolitical events, accidents in natural gas supplying countries will also affect the fluctuation of international natural gas prices. But at this stage, it is mainly the European natural gas market that is affected. The Asian natural gas market, which has not been directly impacted, is relatively less affected, with more stable supply and demand conditions and smaller price increases. Overall, the geopolitical environment of the Asian natural gas market is relatively stable, and the supply of the natural gas market is relatively stable. The geopolitical uncertainty of the European natural gas market is high, with a higher risk of natural gas supply cuts or interruptions.

From the development trend of China's natural gas industry, China's demand for natural gas will further increase under the "dual carbon" policy. As the result of economic developments, the rate of urbanization is accelerating, the level of urbanization is constantly improving, the scale of urban population is expanding, and the demand for residential gas, commercial gas, industrial gas is rapidly growing. China's natural gas resources are quite abundant, and the proven reserves in 2021 ranked sixth in the world, but it is currently in the early stage of exploration and development. Both the proven rate of natural gas resources and the degree of exploitation are relatively low, resulting in a slow growth rate of natural gas production. Over the past year, with the easing of "COVID-19" restrictions in China and the gradual economic recovery, the demand for natural gas used in production and daily life, as well as for transportation, has steadily rebounded. But the growth in industrial gas demand has not met expectations. In 2023, China's natural gas production was approximately 167.16 million tons, a year-on-year increase of 5.6%. The consumption volume was about 283.75 million tons, up 7.6% year-on-year. Natural gas imports amounted to 119.9712 million tons, a year-on-year rise of 9.9%. It is evident that the growth rate of production cannot keep pace with that of consumption. As the economy continues to develop further, China's industrial gas demand is expected to rise steadily, leading to a significant increase in natural gas consumption and a further acceleration in the consumption growth rate. At that time, the growth rate of natural gas production will find it hard to catch up with the consumption growth rate and will struggle to meet the rapidly growing gas demand. Thus, China will have to rely on imported natural gas to satisfy its gas-consuming needs.

In short, under the strategic circumstances of energy transition, China's natural gas imports and dependence on foreign sources are expected to increase in the future. Natural gas imports will still mainly consist of PNG and LNG, with the primary sources being Turkmenistan and Australia, although their shares will gradually decrease, while Russia's proportion will correspondingly increase. In the import structure, the proportion and price of LNG imports will still be higher than those of PNG.

In the long-term perspective, the complex international situation has brought both challenges and opportunities to China's natural gas energy security. The ongoing conflict between Russia and Ukraine has facilitated in-depth cooperation between China and Russia in natural gas exports. The impact of the Israel-Palestine conflict on China's natural gas market is limited, but the existence of potential threats can provide impetus for China's energy construction and will accelerate the development of China's natural gas industry's production, supply, storage and sales system.

In summary, at the current stage, China should strive to diversify its natural gas foreign trade and expand the channels for natural gas imports. This will help reduce the direct structural risks brought about by potential interruptions in natural gas supply and prevent over-reliance on any single major supplier. Additionally, in the face of uncertainties in the international natural gas market, China needs to enhance its analysis and forecasting of the international market. By leveraging the diversification of foreign trade and ensuring energy security, China can maximize the reduction of natural gas import costs.

China's Natural Gas Consumption

In 2023, China's natural gas consumption increased, driven by factors such as the decline in international gas prices and the economic recovery. At present, China's economic development has entered a new normal, with significant adjustments in the economic structure. The economic growth rate has transitioned from high-speed growth to medium-speed growth, and energy demand has also entered a stage of medium-speed growth. In 2024, economic growth, geopolitical factors, financial pressures from international oil prices, and the development of new energy will continue to be the main factors affecting China's natural gas consumption. For a relatively long period in the future, under the influence of high-quality economic development and the energy transition strategy, natural gas, as an important support for China's energy structure optimization and green and low-carbon development, will play an even more important role. China's natural gas industry will also enter a period of development full of opportunities and challenges.

Natural Gas Consumption

(I) China's natural gas consumption in 2023

In 2023, China's apparent consumption of natural gas was 3,945 billion cubic meters, an increase of 7.60% compared with 2022, with an increase of 282 billion cubic meters. The monthly natural gas consumption and year-on-year growth rate for the whole year are shown in Table 23 and Fig. 21.

As can be seen, the Chinese natural gas market is in a period of recovery growth in 2023, with annual consumption returning to positive growth and relatively stable monthly consumption growth. Domestic natural gas consumption in January and February was basically flat compared to previous years. Entering March, influenced by the economic situation and domestic and foreign natural gas prices, the recovery was more obvious, showing a trend of significant year-on-year growth. Affected by the decline in hydropower generation, high summer temperatures, and policy guidance, natural gas power generation grew rapidly, with a year-on-year growth rate reaching a high of 13.1% in June. The year-on-year growth rate in July and August was also at a relatively high level in recent years. Starting from September, the overall consumption of the natural gas market returned to a moderate growth trend. Overall, the natural gas consumption market operated relatively stably throughout 2023, with demand continuing to recover, but with increased volatility.

(II) China's natural gas consumption by industry

According to the "Natural Gas Utilization Policy" (Order No. 15 of 2012 of the National Development and Reform Commission), natural gas utilization can be divided into five categories according to different gas usage purposes: self-use gas, urban gas, power generation, industrial fuel, and the chemical industry. Self-use gas

Table 23 Monthly natural gas consumption in China (2022–2023)

Month	2022 Consumption (hundred million cubic meters)	2023 Consumption (hundred million cubic meters)	Growth rate (%)
1–2	635.2	635.9	0.10
3	297.5	338.9	9.30
4	297.1	317.9	7.30
5	298.6	334.4	12.40
6	288.7	325.2	13.10
7	298.2	324.9	9.60
8	295.8	326.8	11.10
9	279.2	296.1	5.80
10	305.3	319.2	4.50
11	319.5	349.8	9.40
12	347.9	376.5	9.50
Total	3663.0	3945	7.60

Data source "Natural Gas Operation Briefing," National Development and Reform Commission

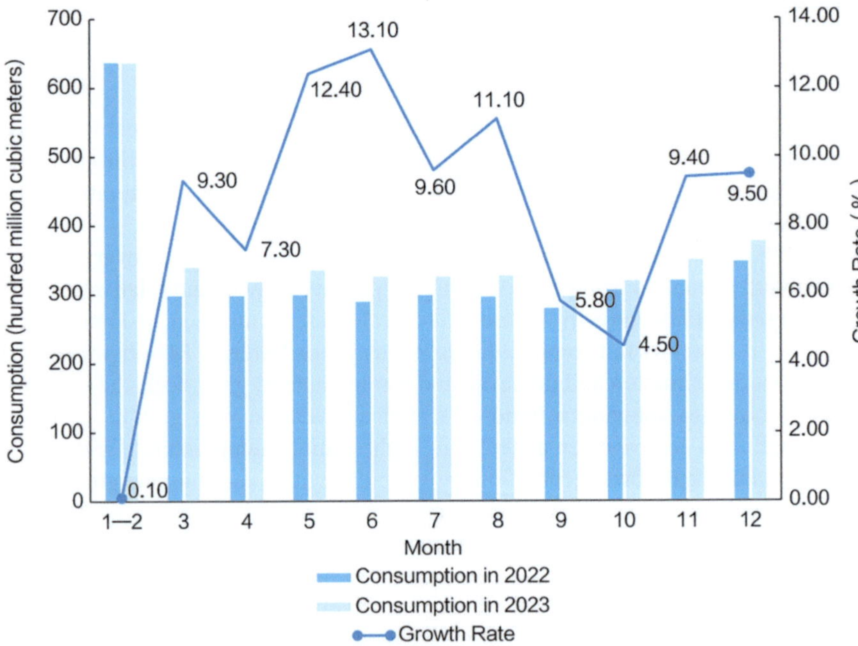

Fig. 21 Monthly natural gas consumption and growth rate in China (2022–2023). *Data source* "Natural Gas Operation Briefing," National Development and Reform Commission

is mainly used in the oil and natural gas extraction industry. Urban gas is mainly used in transportation, storage, and postal services, wholesale and retail, as well as in accommodation, catering services, residential use, and other living consumption sectors. For power generation it is mainly used in the production and supply of electricity and heat. As industrial fuel it is mainly used in industries other than oil and natural gas extraction, the chemical raw materials and chemical products manufacturing industry, and the electricity and heat production and supply industry. In the chemical industry it is mainly used in the manufacturing of chemical raw materials and chemical products.

In 2023, natural gas consumption in China saw a recovery across all four categories outside of self-use gas to varying extents. The restoration of production significantly boosted consumption in the tertiary sector. Natural gas consumption was mainly driven by urban gas and gas for power generation. The increase in urban gas consumption primarily came from the commercial service sector, transportation (with a booming sales of LNG heavy-duty trucks), and heating. The rise in gas consumption for power generation was mainly due to the increase in regional gas-fired power generation capacity and the peak power generation demand during the summer. Industrial users are more sensitive to gas prices, and their demand increase was largely influenced by the gas price trend throughout the year. At the same time, the sluggish market consumption demand was also the main factor that led to the relatively weak increase in industrial user consumption in 2023. In 2023, China's natural gas consumption was dominated by urban gas and industrial fuel, which together accounted for about 74.99%. Specifically, the consumption scale of the four major industries of urban gas, industrial fuel, power generation, and chemical industry all expanded to varying degrees, with year-on-year growth rates of 7.86%, 6.29%, 7.00%, and 2.73% respectively. Among them, the performance of natural gas heavy-duty trucks was eye-catching, with annual sales approaching 160,000 units, a year-on-year increase of 330%, which significantly boosted gas consumption in the transportation sector (Fig. 22).

(III) China's natural gas consumption by region

Due to various factors such as the unevenness in economic development, pipeline construction, and market development, the consumption of natural gas varies from region to region. China's natural gas production areas are mainly concentrated in the east and western regions, while the consumption areas are mainly in the Yangtze River Delta and the Pearl River Delta, leading to a mismatch between supply and demand. The consumption in most provinces, municipalities directly under the central government, and autonomous regions has recovered, but the growth rates vary. Guangdong, Zhejiang, Fujian, Chongqing, Yunnan, and other regions have strong recovery momentum, with growth rates exceeding 10%. Anhui Province's natural gas consumption reached 9.744 billion cubic meters, a year-on-year increase of 24.52%, setting a new record high in both consumption and growth rate. Jiangsu and Hebei have relatively weak recovery momentum and lower growth rates. Jiangsu Province's apparent natural gas consumption was 33.42 billion cubic meters, up

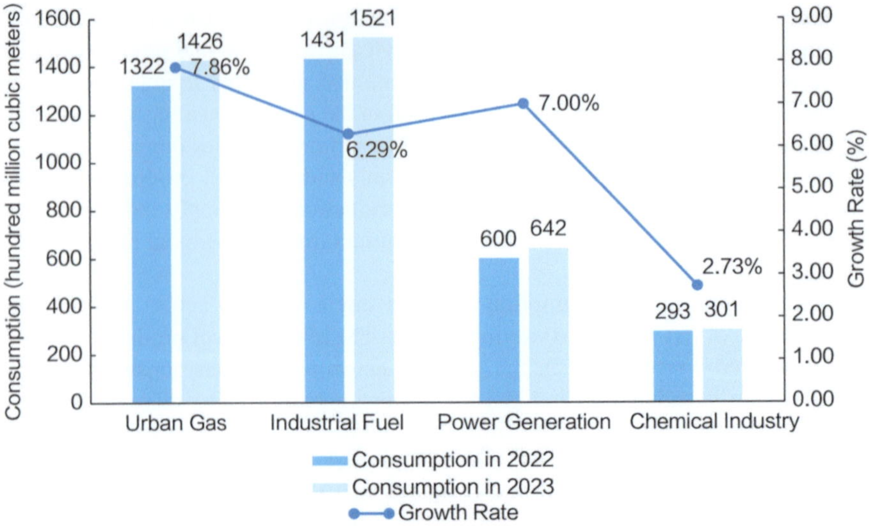

Fig. 22 Natural gas consumption by industry in China in 2023. *Data source* LNG industry information, Chongqing Petroleum and Natural Gas Trading Center

8.8% year-on-year. Jilin, Heilongjiang, and other regions have relatively lagged in the recovery of natural gas consumption and have even experienced negative growth.

(IV) Suggestions for deepening the reform of the natural gas consumption market under the "dual carbon" goals

On October 20, 2023, the National Development and Reform Commission issued the "National Carbon Peak Pilot Construction Plan." The plan proposes to select 100 representative cities and industrial parks nationwide to carry out carbon peak pilot projects. These projects aim to address the bottleneck constraints facing green and low-carbon development. They will explore pathways to carbon peaking of cities and industrial parks with different resource endowments and development foundations, and providing replicable and scalable experiences for the entire nation. Natural gas is not only the most realistic choice for replacing coal and achieving "clean and low-carbon" development, but also a lifelong partner for renewable energy. It is a key energy source for promoting "carbon neutrality." Energy transition has become a fundamental strategy for China's economic and energy development. In developed countries such as the European Union and the United States, the share of natural gas in primary energy consumption is over 20%, while in China it is only 9.1%. Experience from the European Union, the United States, and other countries that have already peaked in carbon emissions shows that vigorously developing natural gas will surely promote the realization of the "dual carbon" goals. Based on this, and considering the current consumption situation and development trend of China's natural gas market, reform suggestions under the "dual carbon" goals are put forward.

1. Adhere to the "gas-determined reform" principle of coal-to-gas conversion to achieve the transformation and upgrading of the energy structure

Adjusting the energy structure is an important measure to achieve the "dual carbon" goals. Promoting the transformation and upgrading of the energy structure by vigorously developing clean energy is an important way to achieve this goal. To achieve the "dual carbon" goals, it is necessary to further increase the proportion of natural gas in primary energy consumption during a certain period and vigorously develop clean energy. It is essential to comprehensively promote the substitution of natural gas, implement the "gas-determined reform," and quickly achieve effective substitution of natural gas for urban gas, industrial fuel, and power generation gas, as well as for industrial waste heat, industrial waste gas, and waste incineration power generation. It is also crucial to ensure orderly growth in gas consumption for residential living, general industry and commerce, and other uses. By 2025, the proportion of natural gas in primary energy consumption should reach around 12%.

2. Increase investment in pipeline construction and improve the efficiency of infrastructure utilization

The construction of natural gas pipeline infrastructure is a prerequisite for deepening the reform of the natural gas consumption market. At present, there are still some problems in the construction of natural gas pipeline infrastructure in China. First, the pipeline transportation capacity is insufficient. The total length of natural gas pipelines in China is more than 130,000 km, only 1/4 of that in the United States, and there is a large gap with developed regions such as Europe. Second, the peak-shaving gas storage capacity is insufficient. There are only 43 gas storage facilities and 32 LNG receiving stations with gas storage capacity, which is far behind the United States and Europe. Third, the construction of pipeline network connectivity in some areas lags behind, and the built pipeline network has not yet formed effective transportation capacity, and there are problems of idle or insufficient pipeline network resources in some areas. In light of this, relevant government departments should further increase investment in pipeline construction, improve the transportation capacity, peak-shaving capacity and transportation efficiency of the pipeline network, and build China into an international natural gas resource allocation center.

3. Accelerate the establishment of the coal-based synthetic natural gas (SNG) market and comprehensively promote green and low-carbon development

Coal-based synthetic natural gas (SNG) is an important part of traditional coal chemical industry. It can not only fill the gap in natural gas supply, but also bring about significant emission reduction benefits, which is of great significance to achieving the "dual carbon" goals. It is suggested to accelerate the establishment of a coal-based SNG market and promote the development of coal-based SNG through adjustments in natural gas prices. On the one hand, coal-based SNG projects can be bundled with wind power and photovoltaic power generation projects. By utilizing their carbon emission offset mechanism and adjusting prices, the consumption of new energy and carbon emission reduction can be realized. On the other hand, the development of coal-based SNG can be promoted in a way that is determined by the availability of gas. On the basis of the existing natural gas prices, appropriate adjustments can be made. By using the price lever, coal-based SNG enterprises can be encouraged to increase investment, expand production capacity and optimize product structure, thus providing strong support for China to achieve the "dual carbon" goals.

Natural Gas Prices

(I) Natural gas pricing mechanisms

Globally, the natural gas pricing mechanisms of various countries can generally be divided into the following four types.

1. Gas-to-gas competition

Prices are formed by the dynamics of supply and demand through competition among gas sources. Natural gas trading takes place in natural gas trading centers, including physical trading centers (such as the Henry Hub in the United States) and virtual trading centers (such as the NBP in the UK). Natural gas trading is mainly based on short-term transactions with fixed prices, and there are also long-term natural gas trading contracts. However, the monthly prices of these contracts are indexed to the price indices of natural gas spot or futures trading.

2. Oil price linkage

The price of natural gas is linked to the price of competitive fuels (usually crude oil, diesel or fuel oil) through a base price and an escalation clause, and sometimes also includes coal prices and electricity prices.

3. Bilateral monopoly

The price is determined through discussion and negotiation by a large buyer and a large seller and remains fixed for a period of time (usually one year). Unlike gas-to-gas competitive bilateral transactions with multiple buyers and sellers, bilateral monopoly pricing has at least one party that is the sole buyer or seller in the market.

4. Regulated pricing

There are three pricing models for regulated pricing: ① Cost-plus pricing, where the price is determined or approved by the relevant government authorities, and the price level includes "cost of service" and a reasonable return on investment; ② The price is set by the government, but there is no explicit pricing method, often based on political and socio-economic factors; ③ The price is below the production and transportation costs.

The reform of natural gas gate station prices in China began with the "Two Guang Pilot" in 2011, and currently over 60% of the gas volume at the gate station level has been liberalized to be formed by the market. In 2011, the National Development and

Reform Commission issued a notice to carry out a pilot reform of the natural gas price mechanism in Guangdong Province and Guangxi Zhuang Autonomous Region. In 2013, the gate station price mechanism reform was rolled out nationwide. In 2015, the Central Committee of the Communist Party of China and the State Council issued "Several Opinions on Promoting Price Mechanism Reform", proposing the overall idea of "regulating the middle and liberalizing the two ends" for China's natural gas price reform. In 2017, the Central Committee of the Communist Party of China and the State Council issued "Several Opinions on Deepening the Reform of the Oil and Natural Gas System," outlining eight tasks for the reform of the oil and gas system. On May 25, 2018, the NDRC issued the "Notice on Rationalizing the Gate Station Price of Residential Natural Gas" (NDRC Price Regulation [2018] No. 794), changing the management of residential gas prices from a maximum gate station price to a benchmark gate station price. The price level was set according to the benchmark gate station price for non-residential gas (including 10% value-added tax). The supply and demand sides could negotiate a specific gate station price based on the benchmark gate station price, within a range of up to a 20% increase and no limit on decrease, thus achieving the unified connection of the price mechanisms for residential and non-residential gas. On March 13, 2020, the "Central Pricing Catalogue" was published, stipulating that the gate station prices for imported pipeline natural gas put into operation after 2015, as well as natural gas in provinces with competitive conditions, would be formed by the market. Other domestic onshore pipeline natural gas and imported pipeline natural gas put into operation before the end of 2014 would temporarily be managed according to the current pricing mechanism, with the prices to be liberalized in a timely manner according to the progress of natural gas market reform.

(II) The current prices of natural gas in China

1. Pipeline transportation prices

On December 5, 2023, in order to implement the decision and plans of the Party Central Committee and the State Council on deepening the reform of the oil and natural gas market system and accelerating the construction of a "national one-network" system, the National Development and Reform Commission issued the "Notice on Determining Cross-provincial Natural Gas Pipeline Transportation Prices" (hereinafter referred to as the "Notice"). For the first time, the NDRC determined the cross-provincial natural gas pipeline transportation prices operated by the National Petroleum and Natural Gas Pipeline Network Group Co., Ltd. on a regional basis. The "Notice" clearly defined the cross-provincial natural gas pipeline transportation rates for the Northwest, Northeast, Central and Eastern, and Southwest price zones. It required the National Petroleum and Natural Gas Pipeline Network Group Co., Ltd. to calculate the specific pipeline transportation prices based on the transportation distance between the natural gas entry and exit points. This pricing is the first since the reform of the natural gas pipeline operation mechanism and is the first time the state has determined cross-provincial natural gas pipeline transport prices on a "one price per zone" basis. After the price determination, the number of cross-provincial natural gas pipeline transport price rates operated by the National Oil and Gas Pipeline Network Group Co., Ltd. was significantly reduced from 20 to 4, creating a relatively unified pricing structure. This move broke the previous segmentation of pipeline operations due to too many price rates, facilitating the interconnection and fair opening of pipeline facilities, accelerating the formation of a "national one-network" system, promoting the free flow of natural gas resources and market competition, and supporting high-quality industry development. The main cross-provincial natural gas pipelines operated by the National Pipeline Network Group are shown in Table 24.

2. Benchmark prices

In accordance with the adjustment of the value-added tax rate for natural gas, it has been decided to correspondingly adjust the benchmark gate station prices for natural gas. Starting from April 1, 2019, the benchmark gate station prices for natural gas in various provinces (regions, municipalities) have been adjusted, as shown in Table 25.

Due to differences among provinces in terms of gas supply links, supply costs, and consumption preferences, there are significant differences in the benchmark prices of natural gas across provinces. The benchmark gate station prices for natural gas in the eastern regions are generally higher than those in the western regions. For example, Shanghai and Guangdong have the highest prices, at 2,040 yuan per thousand cubic meters. In contrast, Xinjiang, as a major natural gas producing region in China, has the lowest benchmark gate station price, at 1,030 yuan per thousand cubic meter. When natural gas production and supply enterprises negotiate specific prices with users, they should fully consider the factor of the reduced value-added tax rate and pass on the benefits of the reduced tax rate to users. Local price regulatory authorities should also take into account the reduced value-added tax rate when arranging natural gas sales prices, ensuring that all the benefits of the value-added tax reform are passed on to users.

3. Consumption prices

(1) Price of LNG.

The China LNG factory price index is exclusively released by the Shanghai Petroleum and Natural Gas Exchange Center, which closely monitors nearly 100 domestic LNG factories and more than 20 receiving stations. The index is calculated based on the transaction data of the exchange center, supplemented by the quotations calculated by the exchange center's shareholder units and partner consulting agencies. It mainly reflects the price trend of the domestic LNG market. The national LNG factory price index for China in 2023 is shown in Fig. 23.

It can be seen that in 2023, domestic LNG prices exhibited a "U"-shaped trend. In the first half of the year, domestic LNG prices gradually declined, with market prices experiencing a downward fluctuation. The primary reason for this was the impact of geopolitical factors in 2022. As international LNG prices returned to a more rational level, and considering the potential for a cold winter due to the La Niña phenomenon, countries and regions around the world replenished their natural

Table 24 Main cross-provincial natural gas pipelines operated by the national pipeline network group

Serial number	Price zone	Tax-inclusive Freight Rate [yuan/(thousand m^3·km)]	Main Trunk Pipeline	Starting Point	End Point	Important Branch Lines and Necessary Liaison Lines
1	Central and Eastern Price zone	0.2783	Shaanxi-Beijing Line 1	Yulin, Shaanxi	Shijingshan, Beijing	Liuyong Branch Line
2			Shaanxi-Beijing Line 2	Yulin, Shaanxi	Tongzhou, Beijing	Renqiu Branch Line
3			Shaanxi-Beijing Line 3	Yulin, Shaanxi	Changping, Beijing	E'an Cang Liaison Line
4			Shaanxi-Beijing Line 4	Yulin, Shaanxi	Shunyi, Beijing	Gaoxi Branch Line, Baoxiangxi Liaison Line, Ma Fang - Xianghe Branch Line, Miyun - Ma Fang Liaison Line, Ying Zhang Branch Line Liaison Line, Zhangbei Branch Line, etc
5			Hongqing Line	Langfang, Hebei	Dagang, Tianjin	Yongjing Branch Line, Hongqing Duplex Line, Hongqing Line 3, Wangzhuangzi Branch Line, Beijing 58 Pipeline, Dagang Gas Storage Pipeline
6			Tangshan LNG External Pipeline	Tangshan, Hebei	Tangshan, Hebei	Fengnan-Ninghe Branch
7			West–East Gas Pipeline Phase I East Section (East of Zhongwei and beyond)	Zhongwei, Ningxia	Qingpu, Shanghai	Nanjing-Wuhu Trunk Line, Changzhou-Changxing Trunk Line, Longchi-Yangzi Yangba Branch, Huaiwu Branch, Dingyuan-Hefei Duplex, Taixing-Rong Branch, etc

(continued)

Table 24 (continued)

Serial number	Price zone	Tax-inclusive Freight Rate [yuan/(thousand m^3·km)]	Main Trunk Pipeline	Starting Point	End Point	Important Branch Lines and Necessary Liaison Lines
8			West–East Gas Pipeline Phase II East Section (East of Zhongwei and beyond)	Zhongwei, Ningxia	Guangzhou, Guangdong	Pingtai Trunk Line (Henan Section), Pingtai Trunk Line (Shandong Section), Nanchang-Shanghai Trunk Line, Guangzhou-Nanning Trunk Line (Guangxi Section), Guangzhou-Nanning Trunk Line (Guangdong Section), Guangzhou-Shenzhen Trunk Line (Guangdong Section), Zhongwei-Jingbian Main Line, etc
9			West–East Gas Pipeline Phase III East Section (East of Zhongwei)	Zhongwei, Ningxia	Fuzhou, Fujian	Zhongjing Connection Line, Minyue Trunk Line (Guangdong Section), Dehua Connection Line, Changsha Branch, Changsha Connection Line, Fuzhou Connection Line
10	Central and Eastern Price Zone	0.2783	Sichuan Gas Eastward	Dazhou, Sichuan	Qingpu, Shanghai	Dahua Special Line, Sichuan-Weiyuan Branch, Nanjing Branch, Jiangxi Branch, Wushi Chemical Branch, West First Line Connection Line, etc
11			Zhongwu Line	Zhong County, Chongqing	Wuhan, Hubei	Jingxiang Branch, Qianxiang Branch, Wuhuang Branch
12			Jiangsu LNG External Pipeline	Yangzhou, Jiangsu	Nantong, Jiangsu	

(continued)

Table 24 (continued)

Serial number	Price zone	Tax-inclusive Freight Rate [yuan/(thousand m^3·km)]	Main Trunk Pipeline	Starting Point	End Point	Important Branch Lines and Necessary Liaison Lines
13			China-Myanmar Line (Guiyang-Guigang Section)	Guiyang, Guizhou	Guigang, Guangxi	Duyun Branch, Hechi Branch, Guilin Branch, Qinzhou Branch, Fangchenggang Branch
14			Yulin-Jinan Line	Yulin, Shaanxi	Dezhou, Shandong	Yuma Branch
15			Guangxi LNG External Pipeline	Beihai, Guangxi	Liuzhou, Guangxi	Liudong Branch, Guangxi West Branch, Guilin Branch
16			Anji Line	Hengshui, Hebei	Dezhou, Shandong	
17			Qingning Line	Qingdao, Shandong	Nanjing, Jiangsu	
18			Zhonggui Connection Line	Ningxia Zhongwei	Guizhou Guiyang	Tianshui branch line, Longxi branch line, Longnan branch line, Zhonggui Jiangjin station to Zhongren branch line, Zhonggui line and Yuanba-Puguang interconnection gas pipeline
19			Ji-Ning Contact Line	Hebei Hengshui	Jiangsu Yangzhou	Linyi branch line, Jining branch line, Dezhou branch line, Wucheng branch line, Liuzhuang storage gas pipeline branch line, Pizhou-Xuzhou branch line, Pizhou-Lianyungang branch line

(continued)

Table 24 (continued)

Serial number	Price zone	Tax-inclusive Freight Rate [yuan/(thousand m^3·km)]	Main Trunk Pipeline	Starting Point	End Point	Important Branch Lines and Necessary Liaison Lines
20			E-An-Cang Line	Hebei Shijiazhuang	Hebei Cangzhou	Puyang main line, Baoding main line, Anping station-Anji line contact line, Nanle contact line, Cangzhou station-Tianjin LNG contact line
21			Changning Line	Shaanxi Yulin	Ningxia Yinchuan	
22	Central-Eastern Price Zone	0.2783	Tianjin LNG Export Pipeline	Tianjin Nangang Industrial Zone	Hebei Tangshan, Shandong Binzhou	
23			Lanyin Line East Section	Ningxia Zhongwei	Ningxia Yinchuan	
24			Xin-Yue-Zhe Line	Hubei Qianjiang	Guangdong Shaoguan	Xin gas pipeline and West Second Line Shaoguan connection pipeline, Guangxi main line
25			Zhongkai Line	Henan Puyang	Henan Kaifeng	Nansi Line, Pu-Fan-Tai pipeline
26			China-Russia East Line South Section	Hebei Langfang	Jiangsu Taizhou	
27			Mongolian Western Pipeline	Tianjin Binhai New Area	Hebei Baoding	Tianjin LNG contact line, Langfang branch line

(continued)

Table 24 (continued)

Serial number	Price zone	Tax-inclusive Freight Rate [yuan/(thousand m^3·km)]	Main Trunk Pipeline	Starting Point	End Point	Important Branch Lines and Necessary Liaison Lines
28			Su-Wan Pipeline	Jiangsu Yancheng	Anhui Hefei	Binhai-Xuyi section, Anhui Tianchang-Hefei section, Funing power plant dedicated line, China-Russia East Line South Section and Jiangsu Binhai LNG export pipeline interconnection pipeline
29	Northwest Price Zone	0.1262	West–East Gas Transmission Line 1 West Section (West of Zhongwei)	Xinjiang Lunan	Ningxia Zhongwei	
30			West–East Gas Transmission Line 2 West Section (West of Zhongwei)	Xinjiang Horgos	Ningxia Zhongwei	Lun Spit Branch Line, Dushi Petrochemical Branch Line, Jiayuguan City Gas Supply Branch Line, Wuwei Branch Line, Jiuquan Gas Supply Branch Line, Zhangye City Gas Supply Branch Line, Jinchang Branch Line
31			West–East Gas Pipeline III West Section (Zhongwei Westward)	Xinjiang Horgos	Zhongwei, Ningxia	Yihuo Line
32			Se Ning Lan Pipeline	Qinghai Golmud	Lanzhou, Gansu	Lianglan Branch Line, Lande Line, Liu Petrochemical Branch Line, Xining Branch Line, Gannan Branch Line, Ganxi South Branch Line, etc
33			Lanyin Line West Section	Lanzhou, Gansu	Baiyin, Gansu	Jingtai Branch Line, Baiyin Branch Line

(continued)

Table 24 (continued)

Serial number	Price zone	Tax-inclusive Freight Rate [yuan/(thousand $m^3 \cdot km$)]	Main Trunk Pipeline	Starting Point	End Point	Important Branch Lines and Necessary Liaison Lines
34			Shanwu Line	Xinjiang Turpan	Xinjiang Urumqi	
35	Northeast Price Zone	0.1828	Sino-Russian East Line (Heihe-Yongqing Section)	Heilongjiang Heihe	Langfang, Hebei	Changling-Changchun Branch Line, Mingha Branch Line, Daqing-Harbin Branch Line, Middle Section Interconnection Pipeline, Wuqing Station Connection Project
36			Qinshen Line	Hebei Qinhuangdao	Shenyang, Liaoning	Huludao Branch Line, Panjin Branch Line, Shenyang Branch Line, Jinxi Petrochemical Branch Line, Jinzhou Petrochemical Branch Line, Yingpan Liaison Line
37			Dashen Line	Dalian, Liaoning	Shenyang, Liaoning	Shenfu Branch Line, Dalian Branch Line
38			Hashen Line	Shenyang, Liaoning	Changchun, Jilin	Pingshan Line Branch Line
39			Yongtangqin Pipeline	Langfang, Hebei	Qinhuangdao, Hebei	
40	Southwest Price Zone	0.3411	Sino-Myanmar Line (Ruili-Guiyang Section)	Dehong, Yunnan	Guiyang, Guizhou	Lijiang Branch Line, Yuxi Branch Line, Western Yunnan Natural Gas Ring Network Ruili to Longchuan Section

Note The pipelines listed above are the main cross-provincial natural gas pipelines operated by the National Oil and Gas Pipeline Network Group, as well as their important branch lines and necessary connecting lines. For other cross-provincial natural gas pipelines and their important branch lines and necessary connecting lines not specified in the table, the transport price rates of the respective price zones shall be applied. After the provincial pipeline networks are integrated into the national pipeline network, the National Oil and Gas Pipeline Network Group is encouraged to optimize and integrate the pipeline transport system based on the "national one-network" concept. Where conditions permit, branch lines and other pipelines shall be included in the unified pricing of the respective price zones

Source National Development and Reform Commission

Table 25 Natural gas benchmark gate station prices in various provinces (Regions, Cities)

Province/City	Base gate station price	Province/City	Base gate station price
Beijing	1860	Hubei	1820
Tianjin	1860	Hunan	1820
Hebei	1840	Guangdong	2040
Shanxi	1770	Guangxi	1870
Inner Mongolia	1220	Hainan	1520
Liaoning	1840	Chongqing	1520
Jilin	1640	Sichuan	1530
Heilongjiang	1640	Guizhou	1590
Shanghai	2040	Yunnan	1590
Jiangsu	2020	Shaanxi	1220
Zhejiang	2030	Gansu	1310
Anhui	1950	Ningxia	1390
Jiangxi	1820	Qinghai	1150
Shandong	1840	Xinjiang	1030
Henan	1870		

Unit: yuan/m^3 (including 9% VAT)
Note The gas delivery point in Shandong is the boundary of Shandong Province
Source National Development and Reform Commission "Notice on Adjusting Natural Gas Benchmark Gate Station Prices"

gas inventories to relatively high levels. After experiencing a "mild winter" in the Northern Hemisphere, regions had ample inventory reserves. The supply and demand contradictions in the international market gradually eased, and buyers' willingness to purchase spot cargoes weakened. This led to the inability of international market LNG prices to maintain their high levels since the beginning of 2023. Consequently, the risk premium for international natural gas supply came to an end, resulting in a continuous downward shift in the cost of domestic imported LNG resources. The interplay of multiple bearish factors caused domestic LNG prices to steadily decline in the first half of the year.

Starting in mid-June, due to the impact of the proposed closure of the Groningen gas field in the international market and the delay in the resumption of production at the Nyhamna liquefaction plant in Norway, coupled with market concerns about the El Niño phenomenon, international natural gas prices have risen, leading to an increase in the cost of domestic LNG spot imports, and therefore domestic LNG prices have stopped falling and rebounded. However, under the relaxed supply and demand situation in the international market, domestic LNG prices are only slightly boosted in the short term. Beginning in mid-August, the LNG market entered its traditional seasonal peak demand period, which led to a continuous upward trend in prices. However, due to the limited progress in the construction of domestic gas storage facilities, the slow recovery of natural gas demand, and the restricted number

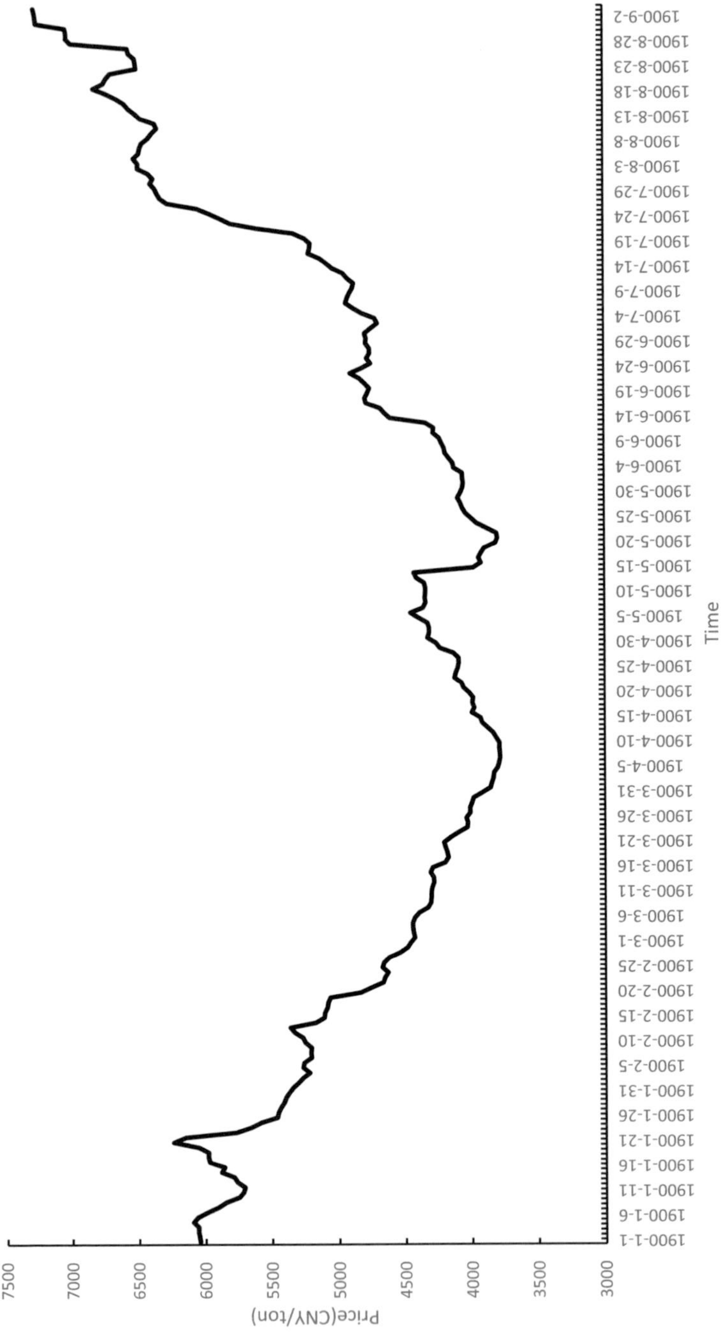

Fig. 23 China's national LNG factory price index for 2023. *Data source* Shanghai Petroleum and Natural Gas Exchange Center

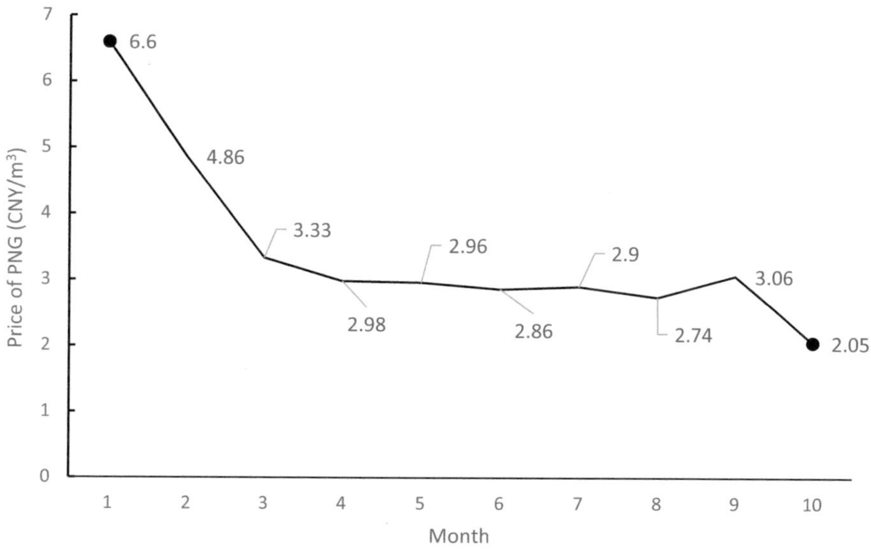

Fig. 24 Average monthly sales price of pipeline natural gas in 2023. *Data source* Shanghai Petroleum and Natural Gas Exchange Center

of new terminal stations, the peak demand for LNG did not result in a significant price surge.

(2) Price of PNG.

The Shanghai Petroleum and Natural Gas Exchange Center announced the average monthly sales price of pipeline natural gas in 2023 (Data for March and December were not announced), as shown in Fig. 24.

As can be seen, the average monthly sales price of pipeline natural gas in 2023 showed a high level in January and then continued to decline steadily. Except for the data for March and December, which were not announced, the PNG monthly average sales price has been falling since May. In November, due to seasonal demand and policy influences, the PNG monthly average sales price briefly rebounded.

(III) Zonal pricing for interprovincial natural gas pipeline transport

With the implementation of the reform of the oil and gas pipeline operation mechanism, the National Petroleum and Natural Gas Pipeline Group Co., Ltd. (hereinafter referred to as the "National Pipeline Group") was established on December 9, 2019, and the inter-provincial natural gas pipeline was transferred from the oil company to the National Pipeline Group in 2020. In 2021, the "Natural Gas Pipeline Transportation Price Management Measures (Interim)" and the "Natural Gas Pipeline Transportation Pricing Cost Audit Measures (Interim)" were introduced. On December 5, 2023, the National Development and Reform Commission announced the first approved transportation rates for the four price zones of China's inter-provincial natural gas pipeline: northwest, northeast, central and eastern, and southwest. This marked the completion of the first cost audit and the first zoning rate approval for the natural gas pipeline operated by the National Pipeline Group. Over the past four years, China's inter-provincial natural gas pipeline has initially formed a "national one-network" in physical terms, and the accompanying price design, cost audit, rate approval and other price management reforms have followed. In the future, as the pipeline operation rules gradually mature, the management of pipeline transportation prices will be further optimized and refined. This will accelerate the formation of a market competition pattern where gas sources flow freely.

1. Transportation rates have risen and fallen

In the "Natural Gas Pipeline Transportation Price Table" section of the National Pipeline Group's information disclosure, the transportation rates of 10 subsidiaries and the names of some pipelines are published. In conjunction with the "Basic Information of Oil and Gas Pipeline Facilities" disclosed at the same time, a comparison has been made of the changes in transportation rates before and after pricing for 28 pipelines under 9 subsidiaries of the National Pipeline Group (see Fig. 25).

As can be seen from the comparison, for specific pipelines, the newly determined "one price per zone" rates have both increased and decreased compared to the original "one price per enterprise" rates. In terms of pricing zones, the Northwest Zone, being the production area for domestic natural gas and the strategic corridor for imported Central Asian natural gas, has pipelines that were built relatively early. The newly determined transportation rate is close to the original rate of the West Line III and is the lowest among the four pricing zones. The Northeast Zone, serving as the strategic corridor for imported Russian natural gas, has a determined pricing zone rate that is essentially the same as the original trial price of the China-Russia East Line North Section, which was the last to be commissioned. The Southwest Zone, acting as the strategic corridor for imported Myanmar natural gas, has the most complex terrain, high construction costs, late pipeline commissioning, and low transportation volumes. The newly determined transportation rate has decreased by about 14.7% compared to the original rate of the China-Myanmar Line and is the highest among the four pricing zones. The Central and Eastern price zone is the main consumption area of natural gas in China, and the newly approved price zone transportation price rate is close to the original transportation price rate of the main cross-provincial pipelines in the price zone, such as Shaanxi-Beijing Line 1, Shaanxi-Beijing Line 2, Shaanxi-Beijing Line 3, and Shaanxi-Beijing Line 4.

2. The zonal transportation rates divide the market into transit-oriented and consumption-oriented segments

The four price zone markets can be divided into transit-oriented type and consumption-oriented segments using three physical hubs: Zhongwei in Ningxia, Guiyang in Guizhou, and Yongqing in Hebei. The northwest, northeast, and southwest transit-oriented pricing zones mainly serve as corridors for imported and domestic piped natural gas. These zones have relatively weaker price-bearing capacities and smaller end-consumption demands for natural gas. The newly determined

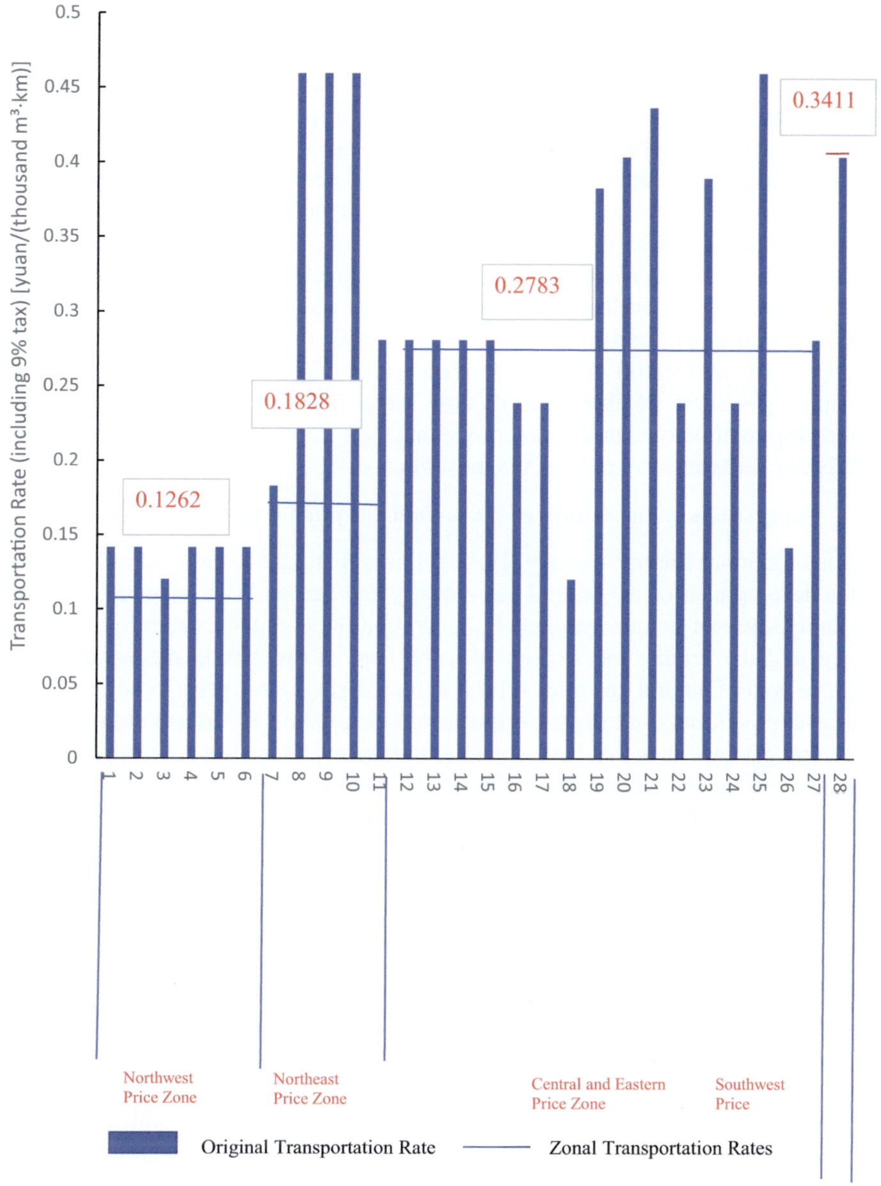

Fig. 25 Changes in transportation rates of major inter-provincial natural gas pipelines. *Data source* National Petroleum and Natural Gas Pipeline Group, National Development and Reform Commission

transportation rates have seen a slight overall decrease, which is conducive to the free flow of gas sources and promotes competition among gas sources from different market entities. This is especially beneficial for the free flow of emerging domestic onshore gas sources, such as coal-to-gas from the Tarim Basin and newly developed gas sources following the granting of exploration rights.

The Central and Eastern price zone, as the main consumption market for natural gas in China, has a strong capacity to bear prices. In addition, China's coastal LNG import channels are mainly concentrated in this region. The regional economic development status and end-market demand determine that the Central and Eastern price zone has the most in-service branch and trunk lines, interconnection lines, and LNG pipelines among the four price zones, with the highest pipeline density. The characteristics of the Central and Eastern Zone, including a more diversified supply of natural gas, more flexible resource flow, larger pipeline turnover volume, and a higher degree of marketization, determine that the competition for pipeline capacity acquisition and route selection is the most intense.

(IV) Suggestions on the reform of the natural gas pricing mechanism

The International Energy Agency (IEA) believes that natural gas pricing is crucial to the development of the natural gas industry and market. The scale and status of a country's use of natural gas largely depend on the country's natural gas pricing policy, and the pricing mechanism is at the core of the market mechanism. The ultimate goal of natural gas price reform is for the gas source price to be formed by market competition, with the government only regulating the transportation prices of natural gas pipelines that have a natural monopoly. China has developed into the world's third-largest natural gas consumer and the largest natural gas importer. The rapid development of the natural gas market has benefited from continuous deepening of price reforms and continuous improvement of the price formation mechanism. For a considerable period of time in the future, China's natural gas market will continue to be in a rapid development stage, and it will still need to continuously deepen price reforms and continuously improve the price formation mechanism.

(1) The reform of pipeline transportation prices should focus on the overall situation and direction of natural gas marketization reform, and accelerate the construction of regional competitive markets. In the future, based on the implementation effects of zonal transportation rates, and in combination with the integration process of national pipelines with provincial pipelines, it is possible to comprehensively consider selecting some provinces in the central and eastern regions with diverse supply entities, large consumption volumes, high pipeline density, and strong downstream price-bearing capacity, such as Guangdong Province and Zhejiang Province, to pilot the "same province, same price" for pipeline transportation fees. Through mechanism design, the impact of pipeline transportation paths on pipeline transportation prices can be eliminated.
(2) Improve the management of gate station prices. Further improve the management of gate station prices, with a focus on refining the policy of combining regulation and deregulation. This includes narrowing the scope of regulated gas

and expanding the scope of non-regulated gas. It is also essential to clarify the scope of supply assurance and the responsibilities and obligations of all parties involved.

(3) Improve the management of infrastructure prices. Further reasonably determine the allowed rate of return. The current method stipulated by the National Development and Reform Commission specifies that the allowed pipeline transportation price and LNG receiving station gasification fees, after tax, shall not exceed 8%. In contrast, mature natural gas markets in Europe and America generally do not exceed 6%. The NDRC's "Notice on the Action Plan for Deepening Price Mechanism Reform During the 14th Five-Year Plan Period" (NDRC Price [2021] No. 689) proposes that cost review should be an important procedure for the government to set and adjust prices, explores the establishment of a standard cost system, and introduces cost incentive and constraint mechanisms.

(4) Improve urban gas price management. The NDRC issued the "Notice on the Action Plan for Deepening Price Mechanism Reform During the 14th Five-Year Plan Period" (NDRC Price [2021] No. 689), which proposes to improve the linkage mechanism between end-user sales prices and procurement costs. It also actively coordinates the fair opening of urban gas distribution networks, reduces the levels of gas distribution, strictly regulates distribution prices, and explores the marketization of end-user sales prices.

Development Trend of Natural Gas in the Later Period of the "14th Five-Year Plan"

In the later period of the "14th Five-Year Plan," China's continued advancement of its energy transformation strategy will drive rigid growth in energy and natural gas demand. At present, natural gas accounts for less than 10% of China's primary energy consumption, which is far lower than the international level of 23%. This indicates substantial room for growth in natural gas consumption. Therefore, the "14th Five-Year Plan" period is a time of rapid development of natural gas in China. The marketization of China's natural gas market will further accelerate, prompting adjustments in the upstream and downstream industries, which will in turn drive the growth of natural gas consumption in China in the later period of the "14th Five-Year Plan."

Developed countries' natural gas consumption development period averages 30 years, which mainly includes two parts: the first 15 years of rapid development and the following 15 years of maturity. If the average annual growth rate of natural gas consumption is between 8 and 10%, it is considered to be in the rapid development stage; if it is below 5%, it indicates that it has entered the maturity stage. According to the experience of developed countries and based on the predicted average annual natural gas consumption growth rate of 6% to 8% in the "14th Five-Year Plan," the following conclusion can be drawn: Although China's natural gas consumption has passed the stage of rapid growth, it is still in a stage of slow development and has not yet entered the maturity stage. It can be predicted that in the next 10 to 15 years, China's natural gas consumption will still maintain steady growth.

In recent years, the task of energy transition remains arduous, and the trend of natural gas serving as a bridge is becoming increasingly evident. Promoting the growth of natural gas consumption remains a primary direction for China's energy transition, maintaining relatively stable growth despite shocks brought by the complex domestic economic situation and changes in the geopolitical environment. Against the backdrop of the "14th Five-Year Plan," predicting future consumption trends based on the basic characteristics of past natural gas consumption trends has a certain degree of objectivity. Therefore, using the annual natural gas consumption data from 2002 to 2023 as shown in Table 26, an ARIMA (Autoregressive Integrated Moving Average) model is established to predict the natural gas consumption trends for the next five years (2024–2028).

Table 26 Natural gas consumption in China (2002–2023)

Year	Consumption (hundred million cubic meters)
2002	291.84
2003	339.08
2004	396.72
2005	446.08
2006	561.41
2007	705.23
2008	812.94
2009	895.20
2010	1073.00
2011	1341.07
2012	1471.00
2013	1692.00
2014	1786.00
2015	1932.00
2016	2058.00
2017	2373.00
2018	2803.00
2019	3064.00
2020	3067.00
2021	3240.00
2022	3663.00
2023	3945.30

Data source "China Natural Gas Development Report," National Development and Reform Commission, National Bureau of Statistics

Table 27 Stationarity test of lnGASt

Test statistic	−3.340464
P-value	0.013164
1% Critical value	−3.809209
5% Critical value	−3.021645
10% Critical value	−2.650713

(I) Selection and preprocessing of sample data

Considering the authority and availability of the data, the time series data selected for this forecast comes from the official data released by the National Development and Reform Commission, and the traceable time range is from 2002 to 2023. Since 2000, China's natural gas consumption has shown a strong growth trend, which can be judged as a non-stationary time series. In order to better achieve fitting and forecasting, the natural gas consumption data (denoted as GASt) needs to be processed logarithmically, and the processed value is denoted as lnGASt. The stationarity test is carried out on lnGASt, and the test results are shown in Table 27. The test statistic in the table is less than the 5% critical value, and it passes the test at the 95% confidence level, meeting the stability requirement, and the p-value is much less than 0.05, proving that it has good stability.

(II) Time series modeling and model evaluation

The ARIMA (p, d, q) model is employed for forecasting analysis of the time series data of natural gas consumption. The values of the model parameters p, d, and q are determined through the BIC (Bayesian Information Criterion) test. After analysis, the ARIMA (2, 0, 0) model is ultimately selected. Subsequently, the ARIMA (2, 0, 0) model is applied to the data lnLNG, with data from 2002 to 2018 serving as the base data and data from 2019 to 2023 used for comparison to verify the credibility and reliability of the model. The differences between the forecasted data and the actual data obtained by this method are shown in Table 28. It can be observed that the error of the forecasting model is at a relatively low level, indicating that the model is credible and can be used for forecasting.

(III) Forecast of natural gas consumption trend in the next five years

Based on this, the lnLNG data for 2024–2028 can be predicted according to the ARIMA model, and it can be restored to obtain the LNG data for 2024–2028. In other

Table 28 Comparison of forecasting results

Year	Absolute error	Relative error
2019	0.042481	0.005292
2020	0.077554	0.009580
2021	0.026435	0.003215
2022	0.094740	0.011545
2023	0.048489	0.005856

Table 29 Forecast of natural gas consumption from 2024 to 2028

Year	Consumption (hundred million cubic meters)
2024	4027
2025	4141
2026	4169
2027	4117
2028	3997

words, the consumption data for 2024–2028 can be predicted based on the natural gas consumption data from 2002 to 2023. First, the value of lnGASt is forecasted, and then it is converted back to the actual consumption data GASt, which serves as the final forecast value. The forecast results are shown in Table 29.

In 2024, the Chinese natural gas market still has uncertainties due to international geopolitics, international LNG prices, and China's energy transformation policies. It is predicted that China's natural gas consumption in 2024 will be 402.7 billion cubic meters, an increase of 8.17 billion cubic meters, an increase of 2.07%.

Outlook

Based on the historical data of the average import prices of natural gas, domestic production, import volumes, apparent consumption volumes, ex-factory prices, market prices, import prices, export prices of liquefied natural gas (LNG), as well as the production, import, export, and apparent consumption volumes of LNG, the main factors affecting the average import price of gaseous natural gas and the ex-factory price, market price, and import price of liquefied natural gas have been analyzed. Moreover, the recent price fluctuations and their future effects have been analyzed, along with a price forecast for 2024, to provide guidance for the optimization and regulation of natural gas prices.

Natural Gas Price Outlook

Introduction to the Outlook Model

(I) Grey Relational Analysis Model

The Grey Relational Analysis (GRA) model is a method that judges the degree of correlation between factors according to the similarity of the geometric shapes of the change curves of each factor. The closer the curve are, the greater the correlation between the sequences; conversely, the smaller the correlation. It calculates the correlation between factors and ranks the correlation degrees to depict the differences in the changing trend between factors. The specific steps are as follows:

(1) Confirm the reference sequence Y and the comparison sequence $X_{i.}$

The reference sequence Y represents the subject of study, while the comparative sequences X_i represent several factors that influence the subject of study. This can be specifically expressed as:

© Petroleum Industry Press 2025

L. Zhang et al., *Annual Report on Operation of China's Natural Gas Industry: 2023*,
https://doi.org/10.1007/978-981-95-2371-9_4

$$Y = \{y(1), y(2), \ldots, y(t), \ldots, y(T)\} \quad (1)$$

$$X_i = \{x_i(1), x_i(2), \ldots, x_i(t), \ldots x_i(T)\} \quad (2)$$

where X_i is the i-th comparative sequence, and T is the number of data points in each sequence.

(2) Perform dimensionless processing of data.

Since different dimensions can lead to sequences being incomparable or even drawing incorrect conclusions, it is necessary to eliminate the dimensional impact of the data. The expression for mean dimensionless processing is as follows:

$$xi = \frac{xi}{\overline{x}} \quad (3)$$

where $\overline{x}$ is the average value of the sequence.

(3) Calculate the difference sequence Δ_i, the maximum difference $\Delta_{\max}$ and the minimum difference $\Delta_{\min}$.

$$\Delta_i = |y(k) - x_i(k)|, \quad k = 1, 2, \ldots, T \quad (4)$$

$$\Delta\max = \max_i \max_k \Delta i \quad (5)$$

$$\Delta\min = \min_i \min_k \Delta i \quad (6)$$

(4) Calculate the relational coefficient λ.

The relational coefficient describes the degree of close connection between two factors, and the specific calculation formula is as follows:

$$\lambda(Y(k), Xi(k)) = \frac{\Delta\min + \rho\Delta\max}{\Delta i(k) + \rho\Delta\max} \quad (7)$$

Here, the value range of λ is [0,1]. $\lambda = 1$ indicates the highest degree of close connection between two factors, while $\lambda = 0$ indicates the lowest degree of close connection. ρ is the distinguishing coefficient, with $0< \rho <1$, and it is usually taken as 0.5.

(5) Calculate the degree of correlation γ.

The degree of correlation is a measure of the extent to which two factors are associated over time, and its calculation formula is as follows:

$$\gamma i = \frac{1}{T}\sum_{k=1}^{T} \lambda(Y(k), Xi(k)) \quad (8)$$

(II) Vector Autoregressive Model (VAR)

VAR is a model that transforms the endogenous variables involved in the system into functions of lagged values to construct the model. It can effectively analyze the impact of random disturbances on economic variables, thereby better understanding the relationships between these variables and effectively controlling and regulating them.

The standard form of the VAR model is as follows:

$$Y_t = A_0 + A_1 Y_{t-1} + A_2 Y_{t-2} + \cdots + A_p Y_{t-p} + e_t \tag{9}$$

where $Y_{t=}$ is a sequence variable of n elements, is a random disturbance term, is a constant term, is a coefficient matrix, and p is a lag term. Missing matrices. 需要添加矩阵

The stationarity of the VAR model is very important. At the beginning of modeling, it is necessary to ensure the stationarity of the sequence. Establishing the characteristic roots of the model is a method to judge the stability of the model. When all the reciprocal of the characteristic roots of the VAR model are within the unit circle, it indicates that the model is stable. For a stable model, the impulse response between variables can be further studied.

The Impulse Response Function (IRF) is used to measure the impact trajectory of a one-standard-deviation shock of a random disturbance term on the current and future values of other variables. It can intuitively depict the dynamic interaction and effect between variables. A VAR model can be represented as an infinite VMA (Vector Moving Average) process:

$$Y_{t+s} = \varepsilon_{t+s} + \varphi_1 \varepsilon_{t+s-1} + \varphi_2 \varepsilon_{t+s-2} + \cdots + \varphi_s \varepsilon_t + \cdots \tag{10}$$

$$\varphi_{t+s} = \frac{\partial Y_{t+s}}{\partial \varepsilon t} \tag{11}$$

Here, Y_t and ε_t are $n \times 1$ vectors, and φ_s is a $n \times n$ matrix. The element in the i-th row and j-th column of φ_s represents the impact on the i-th variable y_{it} at time $t + s$, given that the error term ε_{jt} of the j-th variable y_{jt} receives a one-unit shock at time t, while all other error terms remain unchanged at any period.

Equation (11) is called the impulse response function, which describes the response process of $y_{i,t+s}$ to a one-time shock of ε_{jt}, under the premise that other variables remain unchanged at time t and previous periods.

(III) Deep Learning Model (LSTM)

Long Short-Term Memory (LSTM) is an improved type of Recurrent Neural Network (RNN). Compared to traditional RNNs, it includes four interacting gate layers, namely the forget gate, input gate, cell state update, and output gate. The functionsof each gate layer of the LSTM network and their calculation formulas are as follows:

1. Forget gate

The first step of LSTM is to determine what information to discard from the cell state. The forget gate f_t is responsible for selectively forgetting based on the current input $x_{t.}$ By calculating $f_{t,}$ it controls whether to retain or forget the data from the previous node's state $C_{t\text{-}1.}$ If f_t is 1, it means to retain the data from the previous cell; if f_t is 0, it means to forget the data from the previous cell. The expression of the forget gate f_t is as follows:

$$f_t = \sigma\left(W_f * \left[x_t, h_{t-1}\right] + b_f\right) \tag{12}$$

Here, σ represents the sigmoid activation function, W_f is the weight matrix for the multiplication of the input data *xt* and the output $ht-1$ of the previous cell's hidden layer, and *bf* is the bias of the forget gate.

2. Input gate

The second step is the input gate i_t, which decides what new information to store in the cell state. This gate layer includes two parts. In the first part, σ is used to control the input values that need to be updated, and the output value i_t represents the information to be remembered. The second obtains the candidate memory cell $\tilde{C}_t$ through the activation function tanh. The expressions for $\tilde{C}_t$ and i_t in the input gate are as follows:

$$i_t = \sigma\left(W_i * \left[x_t, h_{t-1}\right] + b_i\right) \tag{13}$$

$$i = \tanh\left(W_c * \left[x_t, h_{t-1}\right] + b_c\right) \tag{14}$$

Here, W_i is the weight matrix for the input data x_t of the input gate i_t, W_c is the weight matrix for the newly generated information at the current time step, and b_i and b_c are the biases for the input data and the newly generated information in the current cell, respectively.

The third step is the cell state update, which updates the old memory state $C_{t\text{-}1}$ to the new memory state $C_{t.}$ This process aims to discard the useless information in the current cell and retain the useful information. The expression of the cell state update process is as follows:

$$C_t = C_{t-1} * f_t + i_t * t \tag{15}$$

3. Output Gate

The final step is the output gate, which imposes the ultimate constraint on the actual output of the cell. First, the sigmoid activation function is used to obtain the initial output o_t. Then, a tanh layer scales the state C_t to be passed to the next time step to the range (−1,1). Finally, the initial output is multiplied to obtain the final output h_t. The expression for the output gate is as follows:

$$o_t = \sigma\left(W_0 * \left[x_t, h_{t-1}\right] + b_o\right) \tag{16}$$

$$h_t = o_t * \tanh + (C_t) \tag{17}$$

where W_o is the weight matrix of the output gate o_t and b_o is the bias.

Gaseous Natural Gas Price Outlook

(I) Analysis of factors affecting the price of gaseous natural gas

Taking the natural gas production and price data from April 2018 to December 2023 as the research object, the factors affecting natural gas prices are analyzed using Grey Relational Analysis (GRA). The average import price of natural gas is used as the reference sequence, while the production, import volume, and apparent consumption of natural gas are used as the comparative sequences to study their correlation with the average import price of natural gas.

The grey relational coefficients obtained are shown in Table 1 and Fig. 1. The grey relational coefficients between the average import price of natural gas and the production, import volume, and apparent consumption of natural gas are 0.6812, 0.6508, and 0.6792, respectively. The results show that the impact on the average import price of natural gas is in the order of: natural gas production > natural gas import volume > natural gas apparent consumption.

As can be seen from Table 1, the average import price of natural gas has the highest correlation with natural gas production, with the largest correlation value. Natural gas production, as the main source of domestic natural gas supply, can ease market tension and thus affect the average import price when it increases. The apparent consumption of natural gas ranks second in correlation with the average import price. An increase in apparent consumption indicates rising domestic demand for natural gas, which can lead to supply shortages and push up prices. The import volume of natural gas has the weakest impact and ranks last. An increase in import volume suggests greater dependence on external natural gas resources and vulnerability to international price fluctuations, which can affect domestic natural gas prices. In summary, all these factors influence the average import price of natural gas by affecting the supply and demand dynamics and price movements in the natural gas market.

Table 1 The grey relational coefficient and ranking of each factor with the average import price of natural gas

Indicator	Grey relational coefficients	Ranking
Natural gas output	0.6812	1
Natural gas import volume	0.6508	3
Apparent consumption of natural gas	0.6792	2

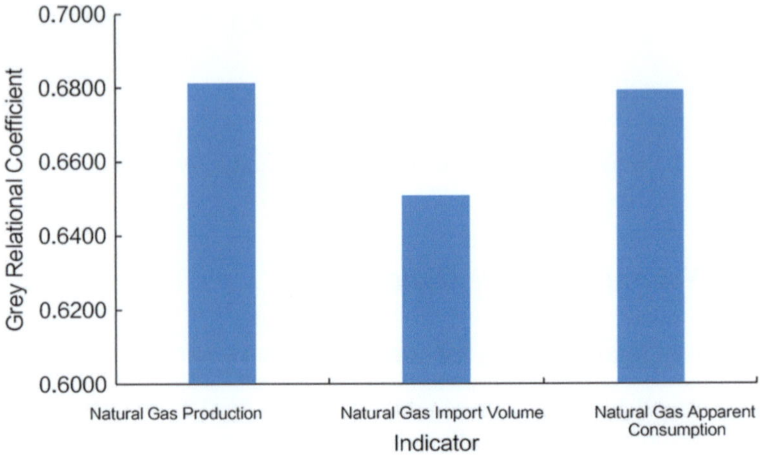

Fig. 1 The results of the grey relational coefficient between each factor and the average import price of natural gas

(II) Analysis of the price trend of gaseous natural gas

To analyze the impact of various factors on the changes in the average import price of natural gas, a Vector Autoregression (VAR) model was established for natural gas production, import volume, apparent consumption, and average import price. The impulse response function obtained from the VAR model was used to analyze the dynamic changes of other variables when the error term of one influencing factor changes.

The prerequisite for establishing a VAR model is that the variables must be stationary. If the vector does not have the characteristics of stationarity, spurious regression may occur. Through the unit root test method, it was found that a stationary time series can be obtained by taking the second-order difference of the original series. And based on the AIC information criterion, the optimal lag order of 3 was selected to establish the VAR(3) model on the basis of the stationary time series. Finally, through a rigorous characteristic root check, as shown in the AR root diagram in Fig. 2, all the reciprocal of the characteristic roots of the model are within the unit circle, indicating that the model has good stability and can be used for impulse response analysis.

After giving a standard deviation shock to natural gas production, import volume, apparent consumption, and average import price respectively, the response changes of the average import price of natural gas are shown in Fig. 3. The horizontal axis represents the lag periods of the shock effect, i.e., each point on the horizontal axis represents the response to the shock at a specific future period. The vertical axis represents the response values of the average import price of natural gas to the shocks of various factors. The blue solid line represents the estimated response of

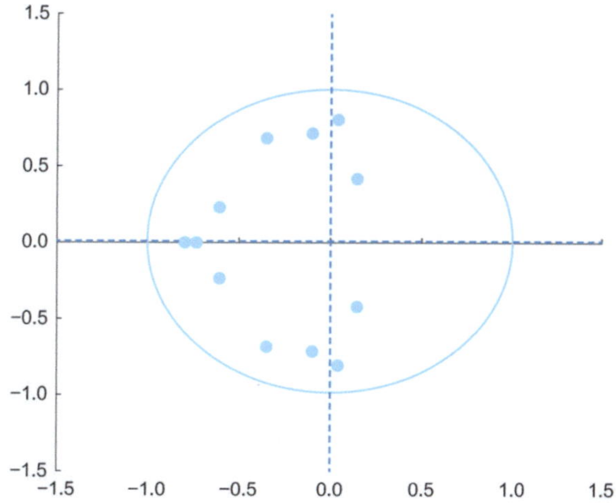

Fig. 2 AR root diagram

the average import price of natural gas to the shocks from other variables at a certain moment, and the black dashed lines represent the two standard errors of the estimated values respectively.

Figure 3a shows the impact of natural gas production on the average import price of natural gas. After a positive shock is given to natural gas production, the average import price of natural gas shows a positive response in the first month, with the strongest degree of response. Starting from the third month, it begins to decline rapidly and shows a negative impact, with an overall cyclical oscillation pattern until the impact disappears after 12 months.

Figure 3b shows the impact of natural gas import volume on the average import price of natural gas. After a positive shock is given to natural gas import volume, the average import price of natural gas shows a relatively weak positive response in the first month, and then gradually weakens with a cyclical oscillation trend, with the impact almost disappearing by the fifth month.

Figure 3c shows the impact of natural gas apparent consumption on the average import price of natural gas. After a positive shock is given to natural gas apparent consumption, the average price of natural gas shows a negative response in the first month, recovers to a positive response in the second month with the strongest degree of response in that period, and then alternates between positive and negative responses, with the impact basically disappearing after 10 months.

Figure 3d shows the impact of the average import price of natural gas on itself. After a positive shock is given to the average import price of natural gas, it shows the strongest positive response in the first month, rapidly declines in the second month to show the strongest negative response, then recovers with slight oscillations, and the impact almost disappears after the fourth month.

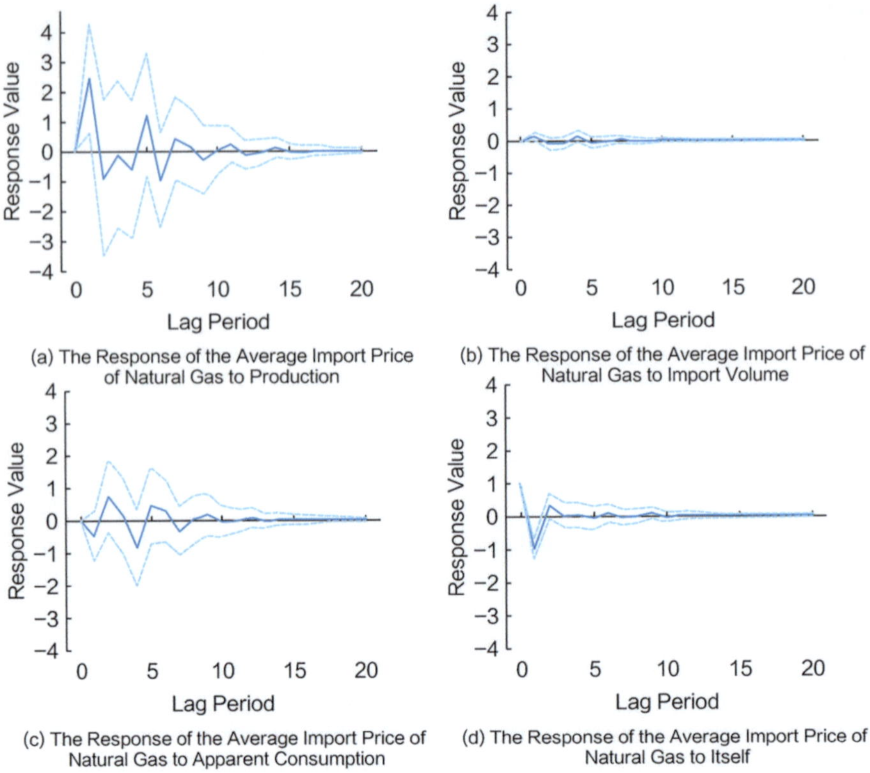

Fig. 3 Impulse response of the average import price of natural gas to shocks from various factors

Overall, natural gas production, import volume, apparent consumption, and average import price all have a certain degree of impact on the average price of natural gas. Among them, the impact of natural gas production is the most significant and lasting. The influence of natural gas apparent consumption is relatively strong and persists until it almost disappears at the 10th month. The impact of the average import price of natural gas itself is relatively weak and almost disappears by the 4th month. The influence of natural gas import volume maintains a weak positive and negative alternation in the first 5 months and almost disappears after 5 months, with its own impact being relatively weak.

(III) Analysis and forecast of gaseous natural gas prices

Regarding the prediction of the average import price of natural gas, a natural gas price prediction model based on the LSTM network was built, with historical natural gas production, import volume, apparent consumption, and average import price as inputs, and the variables at the current moment as outputs. Considering the data sample requirements of machine learning algorithms, the monthly data from April 2018 to May 2023, totaling 62 months, were used as the training set, and the monthly data from June 2023 to December 2023, totaling 7 months, were used as the test set.

Table 2 and Fig. 4 show the prediction results of the average import price of natural gas on the training set and the test set. In terms of prediction performance, the LSTM can be applied to the prediction problem of the average import price of natural gas. The predicted price trend is in line with the actual price and has a good fit on the test set. However, it was found that there is a certain lag at the peaks in January 2022 and January 2023, and the prediction deviation of the peaks is relatively large, which may be due to the influence of external factors that lead to poor fitting performance of the LSTM. But overall, the model has a good fitting effect and is helpful in grasping the changing trend of the natural gas market in the near future.

Table 2 Forecast values of the average import price of gaseous natural gas

Time	Actual value	Forecast value	Time	Actual value	Forecast value
April 2019	468.61	436.6037	September 2021	384.76	368.9502
May 2019	456.72	434.8262	October 2021	399.23	391.8662
June 2019	449.71	430.6232	November 2021	429.01	410.8168
July 2019	443.46	424.9065	December 2021	459.89	440.5660
August 2019	438.65	420.7025	January 2022	769.17	469.6523
September 2019	436.41	417.6271	February 2022	664.67	636.1180
October 2019	435.38	413.7766	March 2022	625.02	606.9906
November 2019	434.41	408.9441	April 2022	603.39	595.4441
December 2019	432.05	409.1299	May 2022	598.65	583.8827
January 2020	413.40	406.1219	June 2022	591.11	585.0290
February 2020	397.59	396.0191	July 2022	585.80	579.0629
March 2020	391.39	379.2799	August 2022	596.79	577.15484
April 2020	387.09	367.3771	September 2022	612.01	577.6392
May 2020	377.90	358.9345	October 2022	621.91	588.2360
June 2020	368.06	353.0487	November 2022	630.45	588.2475
July 2020	363.15	348.9914	December 2022	640.92	610.6438
August 2020	349.27	346.1840	January 2023	688.53	622.3758
September 2020	338.87	344.2044	February 2023	667.34	635.0966
October 2020	332.80	342.8566	March 2023	628.74	617.2437
November 2020	330.05	342.1825	April 2023	604.58	589.8067
December 2020	329.08	341.5282	May 2023	582.93	568.7227
January 2021	345.31	341.2972	Test set		
February 2021	351.61	341.5615	June 2023	565.51	564.3234

(continued)

Table 2 (continued)

Time	Actual value	Forecast value	Time	Actual value	Forecast value
March 2021	351.10	341.6762	July 2023	551.78	551.4588
April 2021	343.12	341.6080	August 2023	541.39	544.2005
May 2021	343.85	341.7799	September 2023	533.90	541.6460
June 2021	351.20	343.3702	October 2023	529.56	534.7067
July 2021	358.18	348.2468	November 2023	529.93	527.5609
August 2021	370.98	356.4419	December 2023	529.9374	533.2796
MAPE	Training set: 4.217%				
	Test set: 0.612%				

Unit: US dollars/ton
Note "MAPE" stands for Mean Absolute Percentage Error

(IV) Price Outlook for Gaseous Natural Gas in 2024

The above shows that the natural gas average import price prediction model based on LSTM is effective. Therefore, a single-step rolling prediction is performed using this model to obtain the estimated values of the average import price of natural gas for each of the 12 months in 2024. The specific values are shown in Table 3.

Figure 5 is a visualization of Table 3, which more clearly and intuitively shows the trend of changes in the average import price of natural gas in 2024. The results show that the average import price of natural gas is highest in January 2024, then begins to decline, with a stable downward trend from February to September, and a more gradual change from October to December. At the beginning of the year, due to the peak in winter gas consumption and the global economic recovery, the demand for natural gas likely reaches its peak, leading to a rise in prices. Subsequently, as temperatures rise and the economic recovery gradually stabilizes, the demand for natural gas will likely gradually decrease, causing prices to fall accordingly.

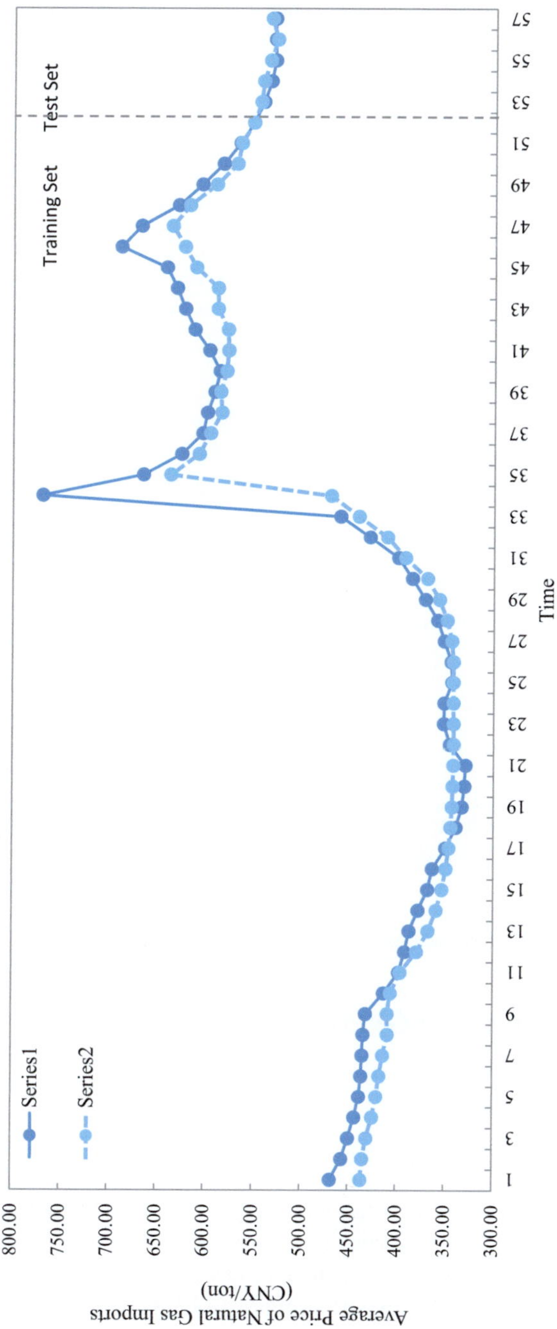

Fig. 4 Results of the prediction model for the average import price of gaseous natural gas

Table 3 Estimated average import prices of gaseous natural gas in 2024

Time	Average import price of natural gas (USD/ton)
January 2024	533.8166
February 2024	496.2208
March 2024	477.1817
April 2024	461.3807
May 2024	446.3370
June 2024	431.2771
July 2024	416.9027
August 2024	402.7792
September 2024	391.1216
October 2024	383.2199
November 2024	378.5863
December 2024	374.8477

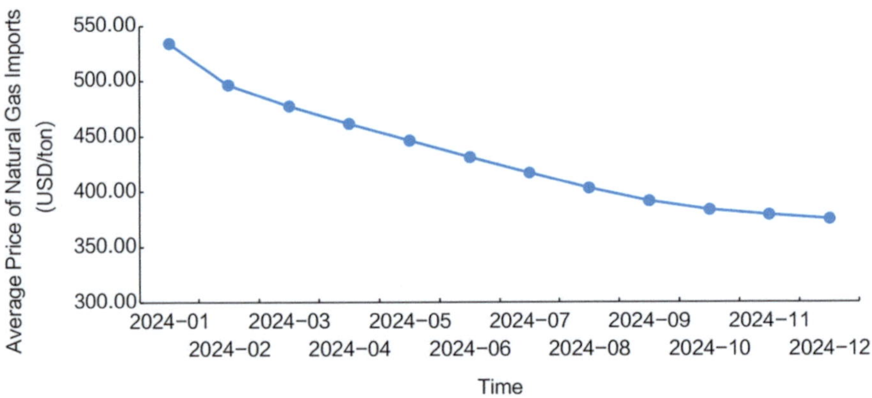

Fig. 5 Estimated average import prices of natural gas in 2024

Liquefied Natural Gas (LNG) Price Outlook

(I) LNG ex-factory price.

1. Analysis of factors affecting LNG ex-factory prices

Similar to natural gas prices, a grey relational analysis was used to explore the factors influencing the LNG ex-factory prices. The study uses LNG data from June 2019 to December 2023, with the LNG ex-factory prices as the reference sequence and LNG production, import volume, export volume, and apparent consumption as the comparative sequences.

The results of the grey relational analysis are shown in Table 4 and Fig. 6. The grey relational coefficients between LNG ex-factory price and LNG production, imports, exports, and apparent consumption are: 0.9218, 0.9182, 0.7766, and 0.9041, respectively. The order of influence is: LNG production > LNG import volume > LNG apparent consumption > LNG export volume.

LNG production, import volume, export volume, and apparent consumption collectively determine the market supply and demand situation, thereby affecting the price of LNG. Among these factors, LNG production ranks first and has the most significant impact on the LNG ex-factory prices. An increase in LNG production may lead to a supply surplus, thereby suppressing the rise in LNG ex-factory prices. The import volume of LNG has a considerable impact on the LNG ex-factory prices. An increase in import volume may increase market supply and affect prices. An increase in LNG apparent consumption indicates a growing market demand for LNG, which in turn affects the LNG ex-factory prices. The impact of LNG export volume on the LNG ex-factory price is relatively small. Changes in export volume may not directly affect domestic ex-factory prices, but if an increase in export volume leads to a decrease in domestic supply, it may indirectly affect prices.

2. Trend analysis of LNG ex-factory prices

To analyze the impact of various factors on the change in LNG ex-factory prices, a VAR model is established for LNG production, import volume, export volume,

Table 4 Grey relational coefficient and ranking of various factors with LNG ex-factory prices

Indicator	Grey relational coefficient	Ranking
LNG production	0.9218	1
LNG import volume	0.9182	2
LNG export volume	0.7766	4
LNG apparent consumption	0.9041	3

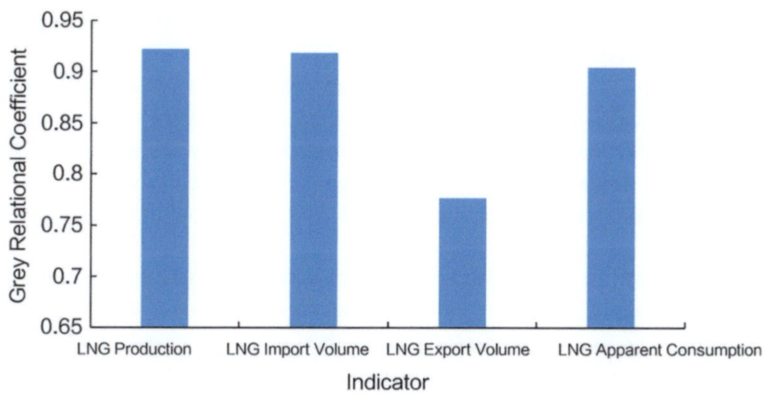

Fig. 6 The Results of the grey relational analysis between the various factors and the LNG ex-factory prices

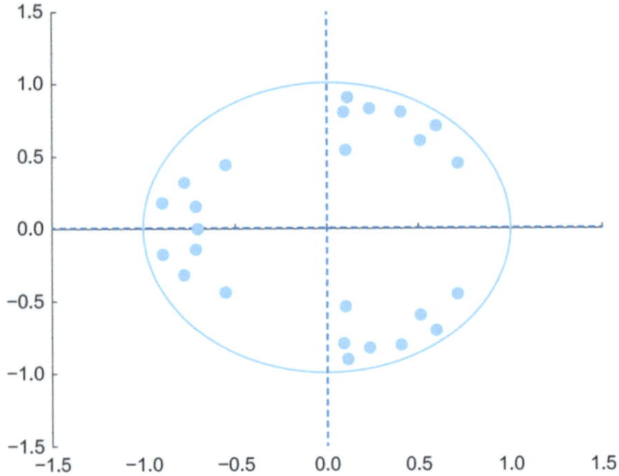

Fig. 7 AR root diagram

apparent consumption, and factory price, and the impulse response relationships are analyzed.

After performing a first-order three-step difference on the original data, a first-order difference is then applied to obtain a stationary time series. Based on the AIC information criterion, the optimal lag of 5 is selected to establish a VAR(5) model. As shown in Fig. 7, the AR roots indicate that all the reciprocals of the model's characteristic roots lie within the unit circle, indicating that the model has good stability.

After applying a one-standard-deviation shock to LNG production, import volume, export volume, apparent consumption, and factory price, the corresponding changes in the LNG ex-factory price are shown in Fig. 8.

Figure 8a shows the impact of LNG production on LNG ex-factory prices. After a positive shock to LNG production, the LNG ex-factory price shows a weak negative response in the first month, then rebounds to produce a positive response. The strongest negative response occurs in the eighth month, after which it rebounds again and the impact gradually disappears.

Figure 8b shows the impact of LNG import volume on LNG ex-factory prices. After giving a positive shock to LNG imports, the LNG ex-factory price experiences a weak negative response in the first month. It then slowly recovers to produce a positive response, drops again to produce a negative response in the fifth month. The overall trend shows a cyclical oscillation trend, and the impact almost disappears in the 20th month.

Figure 8c shows the impact of LNG export volume on LNG ex-factory prices. After a positive shock to LNG export volume, the LNG ex-factory price shows a strong positive response in the first month. It then generates a negative response in the fourth month, which is the strongest response during this period. Subsequently, positive and negative responses alternate, and the impact basically disappears by the 30th month.

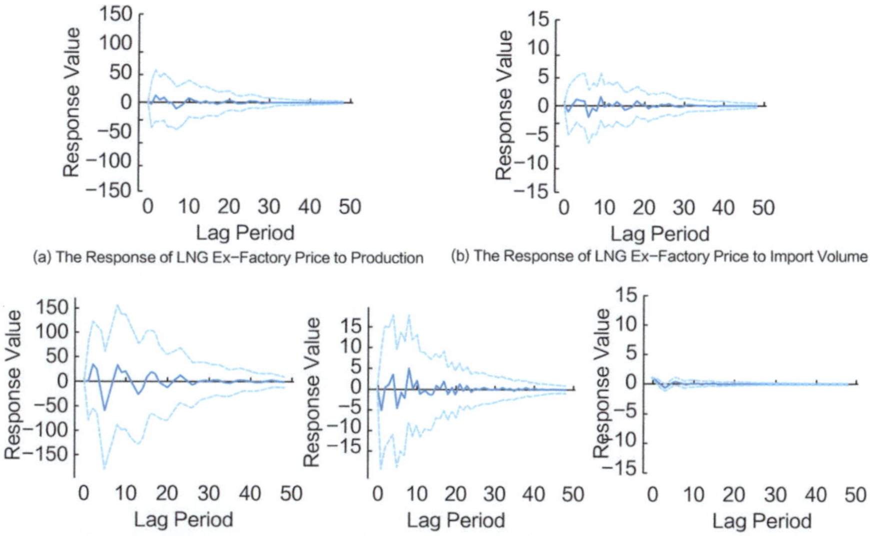

(a) The Response of LNG Ex–Factory Price to Production
(b) The Response of LNG Ex–Factory Price to Import Volume
(c) The Response of LNG Ex–Factory Price to Export Volume
(d) The Response of LNG Ex–Factory Price to Apparent Consumption
(e) The Response of LNG Ex–Factory Price to Itself

Fig. 8 Impulse response of LNG Ex-factory prices to shocks from various factors

Figure 8d shows the impact of LNG apparent consumption on LNG ex-factory prices. After a positive shock to LNG apparent consumption, the LNG ex-factory price exhibits a positive response in the first month. It then slowly recovers to produce a positive response again, with the strongest response occurring in the eighth month. Overall, the response changes display a cyclical oscillation trend, and the impact almost disappears starting from the 24th month.

Figure 8e shows the impact of LNG ex-factory prices on themselves. After a positive shock to the LNG ex-factory price, it exhibits the strongest positive response in the first month. It then drops to produce a negative response in the second month. Overall, the response oscillates weakly between positive and negative values, and the impact almost disappears after the fourth month.

Overall, the production volume, import volume, apparent consumption volume, and ex-factory price of LNG all have a certain degree of influence on the ex-factory price of LNG. Among them, the influence of LNG export volume is the most significant and enduring. The influence of LNG production volume is relatively strong, but it almost disappears by the 12th month. The influence of LNG apparent consumption volume is relatively weak and fluctuates more severely, almost disappearing by the 24th month. The influence of LNG import volume is very weak and disappears after the 20th month. The influence of the ex-factory price of LNG on itself is the weakest, with a lower degree of dependence.

3. LNG ex-factory price forecast analysis

In the research on LNG price forecasting, a prediction model based on the LSTM network in Python is established. In the prediction model for LNG ex-factory prices, historical LNG production, import volume, export volume, apparent consumption, and ex-factory prices are used as inputs, with the variables of the current moment as the output to train the model. The 55 monthly data from June 2019 to May 2023 are used as the training set, and the 7 monthly data from June 2023 to December 2023 are used as the test set.

Table 5 and Fig. 9 show the prediction results of LNG ex-factory price on the training set and test set. It can be seen that the model has a good fitting effect, showing good fitting performance on both the test set and the training set, which proves that the model has good robustness[1]. Although there may be some deviations at the peaks, the model has a high degree of consistency with the changing trend of LNG ex-factory prices. Overall, it indicates that the model is suitable for the prediction of LNG ex-factory prices and helps to explore the future changing trend of LNG ex-factory prices.

4. Outlook for the LNG ex-factory price in 2024

Based on the established LSTM model, the estimated values for the LNG ex-factory prices in 2024, obtained through a single-step rolling prediction method, are shown in Table 6.

Figure 10 is a visualization of Table 6. The results show that the LNG ex-factory price in 2024 shows a monthly downward trend, with the downward trend increasing month by month in the first half of the year and slowing down month by month in the second half of the year. The adjustment of the global energy structure and the strengthening of environmental protection policies may lead to a decrease in the demand for LNG in 2024, resulting in an oversupply in the LNG market and a decrease in prices. In addition, factors such as weather, production strategy, and market competition will also affect LNG prices.

(II) LNG market price.

1. Analysis of factors affecting LNG market price

In studying the factors affecting the LNG market price, the LNG market price is used as the reference sequence, and LNG production, import volume, export volume, and apparent consumption are used as the comparative sequences. The results of the grey relational analysis are shown in Table 7 and Fig. 11. The grey relational coefficients between the LNG market price and LNG production, import volume, export volume, and apparent consumption are 0.9211, 0.9221, 0.7766, and 0.9045, respectively. The order of influence is: LNG import volume > LNG production > LNG apparent consumption > LNG export volume.

The market supply of LNG is directly affected by production and import volumes, which in turn significantly affect market prices. The apparent consumption of LNG affects market demand, and an increase in market demand may lead to a price

[1] In machine learning, robustness refers to a model's ability to resist noise, outliers, and other disturbances in the data. A robust model is typically more reliable in dealing with real-world data situations.

Table 5 Predicted values of LNG ex-factory prices

Time	Actual value	Predicted value	Time	Actual value	Predicted value
June 2020	2528.28	2985.3107	July 2021	4388.66	4281.2084
July 2020	2532.17	2922.1972	August 2021	5522	5624.4130
August 2020	2563.51	2919.1730	September 2021	5995.21	6100.7700
September 2020	2601.96	3036.7282	October 2021	6533.88	6440.7255
October 2020	3167.52	3659.0358	November 2021	7477.9	7237.2919
November 2020	3692.61	4179.0161	December 2021	5420.45	6499.7880
December 2020	5346.72	4753.7167	January 2022	4951.57	6028.4223
January 2021	5374.15	4763.4711	February 2022	6688.25	6412.9853
February 2021	3768.05	4345.1113	March 2022	8179.81	8014.3901
March 2021	3425.6	3648.7397	April 2022	7986.23	7935.1162
April 2021	3376.9	3659.5737	May 2022	7472.9	7361.3818
May 2021	3612.15	4080.3232	June 2022	6543.68	7082.9194
June 2021	3774.38	4139.3266	July 2022	6434.9	6882.6435
August 2022	6900.69	7069.9692	May 2023	4221	4396.3178
September 2022	7066.9	7274.8398	Test set		
October 2022	7359.27	7186.5126	June 2023	4112.23	4132.4052
November 2022	6023.86	5794.7094	July 2023	4135.52	4018.4890
December 2022	7176.09	6863.9877	August 2023	3898	4277.5244
January 2023	6869.82	6610.2153	September 2023	4298.3	4375.6665
February 2023	6471.6	6504.3627	October 2023	4803.43	4681.9833
March 2023	5389.43	5434.9838	November 2023	5401.38	5095.2255
April 2023	4785.35	4931.1015	2023/12	5532.040662	5557.0908
MAPE	Training set: 7.5582%				
	Test set: 3.3580%				

Unit: US dollars/ton

increase, thus having a strong impact on market prices. In contrast, the export volume of LNG has a relatively minor impact on market prices, because changes in export volumes do not directly affect market prices. However, if an increase in export volume leads to a decrease in domestic supply, it may indirectly affect market prices.

2. Trend analysis of LNG market prices

In order to analyze the impact of various factors on the changes of LNG market prices, a VAR model is established for LNG production, import volume, export volume, apparent consumption, and market price, and the impulse response relationships were analyzed.

After performing a first-order three-step difference on the original data, a first-order difference was then applied to obtain a stationary time series. Using the AIC information criterion, the optimal lag order of 5 was selected to establish the VAR(5) model. As shown in Fig. 12, the AR roots indicate that all the reciprocals of the

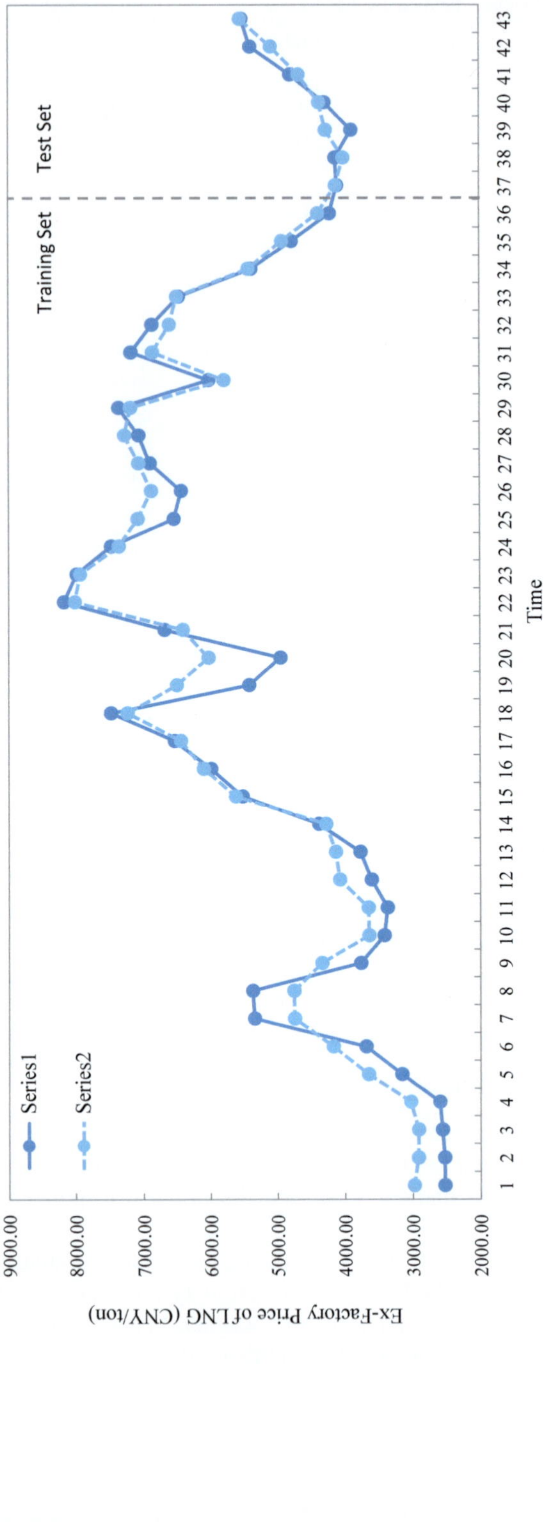

Fig. 9 Results of the LNG Ex-factory price prediction model

Table 6 Estimated values of LNG ex-factory prices in 2024

Time	LNG ex-factory price (USD/ton)
2024/1	6057.3049
2024/2	5792.3234
2024/3	5405.0846
2024/4	5037.8906
2024/5	4519.7971
2024/6	4001.4150
2024/7	3568.0395
2024/8	3225.0249
2024/9	2897.8624
2024/10	2615.6779
2024/11	2354.9398
2024/12	2089.4578

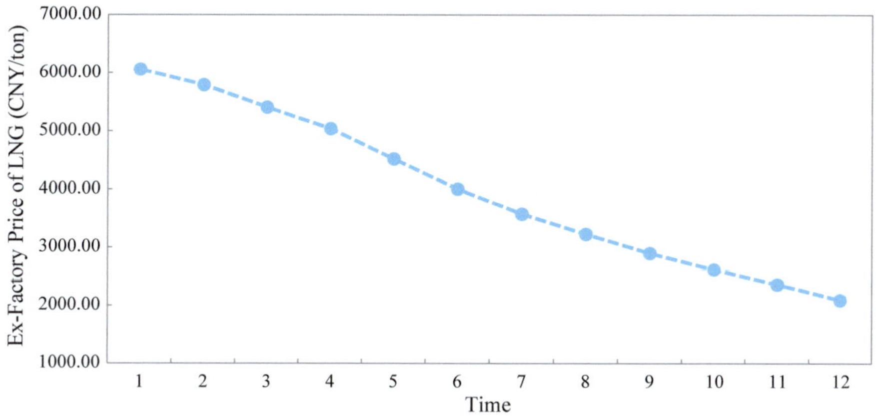

Fig. 10 Trend of the estimated values of LNG Ex-factory prices in 2024

Table 7 Grey relational coefficient and ranking of various factors and the LNG market price

Indicator	Grey relational coefficient	Ranking
LNG production	0.9211	2
LNG import volume	0.9221	1
LNG export volume	0.7766	4
LNG apparent consumption	0.9045	3

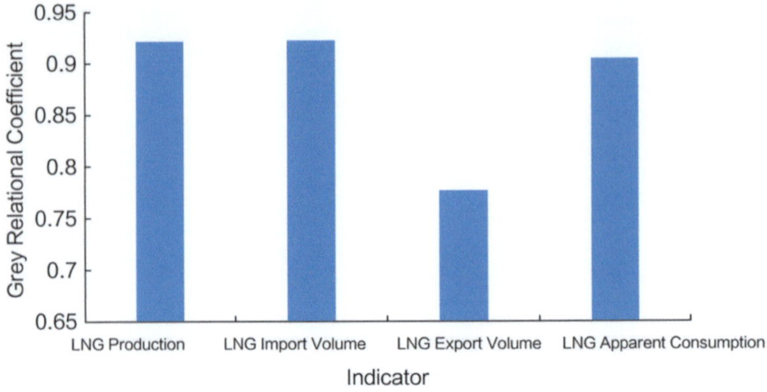

Fig. 11 Results of grey relational analysis between various factors and the LNG market price

model's characteristic roots lie within the unit circle, demonstrating that the model has good stability.

After giving a standard deviation shock to LNG production, import volume, export volume, apparent consumption, and market price, the response changes in LNG market price are shown in Fig. 13.

Figure 13a shows the impact of LNG production on LNG market prices. After a positive shock is given to LNG production, the ex-factory price of LNG shows a slight negative response in the first month, then recovers to produce a slightly stronger positive response. The overall response degree is relatively weak and the impact almost disappears after the 10th month.

Figure 13b shows the impact of LNG import volume on LNG market price. After a positive shock is given to LNG import volume, the LNG market price exhibits a

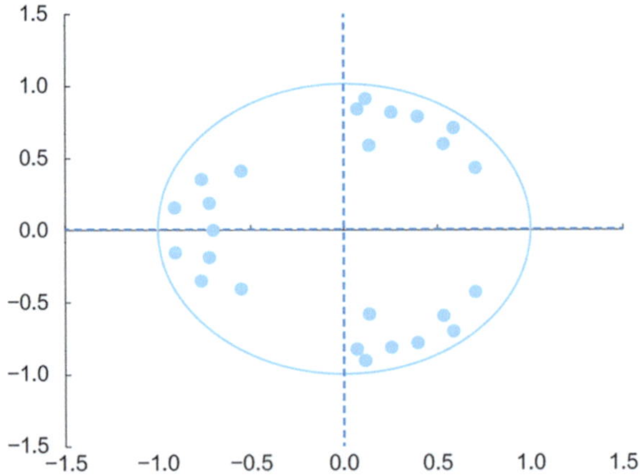

Fig. 12 AR root diagram

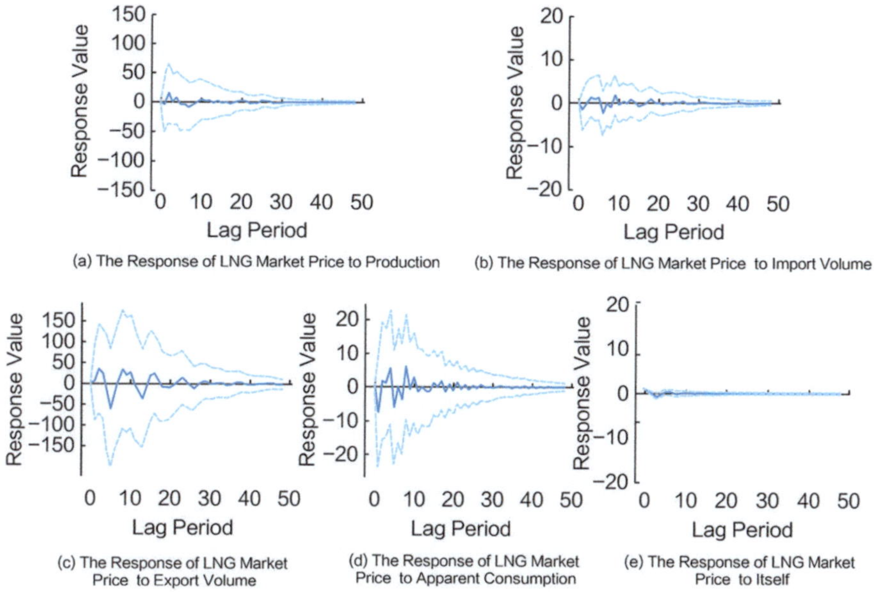

Fig. 13 Impulse responses of LNG market price to shocks from various factors

weak negative response in the first month. It then slowly rebounds to produce a weak positive response, which gradually weakens over time with alternating positive and negative responses. The impact almost disappears after the 20th month.

Figure 13c shows the impact of LNG export volume on LNG market prices. After a positive shock to LNG export volume, the LNG market price shows a strong positive response in the first month. It then decreases, producing the strongest negative response in the fifth month. Subsequently, positive and negative responses alternate, and the impact essentially disappears by the 30th month.

Figure 13d shows the impact of LNG apparent consumption volume on LNG market prices. After a positive shock to LNG apparent consumption volume, the LNG market price exhibits a strong negative response in the first month. It then rebounds to produce a positive response, overall showing a severe cyclical oscillation. The impact almost disappears after the 28th month.

Figure 13e shows the impact of the LNG market price itself on the LNG market price. After a positive shock to the LNG market price, it produces the strongest positive response in the first month. It then decreases to produce a negative response in the second month, rebounds in the third month, and the impact almost disappears thereafter.

Overall, LNG production, import volume, apparent consumption volume, and market price all have a certain degree of impact on the LNG market price. Among them, the impact of LNG export volume is the strongest and most enduring. The impact of LNG production is relatively strong but less enduring. The impact of LNG apparent consumption volume is strong, with dramatic response changes, lasting until the 28th month when the impact almost disappears. The impact of LNG import volume is weaker but more enduring, while the impact of the price itself is the weakest and least enduring.

3. LNG Market Price Forecast Analysis

In the LNG market price prediction model based on LSTM, historical LNG production, import volume, export volume, apparent consumption volume, and market price are used as inputs, with the variables at the current moment serving as outputs to train the model. The monthly data from June 2019 to May 2023, a total of 48 months, are used as the training set, and the monthly data from June 2023 to December 2023, a total of 7 months, are used as the test set.

Table 8 and Fig. 14 display the prediction results of the LNG market price on both the training and test sets. It can be seen that the model shows a good fitting effect. Although the fitting effect at the peak is slightly worse, the model has learned the trend of sudden changes following the peak. This indicates that the model can be used to predict future trends in LNG market prices.

4. Outlook for LNG Market Prices in 2024

Based on the LSTM model established above, the estimated value of the LNG market price in 2024, obtained through a single-step rolling prediction method, are shown in Table 9.

Figure 15 is a visualization of Table 9. The results show that the LNG market price in 2024 will first decrease and then increase, and then maintain a relatively stable trend. At the beginning of 2024, due to weather conditions and industrial production demand, the supply of LNG is relatively tight and the price is high. As production gradually recovers, the supply situation gradually improves and the price begins to decrease. In April, as the weather gradually warms up, market demand begins to rebound and prices also rise.

(III) LNG import prices

1. Analysis of factors affecting LNG import prices

To study the factors affecting LNG import prices, the LNG import price is taken as the reference sequence, while LNG production, import volume, export volume, and apparent consumption volume are used as the comparative sequences. The results of the grey relational analysis are shown in Table 10 and Fig. 16. The grey relational coefficients between the LNG import prices and the LNG production volume, import volume, export volume, and apparent consumption volume are 0.9365, 0.9144, 0.7782, and 0.9021, respectively. The order of influence is: LNG production > LNG import volume > LNG apparent consumption volume > LNG export volume.

The impact of LNG production on LNG import prices is the greatest. An increase in production directly affects the market supply, impacts import demand to a certain

Table 8 Predicted values of LNG market prices

Time	Actual value	Predicted value	Time	Actual value	Predicted value
2020/6	2540.3	2908.1403	September 2021	5953.55	6430.9516
July 2020	2578.74	2822.2082	October 2021	6636.31	6567.9028
August 2020	2639.71	2910.0224	November 2021	7606.36	7081.2993
September 2020	2695.77	3350.0446	December 2021	5397.7	6981.1938
October 2020	3264.63	3838.6708	January 2022	4985.68	6211.2299
November 2020	3863.52	4518.6259	February 2022	6846.19	6493.7260
December 2020	6153.04	6215.6049	March 2022	8273.57	8471.457031
January 2021	6211.9	6308.0048	April 2022	7999.37	7686.9389
February 2021	3820.2	4182.1308	May 2022	7472.26	7222.4409
March 2021	3494.65	3828.7468	June 2022	6665.1	6867.2749
April 2021	3404.29	3525.7836	July 2022	6468.38	6779.8588
May 2021	3662.17	3866.7316	August 2022	6961.43	6997.1225
June 2021	3847.76	4259.3701	September 2022	7050	7506.0498
July 2021	4469.91	4938.0224	October 2022	7365.31	7284.3374
August 2021	5556	5897.7148	November 2022	5983.5	6038.8764
2022/12	7265.41	6974.5390	2023/6	3997.9	3998.8815
2023/1	6452.81	6249.0458	2023/7	4065.57	4053.2248
2023/2	6225.45	6173.4907	2023/8	3833.57	4307.6235
2023/3	5147.87	5082.0336	2023/9	4250.5	4306.0205
2023/4	4528.53	4598.3574	2023/10	4775.24	4598.2524
2023/5	4101.5	4062.9873	2023/11	5281.45	5027.2675
Test set			2023/12	5751.24	5697.8085
MAPE	Training set: 7.342%				
	Test set: 3.3497%				

Unit: yuan/ton

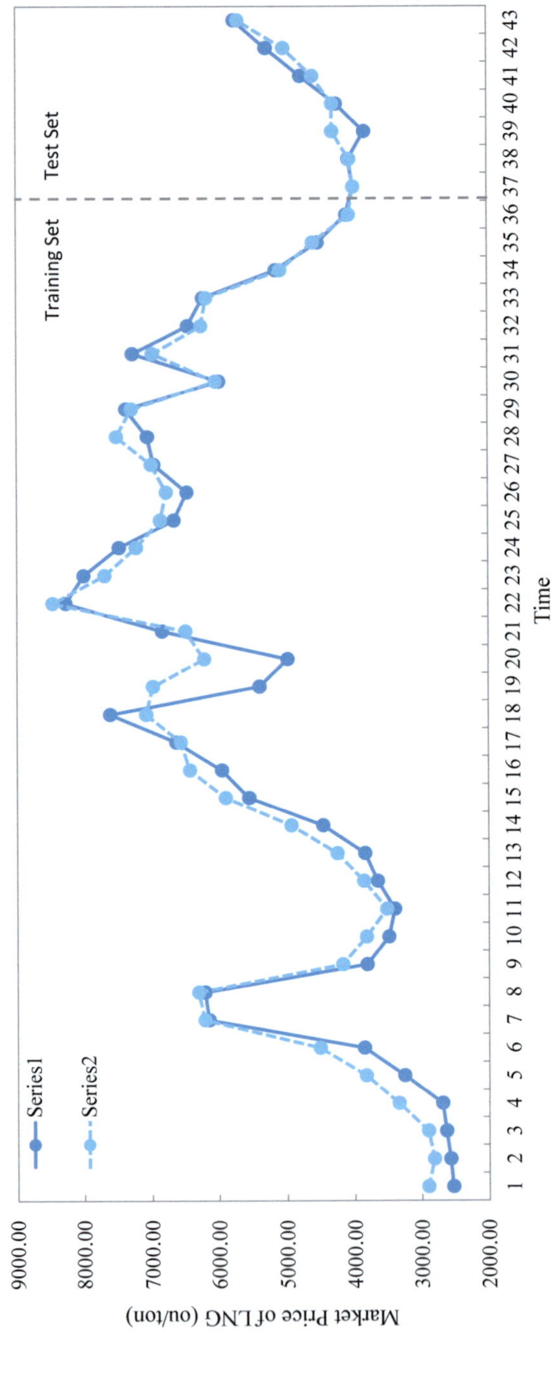

Fig. 14 Results of the LNG market price prediction model

Table 9 Estimated values of LNG market prices in 2024

Time	LNG market price (yuan/ton)
2024/1	5782.5962
2024/2	5474.0684
2024/3	5242.6483
2024/4	5698.1862
2024/5	5759.3203
2024/6	5773.6046
2024/7	5801.9316
2024/8	5837.4540
2024/9	5841.4058
2024/10	5857.7541
2024/11	5882.5117
2024/12	5883.5328

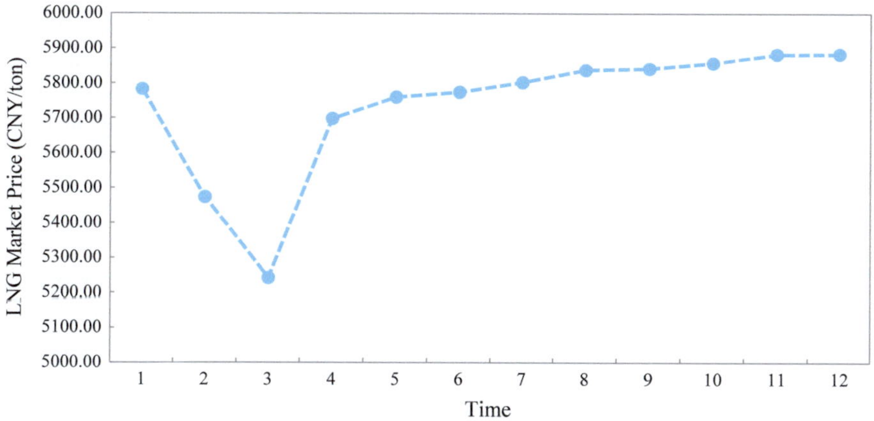

Fig. 15 Trend of estimated values of LNG market prices in 2024

Table 10 Grey relational coefficient and ranking between various factors and LNG import prices

Indicator	Grey relational coefficient	Ranking
LNG production	0.9365	1
LNG import volume	0.9144	2
LNG export volume	0.7782	4
LNG apparent consumption	0.9021	3

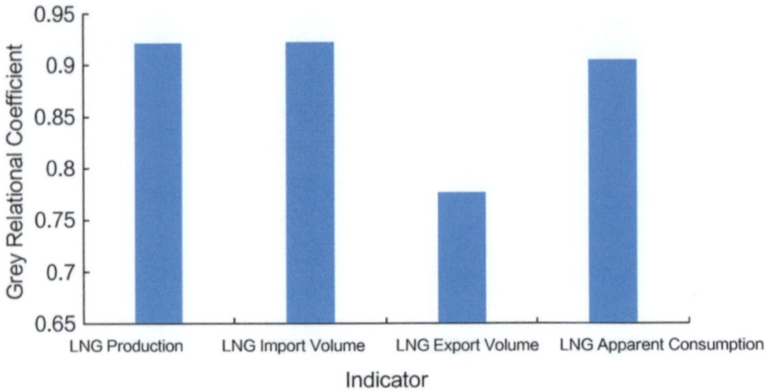

Fig. 16 Results of the grey relational analysis between various factors and LNG import prices

extent, and thereby further affects import prices. The change in LNG import volume is affected by various factors, such as domestic and foreign price differences and international trade policies, which directly affect import costs. The change in LNG apparent consumption volume has a significant impact on import prices by influencing market demand, which in turn affects import prices. The change in LNG export volume has a relatively smaller impact on import prices; fluctuations in export volume do not directly cause changes in prices.

2. Trend analysis of LNG import prices

To analyze the impact of various factors on the change in LNG import prices, a VAR model was established based on LNG production, import volume, export volume, apparent consumption, and import prices. The impulse response relationship was also analyzed.

After applying a first-order three-step difference to the raw data, a first-order difference was taken to obtain a stationary time series. The optimal lag order of 5 was selected based on the AIC information criterion, resulting in a VAR(5) model. As shown in Fig. 17, all the inverse roots of the AR roots are within the unit circle, indicating that the model has good stability.

After applying a one-standard-deviation shock to LNG production, import volume, export volume, apparent consumption, and import price respectively, the corresponding changes in LNG import prices are shown in Fig. 18.

Figure 18a shows the impact of LNG production on LNG import prices. After a positive shock is applied to LNG production, the LNG import price shows a strong positive response in the first month. The overall response alternates between positive and negative changes. The strongest negative impact occurs in the sixth month, and the impact almost disappearing after the 22nd month.

Figure 18b shows the impact of LNG import volume on LNG import prices. After a positive shock is applied to LNG import volume, the LNG import price exhibits a weak positive response in the first month. Negative responses begin in the fifth

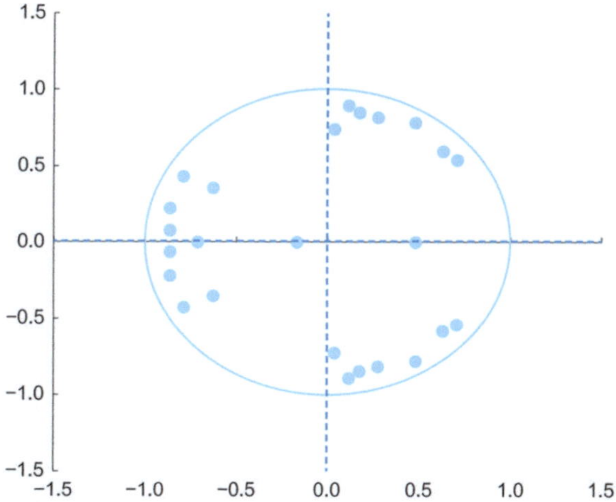

Fig. 17 AR root diagram

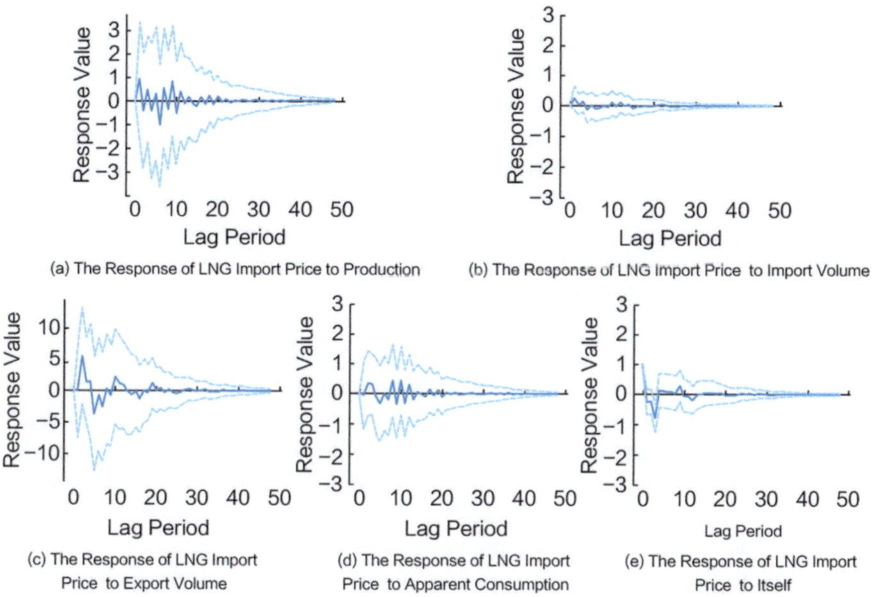

Fig. 18 Impulse response of LNG import prices to shocks from various factors

month, and the influence gradually weakens until it almost disappears by the 15th month.

Figure 18c shows the impact of LNG export volume on LNG import prices. After a positive shock is applied to LNG export volume, the LNG import price shows a positive response in the first month, with the strongest response occuring at this time. It then begins to decline, reaching the strongest negative response in the fourth month. Subsequently, it fluctuates and rises again to produce a positive response. The overall response alternates between positive and negative, with the influence almost disappearing by the 30th month.

Figure 18d shows the impact of LNG apparent consumption on LNG import prices. After a positive shock is applied to LNG apparent consumption, the LNG import price exhibits almost no impact in the first month. It then produces an equally strong positive response in the second and third months, before declining to generate a negative response. Alternating positive and negative responses appear after the ninth month, lasting until the influence almost disappears by the 20th month.

Figure 18e shows the impact of LNG import price itself on LNG import prices. After a positive shock is applied to the LNG import price, it produces the strongest positive response in the first month, declines to generate a negative response in the second month, rebounds to produce the strongest negative response in the third month, then rebounds again to generate a positive response. After the 10th month, it produces a weaker negative response, and the influence almost disappears by the 13th month.

Overall, LNG production, import volume, apparent consumption, and import price all have a certain degree of impact on LNG import prices. Among them, the impact of LNG export volume is relatively strong and persistent, but the impact weakens in the later period. The impact of LNG production is relatively strong and enduring, disappearing after the 22nd month. The impact of LNG apparent consumption is strong, with weaker effects in the early and later periods, the strongest impact in the middle period, and the influence disappearing after the 20th month. The impact of LNG import volume is the weakest, with relatively weak persistence, almost disappearing after the 15th month. Its own impact is weak and lasts until the influence disappears after the 13th month.

3. Forecast analysis of LNG import prices

In the LNG import price forecasting model based on LSTM, historical LNG production, import volume, export volume, apparent consumption, and import price are used as inputs, with the current moment's variables as outputs to train the model. The monthly data from June 2019 to May 2023 are used as the training set, and the monthly data from June 2023 to December 2023 are used as the test set.

Table 11 and Fig. 19 show the prediction results of LNG import prices on the training set and test set. It can be seen from these results that the model has a good fit to the trend of changes in LNG import prices. The sudden change in January 2022 may have been affected by external factors, making the model's fit to its value poor. However, the usual trend of changes is well fitted. Therefore, it can be considered that this model has a certain reference value in predicting the future trend of changes in LNG import prices.

4. Outlook for LNG import prices in 2024

Based on the established LSTM model, the estimated value of LNG import prices in 2024 are obtained using the single-step rolling prediction method, as shown in Table 12.

Table 11 Predicted values of LNG import prices

Time	Actual value	Predicted value	Time	Actual value	Predicted value
June 2020	393.77	394.2642	December 2020	348.48	331.0989
July 2020	386.43	378.91835	January 2021	382.97	344.3147
August 2020	370.07	363.7019	February 2021	404.28	387.5218
September 2020	357.02	352.8147	March 2021	407.19	410.4788
October 2020	349.78	341.7765	April 2021	395.3	388.6920
November 2020	346.97	329.4400	May 2021	392.55	377.1012
June 2021	402.47	374.0000	October 2022	808.1	759.6215
July 2021	414.34	388.0337	November 2022	813.52	729.8137
August 2021	431.76	386.0740	December 2022	822.71	769.5916
2021/9	451.32	416.8912	2023/1	819.09	770.6645
2021/10	471.79	447.7660	2023/2	799.85	749.1834
2021/11	514.79	464.2347	2023/3	742.29	713.2081
2021/12	559.72	511.2987	2023/4	713.35	689.1226
2022/1	969.5	567.0811	2023/5	685.13	653.3270
2022/2	855.76	802.6971	Test set		
2022/3	804.05	773.2708	2023/6	663.54	664.3167
2022/4	783.21	754.4804	2023/7	645.51	647.5237
2022/5	777.92	743.3085	2023/8	633.42	634.6582
2022/6	765.81	744.1889	2023/9	625.5	638.3843
2022/7	758.27	731.6566	2023/10	620.3	623.5654
2022/8	773.03	723.6872	2023/11	620.5325861	617.8875
2022/9	792.45	725.5464	2023/12	620.522183	615.8320
MAPE	Training set: 6.136279%				
	Test set: 0.627553%				

Unit: US dollars/ton

Figure 20 is a visualization of Table 12. The results show that the import price in 2024 generally exhibits a downward trend, with the rate of decrease slowing month by month and beginning to stabilize from September. As temperatures rise and economic recovery gradually stabilizes, the demand for LNG in the market may gradually decline, leading to a corresponding decrease in prices. However, with

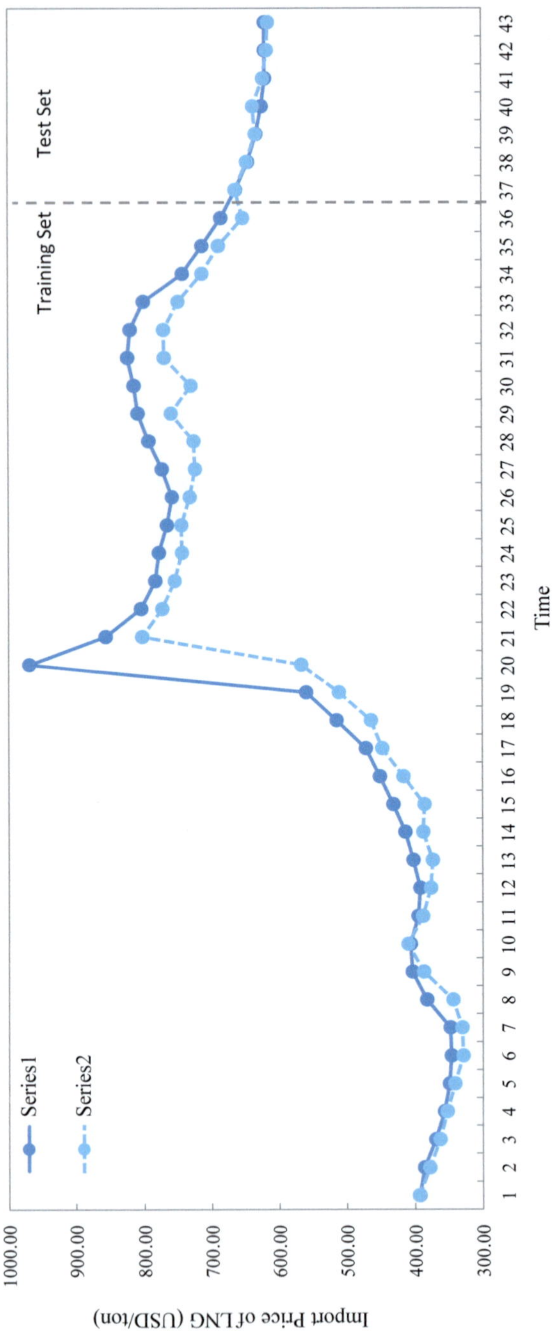

Fig. 19 Results of the LNG import price prediction model

Table 12 Predicted values of LNG import prices in 2024

Time	LNG import price (USD/ton)
2024/1	609.5743
2024/2	575.4534
2024/3	546.9611
2024/4	523.1303
2024/5	502.6526
2024/6	486.9676
2024/7	475.1046
2024/8	466.4055
2024/9	459.8501
2024/10	455.3992
2024/11	452.4188
2024/12	450.1874

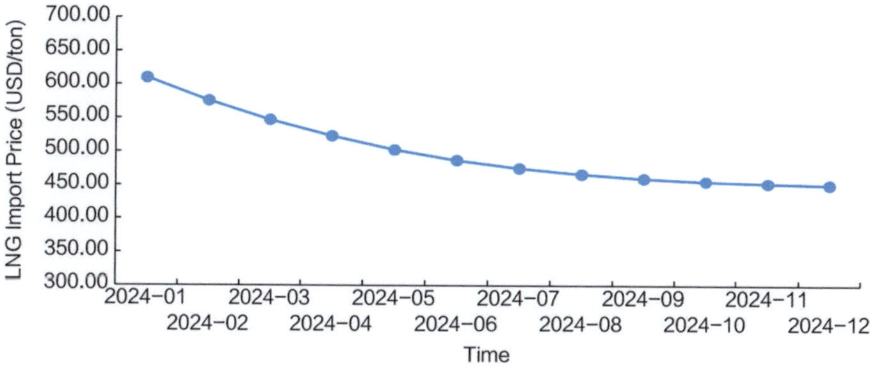

Fig. 20 Trend of estimated values of LNG import prices in 2024

the adjustment of the market supply and demand relationship, the market gradually stabilizes, and the price also gradually stabilizes.

Summary

Based on the historical data of gaseous and liquefied natural gas, the impact of factors such as production, import volume, export volume, and apparent consumption on prices has been studied. The trends of these factors' impacts on prices have been analyzed, and a price forecast for 2024 has been made. The specific conclusions are as follows:

(1) In gaseous natural gas, the degrees of influence of various factors on the average import price of natural gas are as follows: natural gas production > apparent consumption of natural gas > import volume of natural gas. The influence of natural gas production is significant and lasting, while its dependence on itself is relatively small.

(2) In liquefied natural gas, according to the results of grey relational analysis, among the ex-factory price, market price, and import price, production volume, import volume, and apparent consumption have a strong correlation, while the correlation of export volume is weak.

(3) In liquefied natural gas, further trend change studies have found that production volume, import volume, export volume, apparent consumption, and the price itself all have a certain degree of influence. Among them, the influence of the price itself is relatively weak. Production volume, import volume, export volume, and apparent consumption are all reflections of the supply and demand relationship in the natural gas market and have a strong influence on the price of natural gas. However, the degree to which production volume, import volume, export volume, and apparent consumption reflect the balance of supply and demand in the market varies, resulting in different trends and degrees of influence at different prices.

(4) In the 2024 price estimates for gaseous and liquefied natural gas, the import price of gaseous natural gas and the ex-factory price and import price of liquefied natural gas all show a monthly decreasing trend. Factors such as climate temperature and international policies cause different rates of decrease. The overall change in the market price of liquefied natural gas is relatively stable, but the price drops significantly in the second and third months.

Special Topics in the Natural Gas Industry

Natural gas, as a clean and efficient energy source, occupies an important position in the global energy structure. The development of the natural gas industry not only relates to national energy security and sustainable economic development, but also directly affects the quality of life and the ecological environment. Therefore, for some key issues in the development of the natural gas industry, we conduct special discussions from economic, management, environmental, technical and other perspectives, in order to strengthen academic exchanges, promote technological innovation and industrial upgrading, drive high-quality development of the natural gas industry, provide intellectual support for the healthy development of the natural gas industry, and provide a scientific basis for policy making.

Topic 1 is an analysis of the prosperity index of China's natural gas industry, based on the authoritative expert judgment of the three major oil companies, universities, research institutions, and government departments, calculating the prosperity index of China's natural gas industry, natural gas production prosperity index, natural gas sales prosperity index, and using the continuously published 8 years of 32 periods of data, to predict the prosperity index of China's natural gas industry in each quarter of 2024. Topic 2 first analyzes the supply and consumption status of natural gas in the Sichuan and Chongqing regions in 2023, and then makes a forecast for the development and consumption prospects of natural gas in the Sichuan and Chongqing regions in 2024. Topic 3 discusses the issue of digital transformation in the natural gas industry. Overall, the degree of digitalization in the natural gas industry is relatively low, and the returns brought by digitalization are not high. In the future, digital application work can be carried out in several aspects such as the deep application of artificial intelligence in the field of exploration and development, the deep empowerment of integrated intelligent technology for production and operation, and the deep development of natural gas market marketing data resources. Topic 4 carries out research on the full life cycle planning of natural gas development projects, borrowing from the three-dimensional structural system model of American systems engineering scholar Hall, constructing a three-dimensional conceptual model of full

© Petroleum Industry Press 2025

L. Zhang et al., *Annual Report on Operation of China's Natural Gas Industry: 2023*,
https://doi.org/10.1007/978-981-95-2371-9_5

life cycle management of natural gas development projects with the full life cycle of the project, all participants of the project, and project management content as dimensions, and proposes that the core content of the full life cycle planning of natural gas development projects is SSCQ, namely scope planning, schedule planning, cost planning, and quality planning.

Special Topic 1: Analysis of the Prosperity Index of China's Natural Gas Industry

The prosperity index of the natural gas industry is a measure of the operating status and prosperity of the natural gas industry. This indicator can vividly depict the overall operating status of the natural gas industry, reveal the causes of industry fluctuations, and predict short-term, medium-term, and long-term trends in the prosperity of the natural gas industry. In 2015, the research team led by Shen Xilin at Southwest Petroleum University constructed a prosperity index indicator system for China's natural gas industry based on factors affecting the operation of China's natural gas industry. The system consists of 5 secondary indicators and 23 tertiary indicators. The secondary indicators include macroeconomic indicators, expert judgment indicators of the natural gas industry, prosperity indicators of natural gas imports, prosperity indicators of natural gas production, and prosperity indicators of natural gas sales. The tertiary indicators of macroeconomics come from the macroeconomic data released by the state; the natural gas import and export indicators are taken from the relevant materials regularly published by China Customs; the expert judgment indicators of the natural gas industry mainly come from authoritative experts from universities, research institutions, government departments, and the three major oil companies; the relevant indicators for enterprises are obtained through expert surveys, and the survey objects of the indicators are business management experts from production enterprises and sales companies. Based on their grasp of real-time information and objective data of the enterprise, they make judgments on the tertiary indicators, and the index researchers then assign values to the tertiary indicators based on the experts' judgments. The prosperity index of China's natural gas industry has been publicly released since the first quarter of 2016, and as of the fourth quarter of 2023, it has been continuously released for 8 years, a total of 32 periods.

Analysis of the Prosperity Index of China's Natural Gas Industry

The statistical results of the natural gas industry prosperity index in China show that in the fourth quarter of 2023, the natural gas industry prosperity index in China was

292.43, indicating a very prosperous state; the prosperity index of natural gas production companies was 412.81, also indicating a very prosperous state; the prosperity index of natural gas sales companies was 192.67, indicating a prosperous state. The specific data are shown in Table 1 and Fig. 1.

As shown in Table 9, the prosperity index of China's natural gas industry in the fourth quarter of 2023 was 292.43, compared to 245.75 in the same period of 2022, representing an 18.99% year-on-year increase, with significantly improved prosperity status. The prosperity index of natural gas production companies in the fourth quarter of 2023 was 412.81, the same period in 2022 was 332.75, an increase of 24.06% year-on-year, and the prosperity status has significantly increased; the prosperity index of natural gas sales companies in the fourth quarter of 2023 was 192.67, the same period in 2022 was 168.87, showing a year-on-year increase of 14.09%, with a significant rise in the prosperity level.

Analysis of the Prosperity Index of Natural Gas Production Companies

The prosperity index of natural gas production companies is composed of eight tertiary indicators: investment level, capacity change, capacity utilization, company output, price change, operating status, employment, and employee income. In 2023, the prosperity index of natural gas production companies was 412.81, up from 332.75 in the same period of 2022, showing a 24.06% year-on-year increase and significant improvement in prosperity status. Among them: the investment level indicator in 2023 was 329.91, the same period in 2022 was 278.2, an increase of 18.59% year-on-year with significant improvement; the capacity change indicator in 2023 was 548.58, the same period in 2022 was 420.92, an increase of 30.33% year-on-year, and the indicator has significantly increased; the capacity utilization indicator in 2023 was 385.02, the same period in 2022 was 312.04, an increase of 23.39% year-on-year, and the level of capacity utilization has significantly increased; the company output indicator in 2023 was 411.40, the same period in 2022 was 315.33, an increase of 30.47% year-on-year with significant increase; the price change indicator in 2023 was 132.33, the same period in 2022 was 127.06, an increase of 4.15% year-on-year, and the indicator has slightly increased; the operating status indicator in 2023 was 606.59, the same period in 2022 was 485.07, an increase of 25.05% year-on-year, and the prosperity level has significantly increased; the employment indicator in 2023 was 80.89, the same period in 2022 was 82.53, a decrease of 1.99% year-on-year, and there has been a slight decrease; the employee income indicator in 2023 was 151.35, the same period in 2022 was 138.92, an increase of 8.95% year-on-year, and the indicator has slightly increased.

In order to grasp the trend of changes in indicators such as investment level, capacity change, capacity utilization, company output, price change, operating status, employment, and employee income of production companies, the tertiary index data

Table 1 Natural gas industry prosperity index statistics

Serial number	Time (Year. Quarter)	China's natural gas industry prosperity index	Natural gas production enterprises prosperity index	Natural gas sales companies prosperity index
1	2016.1	97.31	94.65	99.33
2	2016.2	95.60	90.57	99.04
3	2016.3	97.79	94.06	101.43
4	2016.4	101.04	97.50	103.82
5	2017.1	103.03	100.73	103.84
6	2017.2	104.54	103.69	104.97
7	2017.3	107.35	106.98	108.31
8	2017.4	114.53	113.82	115.65
9	2018.1	120.02	120.09	119.89
10	2018.2	124.69	126.32	123.77
11	2018.3	131.65	134.73	130.58
12	2018.4	137.18	143.55	131.58
13	2019.1	142.83	149.80	136.05
14	2019.2	147.71	157.84	141.44
15	2019.3	154.18	166.75	146.65
16	2019.4	163.94	179.83	155.28
17	2020.1	162.00	181.42	149.55
18	2020.2	164.14	187.56	148.15
19	2020.3	171.29	199.98	151.59
20	2020.4	178.49	205.28	160.00
21	2021.1	186.54	221.46	159.65
22	2021.2	190.83	230.01	161.27
23	2021.3	202.96	248.96	166.53
24	2021.4	214.77	269.55	170.90
25	2022.1	226.17	287.03	176.61
26	2022.2	233.76	304.55	178.09
27	2022.3	236.77	319.70	164.62
28	2022.4	245.75	332.75	168.87
29	2023.1	257.42	353.73	174.44
30	2023.2	266.31	371.69	176.12
31	2023.3	277.03	390.88	181.91
32	2023.4	292.43	412.81	192.67

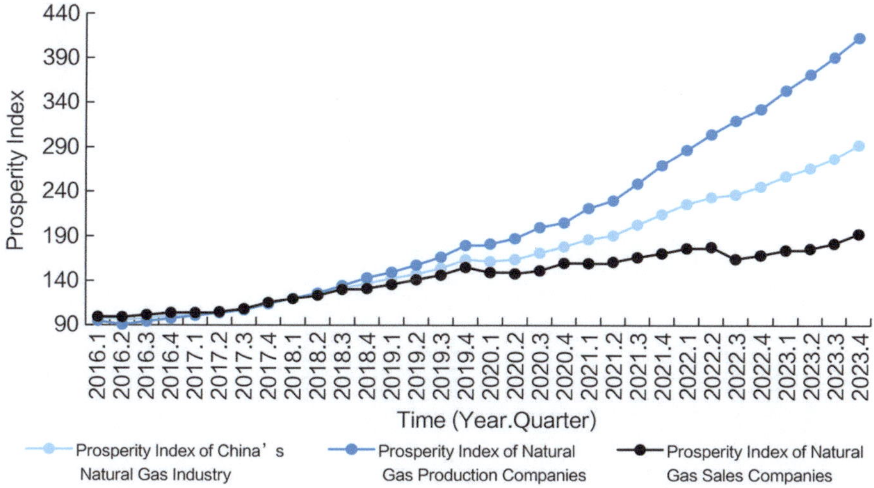

Fig. 1 Prosperity index trend of China's natural gas industry

of the prosperity index of natural gas production from 2019 to 2023 are used for regression analysis, and predictions are made for the tertiary index data of the four quarters of 2024. The specific data are shown in Table 2 and Fig. 2.

The forecast results of the tertiary indicators of the prosperity index of natural gas production companies show: the investment level indicator in 2024 is 347.79, the same period in 2023 is 329.91, an increase of 5.42% year-on-year with slightly improvement; the capacity change indicator in 2024 is 575.54, the same period in 2023 is 548.58, an increase of 4.91% year-on-year, and the indicator has slightly increased; the capacity utilization level indicator in 2024 is 413.89, the same period in 2023 is 385.02, an increase of 7.50% year-on-year, and the level of capacity utilization has slightly increased; the company output indicator in 2024 is 426.33, the same period in 2023 is 411.40, an increase of 3.63% year-on-year, and there has been a slight increase; the price change indicator in 2024 is 140.17, the same period in 2023 is 132.33, an increase of 5.93% year-on-year, and the price indicator has slightly increased; the operating status indicator in 2024 is 640.57, the same period in 2023 is 606.59, an increase of 5.60% year-on-year, and the prosperity level has slightly increased; the employment indicator in 2024 is 83.63, the same period in 2023 is 80.89, an increase of 3.39% year-on-year, and employment has slightly increased; the employee income indicator in 2024 is 153.59, the same period in 2023 is 151.35, an increase of 1.48% year-on-year, and the level of employee income has slightly increased.

According to the forecast values of the tertiary indicators of the prosperity index of natural gas production, the prosperity index of natural gas production is calculated to be: 418.22 in the first quarter of 2024, 423.82 in the second quarter of 2024, 429.42 in the third quarter of 2024, and 435.02 in the fourth quarter of 2024.

Table 2 Forecast of prosperity index tertiary indicators for natural gas production companies

Sequence	Time (Year. Quarter)	Investment level	Capacity Change	Capacity utilization	Output level	Price change	Operating condition	Employment personnel	Employee income
1	2019.1	160.62	165.08	141.93	139.25	106.00	189.88	83.46	98.60
2	2019.2	165.97	176.08	144.29	146.21	106.00	205.70	83.46	101.89
3	2019.3	174.27	190.76	149.10	153.52	104.23	222.85	79.29	105.28
4	2019.4	181.24	209.83	158.05	165.80	104.23	245.13	80.87	111.60
5	2020.1	177.62	205.64	164.37	162.49	102.15	250.03	80.87	109.37
6	2020.2	173.18	210.78	168.48	166.55	102.15	262.54	80.87	112.10
7	2020.3	164.52	231.85	185.33	183.20	99.59	282.23	78.85	106.50
8	2020.4	176.27	248.42	201.21	196.29	103.86	278.19	81.10	108.02
9	2021.1	191.69	273.26	218.82	215.92	106.46	299.06	82.12	114.77
10	2021.2	201.28	282.37	224.29	215.92	105.57	319.00	82.80	115.73
11	2021.3	210.22	307.47	244.22	235.11	112.61	347.35	82.80	118.30
12	2021.4	219.57	334.80	265.93	256.01	120.12	378.23	82.80	120.93
13	2022.1	239.33	358.23	281.89	273.93	124.92	404.70	81.97	124.56
14	2022.2	261.67	382.11	293.16	290.36	129.08	434.38	82.52	127.88
15	2022.3	280.16	415.10	309.87	310.98	130.85	452.92	83.70	129.71
16	2022.4	278.20	420.92	312.04	315.33	127.06	485.07	82.53	138.92
17	2023.1	297.39	453.33	331.39	337.09	129.96	518.54	83.19	140.03
18	2023.2	312.26	487.33	345.31	362.37	129.99	544.47	83.86	142.41
19	2023.3	326.32	514.13	360.85	385.57	132.33	574.42	84.61	144.97
20	2023.4	329.91	548.58	385.02	411.40	132.33	606.59	80.89	151.35

(continued)

Table 2 (continued)

Sequence	Time (Year. Quarter)	Investment level	Capacity Change	Capacity utilization	Output level	Price change	Operating condition	Employment personnel	Employee income
21	2024.1 Predicted value	336.20	551.28	397.73	409.17	137.82	614.33	83.49	150.50
22	2024.2 Predicted value	340.07	559.37	403.12	414.89	138.60	623.08	83.54	151.53
23	2024.3 Predicted value	343.93	567.46	408.50	420.61	139.38	631.82	83.59	152.56
24	2024.4 Predicted value	347.79	575.54	413.89	426.33	140.17	640.57	83.63	153.59

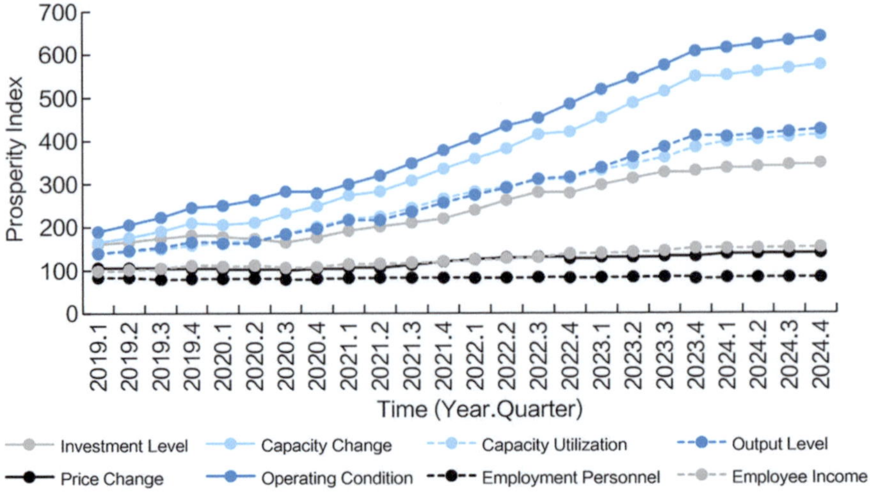

Fig. 2 Forecast of prosperity index tertiary indicators for natural gas production companies

Analysis of the Prosperity Index of Natural Gas Sales Companies

The prosperity index of natural gas sales companies is composed of six tertiary indicators: product sales volume, supply and demand status, price changes, operating conditions, employment personnel, and employee income. In 2023, the prosperity index of natural gas sales companies was 192.67, representing a 5.92% year-on-year increase with a slight increase. Among them: the product sales volume index in 2023 was 330.60, the same period in 2022 was 265.04, a year-on-year increase of 24.74%, the index significantly increased; the supply and demand status index in 2023 was 97.56, the same period in 2022 was 102.45, reflecting a 4.77% year-on-year decrease with slight decline; the price change index in 2023 was 127.37, the same period in 2022 was 118.18, a year-on-year increase of 7.78%, a slight increase; the operating condition index in 2023 was 215.18, the same period in 2022 was 187.77, a year-on-year increase of 14.60% with moderate improvement; the employment personnel index in 2023 was 119.99, the same period in 2022 was 124.08, a year-on-year decrease of 3.30% with slight decline; and the employee income index in 2023 was 129.85, the same period in 2022 was 116.25, a year-on-year increase of 11.70%, a certain increase.

The following is a regression analysis of the tertiary index data of the prosperity index of natural gas sales companies for the five years from 2019 to 2023, and a forecast of the tertiary index for the four quarters of 2024. The specific data are shown in Table 3 and Fig. 3.

The forecast results for the tertiary indicators of the prosperity index of natural gas sales show: the product sales volume index in 2024 is 335.45, the same period in 2023

Table 3 Forecast of prosperity index tertiary indicators for natural gas sales companies

Serial number	Time (Year. Quarter)	Sales volume	Supply and demand conditions	Price changes	Operating status	Employment personnel	Employee income
1	2019.1	179.17	93.18	101.18	153.16	148.92	104.06
2	2019.2	186.85	95.84	98.29	161.91	151.05	108.51
3	2019.3	191.52	98.24	99.52	172.03	151.05	109.87
4	2019.4	210.67	103.15	105.73	180.63	149.16	112.62
5	2020.1	196.63	95.13	97.51	178.63	150.82	113.87
6	2020.2	208.92	89.18	90.20	174.16	148.94	111.02
7	2020.3	227.72	85.61	88.39	175.90	150.43	109.91
8	2020.4	244.80	90.61	94.29	184.70	151.68	112.66
9	2021.1	247.86	89.48	93.11	182.39	149.78	115.48
10	2021.2	253.37	91.46	97.25	182.39	146.45	114.19
11	2021.3	266.03	96.04	102.11	185.43	144.01	118.00
12	2021.4	276.01	100.84	108.49	187.75	140.41	119.47
13	2022.1	286.36	107.14	115.27	192.44	138.66	120.97
14	2022.2	282.27	108.67	117.74	196.56	137.67	124.42
15	2022.3	254.12	102.45	111.81	183.91	125.85	114.64
16	2022.4	265.04	102.45	118.18	187.77	124.08	116.25
17	2023.1	280.42	102.45	120.19	193.97	120.98	119.15
18	2023.2	287.99	98.76	118.03	197.46	118.80	119.15
19	2023.3	305.27	96.79	120.39	203.38	119.99	122.73
20	2023.4	330.60	97.56	127.37	215.18	119.99	129.85
21	2024.1 Predicted value	327.23	102.30	124.60	208.60	116.73	125.96
22	2024.2 Predicted value	329.97	102.48	125.25	209.45	115.94	126.32
23	2024.3 Predicted value	332.71	102.66	125.89	210.29	115.14	126.67
24	2024.4 Predicted value	335.45	102.83	126.54	211.14	114.35	127.03

is 330.60, representing a 1.47% year-on-year increase with slight improvement; the supply and demand status index in 2024 is 102.83, the same period in 2023 is 97.56, a year-on-year increase of 5.41%, a slight increase; the price change index in 2024 is 126.54, the same period in 2023 is 127.37, a year-on-year decrease of 0.65% with possible slight decrease; the operating condition index in 2024 is 211.14, the same

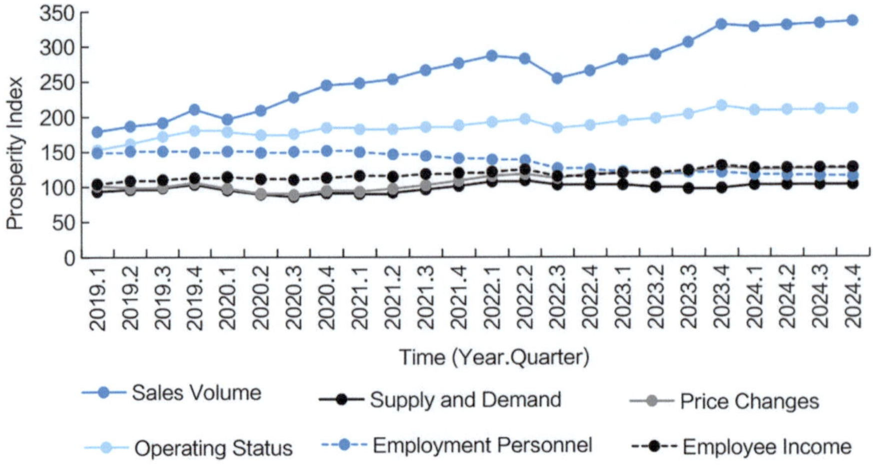

Fig. 3 Natural gas sales prosperity index tertiary indicator forecast

period in 2023 is 215.18, a year-on-year decrease of 1.88%, the index may slightly decrease; the employment personnel index in 2024 is 114.35, the same period in 2023 is 119.99, a year-on-year decrease of 4.7%, the index may slightly decrease; the employee income index in 2024 is 127.03, the same period in 2023 is 129.85, a year-on-year decrease of 2.17%, a slight decrease may occur.

According to the forecast values of the tertiary indicators of the prosperity index of natural gas sales, the prosperity index of natural gas production is calculated to be: 189.81 in the first quarter of 2024, 190.75 in the second quarter of 2024, 191.69 in the third quarter of 2024, and 192.63 in the fourth quarter of 2024.

Conclusion

Based on the forecast values of the prosperity index of natural gas production, the prosperity index of natural gas sales, and the forecast values of natural gas import indicators, the forecast values of the prosperity index of China's natural gas industry for each quarter of 2024 are calculated as follows: 294.70 in the first quarter of 2024, 297.82 in the second quarter of 2024, 300.94 in the third quarter of 2024, and 304.06 in the fourth quarter of 2024.

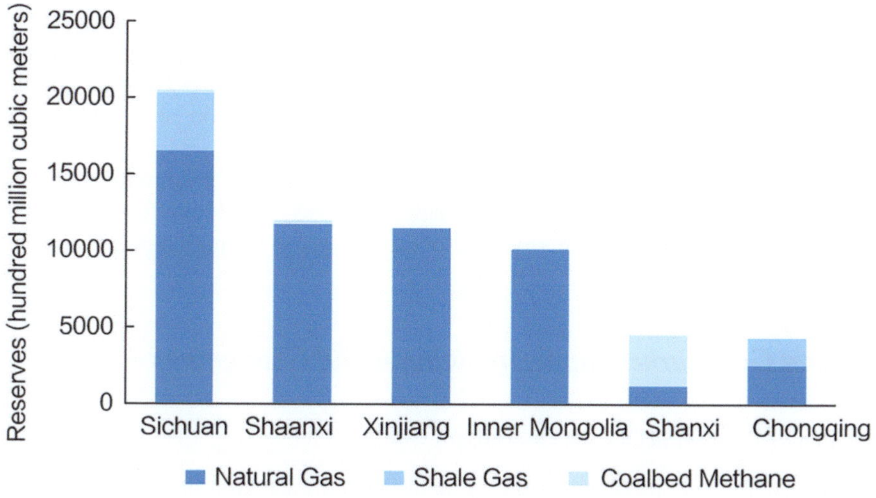

Fig. 4 Energy and mineral reserves in selected Chinese provinces and cities in 2022. *Data source* ministry of natural resources "2022 national mineral resources reserve statistics table"

Special Topic 2: Natural Gas Supply and Consumption in Sichuan and Chongqing in 2023 and Outlook for 2024

The Sichuan-Chongqing region is rich in natural gas resources and is one of the three main gas regions in China. The region boasts a well-developed pipeline network and other infrastructure, along with widespread gas usage range and a high gasification rate. The total natural gas resources in the Sichuan Basin are estimated to be around 40 trillion cubic meters, which is approximately equivalent to the combined natural gas resources in the Ordos, Tarim, and Qaidam Basins, ranking first in China. According to the data from the Ministry of Natural Resources' "2022 National Mineral Resource Reserves Statistical Table," the combined conventional and unconventional gas reserves[1] in Sichuan and Chongqing provinces amount to 2.47 trillion cubic meters, accounting for 32.9% of the national total. Among these, natural gas reserves stand at 1.91 trillion cubic meters, representing 29.1% of the national total, while shale gas reserves amount to 0.55 trillion cubic meters, accounting for as high as 98.7% of the national total. The natural gas reserves in China's major natural gas-producing provinces in 2022 are shown in Fig. 4.

According to relevant materials, the current exploration rate of the Sichuan Basin is only 9%, and the discovery rate is only 14%. Compared with the 60% exploration rate of the Permian Basin in the United States, the 43% exploration rate of the Timan-Pechora Basin in the Soviet Union, and the 28% exploration rate of the Alberta Basin in Canada. These comparisons clearly indicate that the Sichuan Basin remains

[1] The reserves listed in the "2022 National Mineral Resources Reserves Statistics Table" are all retained reserves.

in its early and middle stages of development, and the future development potential is huge. The Sichuan-Chongqing region has positioned shale gas as a strategic pillar industry, actively promoting the construction of China's "Gas Daqing," continuously making new discoveries and breakthroughs in exploration, and significantly enhancing resource production capacity.

This special topic mainly analyzes the development and consumption status of natural gas in the Sichuan-Chongqing region in 2023, and looks forward to the development and consumption prospects of natural gas in the region in 2024.

Natural Gas Market Supply in Sichuan and Chongqing in 2023

(I) Natural gas supply entities and supply volume

The main natural gas producers in the Sichuan-Chongqing region are two state-owned giants, China National Petroleum Corporation (CNPC) and China Petrochemical Corporation (Sinopec). In 2023, these companies continued to focus on strategic breakthroughs and scale-up in natural gas, producing 71.7 billion cubic meters of natural gas in the Sichuan-Chongqing region. Of this, China National Petroleum Corporation produced 45 billion cubic meters and China Petrochemical Corporation produced 26.7 billion cubic meters, accounting for 62.8% and 37.2% respectively.

As one of the main natural gas producing areas in China, the Sichuan-Chongqing region has seen a yearly increase in surplus gas, with some natural gas resources being exported to other provinces and cities in the country. In 2023, the total supply of natural gas in the region was 42.3 billion cubic meters, accounting for only 59.0% of the total natural gas production in the Sichuan-Chongqing region. The export volume was about 29.4 billion cubic meters, accounting for 41.0%, an increase of 4% points from the same period in 2022. The supply breakdown reveals China National Petroleum Corporation supplied 31.5 billion cubic meters of natural gas, and China Petrochemical Corporation supplied 10.8 billion cubic meters, accounting for 76.5% and 25.5% respectively. See Table 4 for details.

In addition, LNG plants in the Sichuan-Chongqing region also provide resource security for the stable operation of the local natural gas market. In 2023, there were 26 operational LNG plants in the Sichuan-Chongqing region, including 17 in Sichuan and 9 in Chongqing, with a designed daily production capacity of 13.58 million

Table 4 Natural gas supply by PetroChina and Sinopec in Sichuan and Chongqing in 2023

Supplier	2023 Supply volume (hundred million cubic meters)	Share (%)	Year-on-year growth rate (%)
PetroChina	315	76.5	4.2
Sinopec	108	25.5	7.2
Total	423	100	6.3

Data source Chongqing Economic and Information Technology Committee and online materials

Table 5 Natural gas supply in Sichuan and Chongqing in 2023

Region	2023 Supply volume (hundred million cubic meters)	Share (%)	Year-on-Year growth rate (%)
Sichuan	278	65.7	4.5
Chongqing	145	34.3	10.7
Total	423	100	6.3

Data source Chongqing Economic and Information Technology Committee and online materials

cubic meters; there were 8 planned LNG plants with a daily production capacity of 7.1 million cubic meters. In 2023, the start-up load rate of the Sichuan-Chongqing liquid plant was about 80%, higher than the national average (55%), with an annual LNG production of about 3.3 billion cubic meters, an increase of 13% year-on-year. In 2023, the LNG resources in the Sichuan-Chongqing region were mainly consumed within the region, but the outflow volume increased, extending to Hunan, Hubei, Jiangxi, Guangxi, and Guangdong. Due to the increase in coastal imported LNG resources and the intensified delivery efforts of receiving stations, the inflow of Sichuan-Chongqing resources to South China decreased compared with 2022.

(II) Regional natural gas supply

In 2023, the natural gas supply in Sichuan was 27.8 billion cubic meters, with China National Petroleum Corporation supplying 22 billion cubic meters and China Petrochemical Corporation supplying 5.8 billion cubic meters; the natural gas supply in Chongqing was 14.5 billion cubic meters, with China National Petroleum Corporation supplying 9.5 billion cubic meters and China Petrochemical Corporation supplying 5 billion cubic meters. See Table 5 for details.

In 2023, the LNG plant in Sichuan Province produced about 2.4 billion cubic meters, an increase of 15% year-on-year, and the LNG plant in Chongqing City produced about 900 million cubic meters, an increase of 7% year-on-year. See Fig. 5 for details.

Natural Gas Market Consumption in Sichuan and Chongqing in 2023

(I) Natural gas consumption in the Sichuan-Chongqing region

The Sichuan-Chongqing region has developed into one of the most mature natural gas markets in the country, with a leading scale of consumption, characterized by a close combination of market and gas sources, a complete downstream industrial chain, diverse utilization methods, and a large number of direct supply customers. In 2023, the natural gas consumption in the Sichuan-Chongqing region was about 42.3 billion cubic meters, an increase of 6.3% year-on-year. Of this, Sichuan Province consumed 27.8 billion cubic meters of natural gas, an increase of 4.5% year-on-year; Chongqing

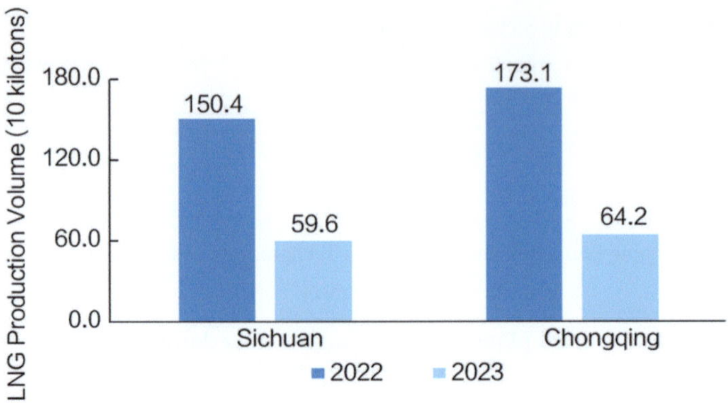

Fig. 5 LNG production comparison in the Sichuan-Chongqing region (2022–2023). *Data source* Longzhong Consulting

City consumed 14.5 billion cubic meters of natural gas, an increase of 10.7% year-on-year. The natural gas consumption of Sichuan Province and Chongqing City ranked 3rd and 10th respectively among the 31 provinces, autonomous regions, and municipalities in China (excluding Hong Kong, Macao, and Taiwan). See Fig. 6 for details.

In 2023, the total consumption of LNG in the Sichuan-Chongqing region was about 2.5 billion cubic meters, an increase of 57% year-on-year. Of this, Sichuan

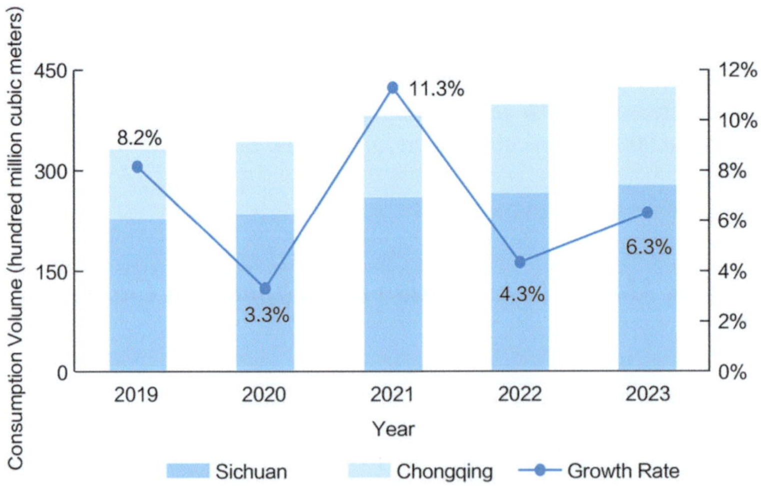

Fig. 6 Trend of natural gas consumption in the Sichuan-Chongqing region (2019–2023). *Data source* Chongqing Economic and Information Technology Committee and online materials

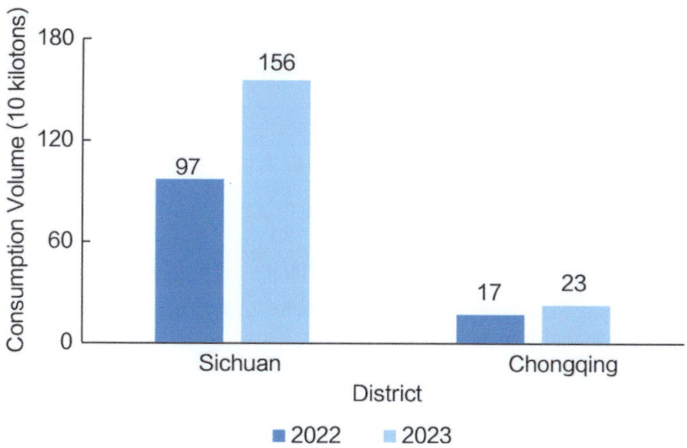

Fig. 7 LNG consumption comparison in the Sichuan-Chongqing region (2022–2023). *Data source* Longzhong Consulting

Province consumed about 2.2 billion cubic meters of LNG, an increase of 61% year-on-year, and Chongqing City consumed about 300 million cubic meters of LNG, an increase of 35% year-on-year. See Fig. 7 for details.

(II) Natural gas consumption structure in the Sichuan-Chongqing region

In 2023, under the comprehensive effects of macroeconomics, energy prices, new energy development, and temperature, the natural gas demand in major consumption areas such as residents, industrial fuel, and natural gas power generation showed different performances, with large differences in consumption increments.

The consumption of natural gas in the Sichuan-Chongqing region is concentrated in the residential and industrial sectors, with their combined consumption accounting for 88.4% of the total regional consumption. Among them, factors such as economic recovery and the concerted efforts of Sichuan and Chongqing to build advanced manufacturing clusters have driven an increase in industrial natural gas consumption by 1.8 billion cubic meters, accounting for 70% of the total increase in consumption. Secondly, as socio-economic activities gradually recover and the economy continues to improve, the revival of tourism and commerce, the rapid growth of commercial gas consumption, and the continuous expansion of the interconnected and coverage of the gas network have driven a steady increase in residential gas demand, with an increase in consumption of about 900 million cubic meters, accounting for 9% of the total increase in consumption. Due to the extreme high temperatures in the Sichuan-Chongqing region in 2022, the amount of natural gas used for peak-shaving power generation increased, and with a large base, the consumption of natural gas in the power generation sector decreased by 2% year-on-year in 2023. For details on the consumption of natural gas for power generation in the Sichuan-Chongqing region in 2023, see Table 6 and Fig. 8.

Table 6 Natural gas consumption in the Sichuan-Chongqing region in 2023

Category	Consumption (hundred million cubic meters)	Share (%)	Year-on-Year growth rate (%)
Residential gas	195	46.1	4.6
Industrial gas	179	42.3	11.1
Power generation gas	11	2.5	−2.0
Other gas	39	9.1	0.1
Total	423	100.0	6.3

Data source Chongqing Economic and Information Technology Commission and online materials. Other gas uses include fertilizer and CNG

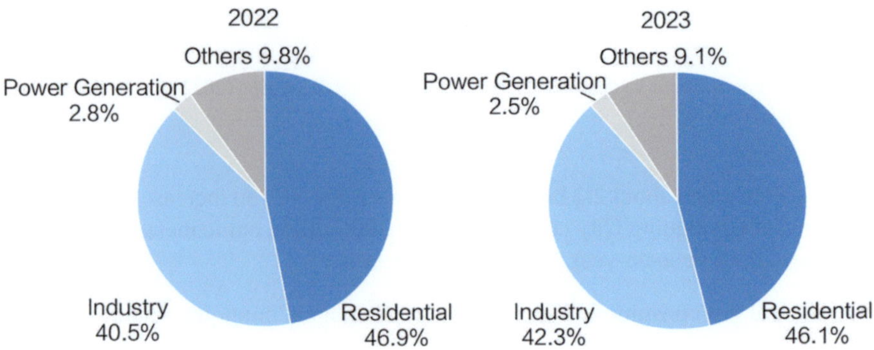

Fig. 8 Comparison of natural gas consumption structure in the Sichuan-Chongqing Region (2022–2023). *Data source* Chongqing Economic and Information Technology Commission and online materials

In 2023, the proportion of natural gas consumption in various fields in Sichuan Province remained relatively stable. The most significant change was in natural gas for fertilizer production. Due to weak downstream market demand, high gas source costs, and reduced operating load rates during the Universiade when some enterprises suspended operations, the proportion of natural gas consumption for fertilizer production decreased from 12.29% in 2022 to 11.78%. For more details, see Fig. 9.

The economy of Chongqing continues to recover and improve, industrial production is steadily recovering, and the amount of gas used in the industrial sector has grown significantly, with a net increase of 1.19 billion cubic meters, a year-on-year growth rate of 14.8%, and natural gas consumption accounting for 63.7%, an increase of 2.3% points from 2022. Residential gas consumption is steadily increasing, with a year-on-year growth rate of 6.9%. See Fig. 10 for details.

In addition, after the transition of epidemic prevention and control, urban socioeconomic activities have returned to normal, leading to a significant increase in the flow of people and goods. Trade and the service industry are thriving, however, due to the continuous promotion of electricity substitution in the terminal consumption link in the Sichuan-Chongqing region and the rapid development of new energy vehicles,

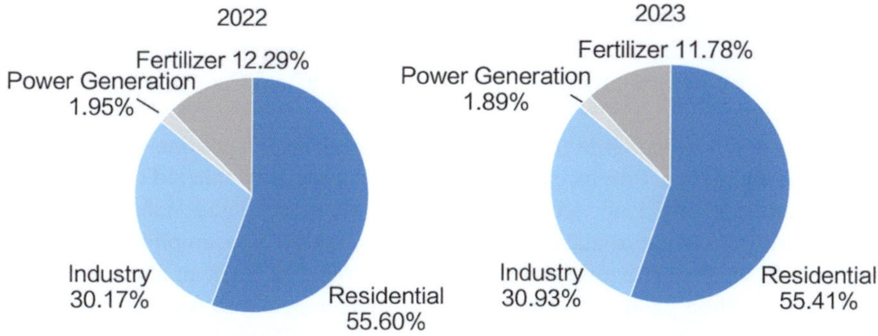

Fig. 9 Comparison of natural gas consumption structure in Sichuan Province (2022–2023). *Data source* online materials

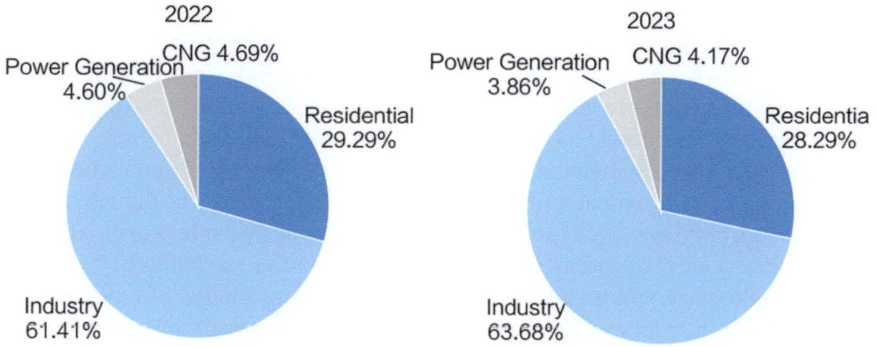

Fig. 10 Comparison of natural gas consumption structure in Chongqing (2022–2023). *Data source* Chongqing Economic and Information Technology Commission

the number of CNG vehicles has decreased, and gas demand continues to shrink. However, the development of the LNG heavy truck market is eye-catching, and the demand for gas has great growth potential.

Natural Gas Market Outlook in Sichuan and Chongqing in 2024

According to the "14th Five-Year" energy development plan of Sichuan and Chongqing, adhering to green development and focusing on resource security, the upstream is mainly to increase production supply. By 2025, the natural gas (including shale gas) production in Sichuan Province will reach 63 billion cubic meters, while the natural gas (including shale gas) production in Chongqing will reach 18.5 billion cubic meters. It is expected that by 2035, the Sichuan-Chongqing region will become

China's first 100 billion-level natural gas production base. At the same time, the development of LNG plants and wellhead gas liquefaction projects in the this region is rapid. The natural gas production targets for Sichuan and Chongqing in 2025 are shown in Table 7.

The economic situation in the Sichuan-Chongqing region continues to improve. In 2023, both Sichuan Province and Chongqing Municipality adhered to the general government's principle of "seeking progress while maintaining stability," implementing all-out efforts to stabilize investment, stimulate consumption, and stabilize market entities. The total annual production value of Sichuan and Chongqing exceeded 9 trillion yuan, of which Sichuan Province achieved 6 trillion yuan and Chongqing City achieved 3 trillion yuan. The annual GDP growth rates were 6% and 6.1% respectively, higher than the national average of 5.2%. In 2024, the government work reports of both places clearly proposed an expected economic growth target of 6%, higher than the expected national economic growth rate, and sought new efforts in areas such as deepening the construction of the Chengdu-Chongqing dual-city economic circle, boosting consumption and expanding investment, accelerating the construction of a modern industrial system, and undertaking industrial transfer in stages. The steadily improving macro economy will escort the steady and progressive development of the natural gas industry in the Sichuan-Chongqing region.

At the same time, Sichuan and Chongqing are accelerating the construction of a new energy system, which will further boost the development of the natural gas industry. Both provinces held energy work conferences at the beginning of 2024 and deployed key content of energy work. The main contents are as follows.

Sichuan Province will accelerate the planning and construction of a new energy and power systems, promoting the complementary development of multiple energies such as hydro, wind, light, hydrogen, and natural gas. The key work includes striving to build a natural gas (shale gas) production base with a capacity of 100 billion cubic meters, and continuously promoting the increase in natural gas reserves and production. Increase the exploration and development of natural gas and constantly optimize the natural gas pipeline network. Strive to optimize energy production and consumption structure, and accelerate the promotion of green and low-carbon energy transformation. Promote the green, safe, and sustainable development of coal, tap the potential of electricity substitution, and accelerate the construction of charging infrastructure. Strive to strengthen the construction of energy reserve facilities and

Table 7 Natural gas production plan for the Sichuan-Chongqing region in 2025

Region	Planned production in 2025 (hundred million cubic meters)	Increase in production compared to 2020 (hundred million cubic meters)
Sichuan	630	192
Chongqing	185	79
Total	815	271

Data source "14th Five-Year Energy Development Plan" of Sichuan Province and "14th Five-Year Energy Development Plan" of Chongqing City (2021–2025)

people's livelihood projects and comprehensively improve the ability of energy security and universal service. Accelerate the construction of natural gas reserve capacity in the province, continuously improve the level of coal reserve dispatch and emergency guarantee, and deeply implement projects such as consolidating and improving of rural power grids.

Chongqing will deepen energy co-protection and mutual aid and accelerate the construction of the Chengdu-Chongqing "Hydrogen Corridor" and "Electricity Corridor." Build new material and new energy industry clusters, accelerate the promotion and application of hydrogen fuel cell commercial vehicles, and promote the integration and interaction of new energy vehicles and the power grid. Accelerate the construction of projects such as the second phase of the Two Rivers Combustion Engine, the Sichuan-Chongqing Natural Gas 100 Billion Cubic Meters Production Base, and the LNG Processing Base. Implement green and low-carbon transformation actions. Promote the carbon peak action plan in key areas such as industry, urban and rural areas, and transportation; accelerate the circularization and energy-saving and carbon-reducing transformation of industrial parks; build green and low-carbon supply chains; promote low-carbon digital energy pilot demonstrations; vigorously develop green manufacturing and assembly-type buildings, and promote intelligent construction Pilot city construction. Encourage the consumption of green and low-carbon products and advocate a green and low-carbon lifestyle.

With the continuous advancement of clean and low-carbon energy transformation in the Sichuan-Chongqing region and the joint action of the "seven-year action plan" for oil and gas reserve increase and production, major oil companies in the region aim to find large and medium-sized gas fields and fully promote natural gas exploration in large gas basins in Sichuan. The natural gas in the Sichuan Basin, especially the exploration of shale gas, is expected to continue to play a major role in the newly discovered reserves. The momentum of a substantial increase in natural gas production remains unabated, and the natural gas production in the basin may exceed 75 billion cubic meters, an increase of about 5% year-on-year. Under the environment of steady economic operation, improved resource supply capacity, and active promotion of terminal gas price reform by the state, the fundamentals of the main gas-using industries are good, and the demand for natural gas in the Sichuan-Chongqing region is further released.

Special Topic 3: Research and Practice on the Digital Transformation of the Natural Gas Industry

Background and Significance of Digital Transformation in the Natural Gas Industry

(I) Background of digital transformation in the natural gas industry.

General Secretary Xi Jinping pointed out, "We must stand at the overall strategic level of the great rejuvenation of the Chinese nation and the world's unprecedented changes in a century, coordinate the two major domestic and international situations, and develop and secure two major events, fully leverage the advantages of massive data and rich application scenarios, promote the deep integration of digital technology and the real economy, empower the transformation and upgrading of traditional industries, foster new industries, new formats and new models, and continuously strengthen and optimize our digital economy," "We must grasp the direction of digitalization, networking, and intelligence, promote the digitalization of industries such as manufacturing, services, and agriculture, use new Internet technologies to transform traditional industries in all aspects and the entire chain, improve total factor productivity, and play the amplification, superposition, and multiplication of digital technology on economic development effect."

As the world's largest energy consumer, China's energy transformation is imminent. In 2016, China issued the "Guiding Opinions on Promoting the Development of 'Internet+' Smart Energy" (Development and Reform Energy [2016] No. 392), aiming to promote the deep integration of energy and modern information technology, promote the transformation of energy production management and marketing models, and reshape the energy industry chain, supply chain, and value chain. Using digital technology and intelligent management models enables distributed energy production, transmission, trading, consumption, and other links to operate efficiently and achieve the healthy and orderly development of clean energy such as photovoltaics, wind power, and nuclear power. In January 2024, the "Implementation Guide for Digital Transformation of Petrochemical Industry" (hereinafter referred to as the "Guide") was officially released. The "Guide" is a specific measure to grasp the direction of digitalization, networking, and intelligence and implement the important instructions of the Party Central Committee and the State Council on new industrialization. Petrochemicals are an important pillar industry of the national economy. It is a typical process manufacturing, with many product types, long process flow, miscellaneous material properties, harsh working conditions, and different production scales in different fields. There are many problems, such as a large number of equipment types, large safety and environmental protection risks, and uneven digital foundation. It is urgent to adapt to local conditions, adhere to problem orientation, pay attention to practical results according to the transformation pain points and development needs of different sub-fields, based on the development direction of high-end, intelligent, green, and safe, clarify the implementation path of deepening integration with new generation information technologies such as 5G, industrial Internet, big data, artificial intelligence, virtual reality, and achieve transformation and upgrading.

In the new situation where new technologies such as artificial intelligence, big data, and mobile internet are accelerating penetration, resource and environmental constraints are continuously strengthening. The task of green and safe development is becoming more urgent, the digital transformation of the natural gas industry is an important means to consolidate the bottom line of energy security, and also an inevitable choice to improve total factor productivity and create new competitive advantages. Under the current wave of digital transformation, the natural gas industry is still relatively lagging compared to other process manufacturing industries. Limited by its large scale and complexity, the digital transformation process of most natural gas companies is relatively slow. Accelerating the digitalization of the natural gas industry, empowering industrial transformation and upgrading with digital technology, achieving deep integration of traditional industries and digital technology, promoting China's energy industry to move towards the middle and high end, accelerating the move towards comprehensive digitalization and intelligence of energy, and promoting the integration of the digital economy and the real economy are of great significance for promoting high-quality economic development in China.

(II) The necessity of digital transformation in the natural gas industry

With the rapid development of technology, digital transformation has become an inevitable trend in all industries, and the natural gas industry is no exception. Digital transformation is an important way to achieve high-quality development in the natural gas industry with the help of digital technology and it is an important support for low-carbon development.

1. Key measures for oil and gas system reform

For a long time, China's natural gas industry has adopted an integrated operation mode of production, transportation, and sales. The oil and gas resources produced by the domestic upstream exploration and development link and overseas LNG resources, the midstream trunk pipeline transportation link is mainly dominated by the three major oil and gas companies, and the three major oil and gas companies' gas transmission capacity and pipeline mileage are in a monopoly position; the downstream natural gas terminal retail link presents a diversified competition pattern. With the official operation of the national pipeline network, natural gas achieves fair and open operation of infrastructure such as pipelines and receiving stations; large urban gas companies gradually expand upstream, the original boundaries between B2B (economic organizations to economic organizations, i.e., wholesale) and B2C (economic organizations to end users, i.e., retail) are gradually blurred, more companies adopt an integrated wholesale and retail model, bringing new changes to the pattern. The reform of the natural gas system brings new problems and challenges to urban gas companies in resource organization and internal management. The three major oil and gas companies lose the monopoly advantage of infrastructure, and how to carry out industry production, supply, and sales under the new conditions becomes a new important research topic. At the same time, facing the state to strengthen the construction of natural gas reserve capacity, alleviating the construction cost of gas storage facilities has become a common problem in front of the main body of the natural gas market.

With the continuous growth of China's urban development, gas companies' service scale and service content are also constantly expanding. This process must inevitably apply many advanced digital technologies and equipment to assist. For example, traditional enterprise operations often need to use manpower statistics and other methods to investigate the market, which is not only inefficient but also not accurate in investigating the market environment. This may cause gas companies to not accurately understand user needs and cannot improve their self-service according to user needs. Instead, they still use the original sales and operation methods, which not only require a lot of manpower and material resources but also cannot serve according to user characteristics in the service process, resulting in poor user experience, low user stickiness, and ultimately leading to the company's market share is not high, which is not conducive to the long-term development of gas companies. If gas companies cannot carry out digital transformation in time. In that case, it will be difficult to match advanced digital technology with the company's production and operation in the production service process, which will hinder the further development of the company. Therefore, in the modern market environment, the digital transformation of gas companies is the only way to achieve sustainable development in the future.

2. Response measures to changes in the natural gas market

The changes in the natural gas market environment are mainly reflected in the supply and demand environment, business model, profit model, and price stability. Natural gas resources have been in short supply for a long time; there are many links in the terminal market, the barriers are high, and the market segmentation is not enough, resulting in no way to reflect the marketing value. With the strengthening of domestic and foreign natural gas supply capacity, the construction of the production, supply, storage, and sales system continues to advance, gradually showing a loose supply situation; the supply and demand environment and fair opening of infrastructure give gas companies a greater say.

In the face of market changes, digital means can effectively perceive market supply and demand changes and predict price trends. Through digital means to improve the management ability and market development level of segmented markets, reduce the operation and management costs of customers, and develop more new business models like integrated supply and use of gas. Through digital means to open up more value-added services and profit channels, promote better integration of natural gas and new energy. Transformation can help natural gas companies improve quality and efficiency, gain stronger management capabilities, open up higher profit space, and has become an important means to respond to market environment changes.

3. An effective path for achieving high-quality development in the natural gas industry

"The 14th Five-Year Plan for National Economic and Social Development and the Outline of the Long-Range Objectives Through the Year 2035 of the People's Republic of China" proposes to transform and upgrade traditional industries and promote the optimization of the layout and structural adjustment of the petrochemical industry. The "14th Five-Year Plan for the Integration of Informatization and Industrialization" issued by the Ministry of Industry and Information Technology clearly proposes that by 2025, informatization and industrialization will achieve integrated development at a wider range, deeper level, and higher level. The "Guiding Opinions on Promoting High-Quality Development of the Petrochemical Industry during the 14th Five-Year Plan" proposes accelerating the collaborative innovation and application of new technologies and new models, promoting demonstration and leadership, and strengthening the empowerment of the industrial Internet. As a traditional industrial industry, the oil and gas industry, facing the new situation and trend of accelerating energy revolution and energy transformation, must effectively use digital technologies represented by cloud computing, Internet of Things, 5G, big data, artificial intelligence, etc., to drive business model restructuring, management model transformation, business model innovation and core capability enhancement, and achieve industrial transformation and upgrading and value growth.

Natural gas is a high-quality, efficient, green, clean, low-carbon energy source. Promoting the high-quality development of urban natural gas business is related to people's livelihood and green development. The natural gas industry is an industry that includes multiple links, such as resource exploration, development, processing, production, storage, transportation, wholesale, and retail. The industry chain is long and mainly includes upstream production, midstream storage and transportation, and downstream consumption. Implementing digital transformation can help improve the management efficiency and operation quality of the entire chain in all aspects. Implementing digitalization in the upstream exploration and development production link can improve exploration and development efficiency, reduce operation costs, and promote resource reserves and production; implementing digitalization in the midstream storage and transportation link can create digital pipelines, improve resource allocation efficiency, optimize transportation methods and scientific scheduling, and ensure the safe and stable operation of the pipeline network. At the same time, it can improve the ability to track and judge market supply and demand changes, reasonably arrange the production plan of gas storage facilities, improve storage emergency response capabilities and economic benefits; implementing digitalization in the downstream consumption link can accurately predict market supply and demand conditions and gas usage fluctuations, improve resource utilization efficiency, and also optimize and improve the user database, deeply mine user needs, build a digital platform, carry out online and offline services, improve user experience, and expand value-added services.

4. The inevitable choice for green and low-carbon development

In order to better achieve carbon neutrality goals, it is necessary to actively transform and accelerate the integrated development of natural gas and new energy, especially to play the supporting role of digital technology in the development of the natural gas industry, such as in the fields of smart microgrids, integrated energy services, energy Internet, etc., actively explore new ideas for integrated development, and improve the overall efficiency of the natural gas industry. The country proposes a strategic plan to build a new type of power system, the significant feature of which is to increase the consumption proportion of renewable energy, accelerate the integrated development of natural gas and new energy, and the market space of natural gas in the field of power generation will continue to grow. Gas companies undergo digital transformation, can play a role in hydrogen fuel cells, integrated energy services, etc., enhance the strategic position of natural gas in the new power system, strengthen environmental protection, and meet the development needs of green and low carbon.

While strengthening the integrated development of natural gas and new energy, it is necessary to strengthen the development of digital technology to cope with new business areas and fullly play to the supporting role of digital technology. Due to the many terminal links in the past natural gas market and the market segmentation not being perfect, the supply and storage value of natural gas is not reflected. However, through the improvement of natural gas business capabilities and digital transformation, the market can be effectively sorted out, and stable development can be achieved in terms of supply and demand relationships, business models, etc., promoting the development of natural gas in terms of supply and demand, thereby promoting the effective use of energy, enhancing the dominant position of natural gas and other energy sources, and reducing the pollution caused by the use of traditional energy to the environment.

(III) The significance of digital transformation in the natural gas industry

Digital transformation refers to the process of transforming and upgrading traditional business and operation models by introducing digital technology and data analysis methods. Digital transformation is the core path to promote the high-quality development of the natural gas industry. By improving efficiency and productivity, improving quality and reliability, and promoting innovation and upgrading, digital transformation can bring huge opportunities and advantages to the natural gas industry, and drive it towards a more efficient, intelligent and sustainable development direction.

1. It is conducive to improving the efficiency and productivity of the natural gas industry

By introducing advanced information technology and automation systems, natural gas companies can achieve high-level production, operations, and management intelligence. For example, through real-time monitoring and control systems, precise monitoring and control can be carried out on the collection, transportation, and distribution of natural gas, improving production efficiency and safety. In addition, digital transformation can also help natural gas companies carry out refined management, optimize resource allocation, and reduce energy waste through data analysis and prediction models, thereby achieving higher levels of production benefits.

2. It is conducive to improving the quality and reliability of the natural gas industry

Natural gas is an important source of energy supply, and its quality and reliability are crucial to socioeconomic development. Through digital transformation, natural gas companies can establish a comprehensive data monitoring and analysis system, monitor the operation of gas sources, pipelines, and equipment in real-time, discover problems in a timely manner, and carry out early warnings and repairs. In addition, digital transformation can also enhance the traceability of the natural gas supply chain, ensure that the quality of natural gas meets relevant standards and requirements, and guarantee the safety and satisfaction of users.

3. Conducive to promoting innovation and upgrading in the natural gas industry

Through the application of digital technology, natural gas companies can carry out more refined data analysis and prediction, deeply understand market demand and user behavior, and provide strong support for product innovation and service upgrading. For example, the precise calculation of gas consumption can be achieved through intelligent metering systems, providing users with personalized energy management solutions, using big data analysis and artificial intelligence technology, predicting and optimizing supply chain demand, and reducing costs and energy consumption. Digital transformation can also promote the integration and innovation of the natural gas industry with other fields, such as combining with renewable energy, energy Internet, etc., to promote the green transformation and sustainable development of the energy industry.

Current Status and Trends of Digital Development in the Natural Gas Industry

(I) Defining digital transformation in the natural gas industry

1. Concept of digitalization

Digitalization is an advanced stage of information technology development and the main driving force of the digital economy. With the rapid development of a new generation of digital technology, all walks of life have created more and more value using digital technology, accelerating the digital transformation of various industries. Informatization can improve work efficiency, precipitate business data, and provide necessary data for digital input, laying a good foundation. At the same time, digitalization uses data integration, connection, and analysis during the informatization stage to gain business insight, optimize operations and decision-making, and also force the optimization and improvement of information systems.

In a narrow sense, digitalization refers to the use of information systems, various sensors, machine vision, and other information and communication technologies to transform complex and variable data, information, and knowledge in the physical world into a series of binary codes, introduce them into the computer, form recognizable, storable, and computable numbers and data, and then establish related data models with these numbers and data for unified processing, analysis, and application.

In a broad sense, digitalization is the use of a new generation of information and communication technologies, such as the Internet, big data, artificial intelligence, cloud computing, blockchain, artificial intelligence, etc., to carry out systematic and comprehensive changes to the strategy, architecture, operation, management, production, marketing, etc., of enterprises and governments and other subjects. It emphasizes the reshaping of the entire organization by digital technology. Digital technology capabilities are no longer just about solving cost reduction and efficiency enhancement problems but have become the core force for enabling model innovation and business breakthroughs.

2. Digital transformation

Digital transformation is a broader concept involving using digital technology at all levels of the organization to reform existing business models and processes to improve efficiency, reduce costs, and increase competitiveness. Digital transformation includes digitalization, but its scope is broader, including changes to organizational culture, work methods, and customer experience. The core of digital transformation is to leverage innovative technologies such as big data, cloud computing, artificial intelligence, the Internet of Things (IoT), and industrial automation to create new value and business opportunities. Specifically, IoT and industrial automation technologies enable the automation of production lines, automatic data collection, and remote monitoring. Artificial intelligence-based automatic recognition technology is used for quality monitoring. Predictive maintenance of equipment is achieved through big data analytics, which helps reduce equipment failures and downtime. Digital procurement and supply chain management improve logistics efficiency while lowering inventory costs. Many manufacturing companies have implemented digital transformation to enhance production efficiency and reduce production costs.

3. The connotation of digital transformation in the natural gas industry

The digital transformation of the natural gas industry refers to the use of advanced digital technologies and solutions by various market entities in the industry to improve the operation and management process of the natural gas industry, improve efficiency, reduce costs, enhance safety, and achieve sustainable development. This transformation includes the application of information and communication technologies such as the Internet of Things, big data analysis, artificial intelligence, cloud computing, blockchain, etc., to achieve real-time monitoring, analysis, and prediction of data, optimize resource utilization, improve production efficiency, reduce environmental impact, improve the production efficiency of enterprises, reduce costs, optimize resource allocation, further enhance the core competitiveness and market competitiveness of enterprises, and then achieve enterprise transformation and upgrading and high-quality development.

(II) Key digital technologies in the natural gas industry

The digital transformation of the natural gas industry relies on several new information technologies to improve efficiency, optimize operations, and promote sustainable development. The key technologies of digitalization in the natural gas industry mainly include the Internet of Things, big data, blockchain, digital twin, and artificial intelligence technology, etc.

1. Internet of Things

Internet of Things (IoT), also known as "the Internet of everything," refers to a new type of technology that uses information sensors, radio frequency identification technology, smart devices, cloud computing, and other devices and technologies to collect real-time information from any object or process that needs to be monitored, connected, and interacted with. It collects necessary information such as sound, light, heat, electricity, mechanics, chemistry, biology, and location. It accesses the network to achieve ubiquitous connections between things and people, thereby realizing intelligent perception, identification, and management of items and processes. IoT technology can achieve interconnection and communication between objects, realizing intelligent, automated, and visual information.

The application of IoT technology in the natural gas industry covers gas well fields, pipeline transportation, gas storage, and other links. These projects use sensors, monitoring equipment, and intelligent control systems to monitor equipment status, predict failures, and operate remotely, improving production efficiency and safety. Its typical application is the management of long-distance pipeline production and operation, which can effectively solve many problems in the management of long-distance pipeline production and operation, such as wide distribution range, large number of pipelines, and long spatial distance. It can effectively avoid mischecks and omissions caused by human factors or single detection equipment in traditional inspection modes, and timely and early detect problems such as aging, blockage, damage, and leakage that occur during pipeline operation. It has a wide range of application prospects in actual scenarios such as pipeline temperature detection, pipeline pressure detection and control, pipeline flow detection, pressurization station control, predictive analysis, and intelligent management.

2. Big data

Big data is a series of processing methods for storing, calculating, statistically analyzing, and processing massive amounts of data. The amount of data processed is usually at the TB level, or even PB or EB level, which is impossible for traditional data processing methods. With the elementization of data production, the continuous development of data science and data technology, and the deep mining and application of data value, a big data revolution is underway. It will drive innovation and change in various fields, such as national strategy and regional economic development, smart city construction, enterprise transformation and upgrading, social management, and personal work and life.

If big data is regarded as an industry, the key to profitability in this industry is to improve the "processing ability" of data and realize the "value-added" of data through "processing". The typical applications of big data in the natural gas industry are production and market decision-making scenarios, that is, using big data analysis technology to predict natural gas demand and market prices, optimize the supply chain and adjust production plans.

3. Blockchain

Blockchain is a type of distributed ledger. A distributed ledger is a shared, encrypted signature, irrevocable transaction record, and extended list arranged in chronological order by all participants in the network. Each record contains a timestamp and a reference link to the previous transaction. Blockchain technology is a distributed ledger technology based on cryptographic theory. Unlike traditional centralized data management methods, blockchain technology has advantages such as decentralization, transparency, immutability, and anonymity. The data management method of blockchain technology brings new opportunities for information management and data exchange in the natural gas industry.

Blockchain is increasingly penetrating various oil and natural gas businesses. The typical scenarios in the natural gas industry are applications in supply chain management and transaction settlement. Blockchain technology can be used to build a natural gas trading platform, realize transparent and traceable transaction data, and apply smart contract technology to achieve automated transaction settlement and contract management.

4. Digital twin

A digital twin is a digital representation technology of real-world entities or systems. In addition to receiving feedback from real-world entities, it can also download data, software, and model updates to real-world entities or nearby systems. Digital twins focus on creating appropriate digital representations of entity operations in the real world, which can be as simple as a dashboard with key performance indicators. Digital twins can also be very complex, existing in physical things, personnel, processes, organizations, and environments, combining inputs from many sensors and external data to support advanced simulations. A digital twin is a structured representation of an entity, including data structures, metadata, physical properties, and key variables. More complex composite digital twins can be assembled from simpler atomic digital twins to form models.

Digital twin technology can achieve comprehensive simulation and monitoring of pipeline systems by establishing mathematical models of the real world, thereby improving the efficiency and safety of pipeline systems. It can also be combined with other digital technologies to form digital operation solutions for the natural gas industry, further improving management levels and service quality.

5. Artificial intelligence

Artificial intelligence is a technical science that researches and develops theories, methods, technologies, and application systems for simulating, extending, and expanding human intelligence. It tries to understand the essence of human intelligence and produce a new kind of intelligent machine that can respond in a way similar to human intelligence. It mainly includes research in areas such as robotics, language recognition, image recognition, natural language processing, and expert systems. Since its birth, the theory and technology of artificial intelligence have become increasingly mature, and the application field has been continuously expanded. Artificial intelligence has received more and more attention in the field of computers and has been applied in robots, economic decision-making, control systems, and simulation systems.

The natural gas industry in China is gradually applying artificial intelligence technology to improve production efficiency and operational management levels. Artificial intelligence supports decision-making through insights from predictive, normative, and cognitive analysis. In this way, it helps the oil and natural gas industry discover and implement new exploration and production concepts to improve the return on investment. For example, Sinopec uses artificial intelligence technology to intelligently monitor gas storage facilities, achieving equipment fault early warning and optimized maintenance plans. In addition, PetroChina has also adopted artificial intelligence technology in production optimization and equipment fault diagnosis to improve the accuracy of equipment fault diagnosis.

(III) Developments in the digitalization of the natural gas industry

With the increasing demand of the natural gas industry to improve efficiency, reduce costs, and enhance safety, based on the rapid development of technology, competitive pressure in the energy industry, requirements for environmental sustainability, and government policy support, the natural gas industry has begun to adopt digital solutions on a large scale to improve business operations, operational efficiency, and decision-making. Digital technology is widely used in natural gas exploration and development, storage and transportation, and the consumer market.

1. Developments in the digitalization of natural gas exploration and development

Since the 1980s, after nearly 40 years of construction, the unified and self-built application systems of various business points in the exploration and development of the natural gas industry have become quite rich. They form an information support system from the operation area, oil and gas factory, oil and gas field company, to the head office, which is fully applied in the exploration and production business, achieving certain application effects.

(1) Oil and gas field exploration includes geophysical exploration, drilling, logging, geology, well logging, and comprehensive supporting operations. The application of digital technology has promoted the transformation of exploration and development production operation methods, reduced on-site personnel, and

improved operation efficiency and safety levels. The digital services and applications of exploration and development research projects have improved the timeliness, accuracy, completeness, and consistency of production data, as well as the operation efficiency of oil and gas wells.

(2) Drilling, well logging, and logging data on the drilling site play an important supporting role in the entire operation system of exploration engineering. Real-time intelligent identification, positioning, tracking, monitoring, and management of the well site, including real-time data collection, analysis, sending, transmission, and display of on-site data, real-time data transmission to the unified application system of production operation management, have improved the level of oil and gas production management and decision-making, and have become the daily work platform for production management and scientific research personnel at different levels; it has improved the efficiency and level of comprehensive research and exploration business management.

(3) In order to improve the ability to prevent operational risks, the digital transformation of real-time monitoring and safety management and early warning-related technologies of well site production dynamics can help drilling experts use more comprehensive integrated data for analysis, improve the level of oil and gas production management decision-making and safety management, and promote the transformation of drilling operation management methods. Remote technology connects the applications of oil and gas fields and headquarters organically, coordinates related work, and provides data and technical support for decision-making by the leadership, operations by the management, and execution by the professional level.

(4) Oil and gas field surface engineering has characteristics such as complex and diverse construction environments, numerous professional categories, complex processes, a variety of materials and equipment, and high safety risks. Relying on digital technology to achieve a safe and efficient production mode, gradually realizing digital factories, digital collection, digital inspections, digital dispatching, digital monitoring, and other core functions, provides strong support for the digital application function design of the entire process of surface pre-construction, surface construction, surface production, and integrity management.

(5) In the work of oil and gas reserve management, facing many problems such as time-consuming and laborious reserve management methods, large amounts of data and information sorting work, and complicated archiving, the application of digital technology can effectively carry out work such as proven reserve increase, recalculation, accounting, settlement, new use, and recoverable reserve calibration and manage multiple elements such as development reserves and oil and gas wells, production, mining rights, administrative divisions, and oil and gas mines, enhancing the decision-making ability to increase reserves, increase production, and increase efficiency, and achieving management of development reserve results, comprehensive business applications, and full life cycle management of workflows.

2. Developments in the digitalization of natural gas storage and transportation

With the continuous expansion of the scale of natural gas storage and transportation facilities, the potential safety incidents, accident losses, and asset integrity risks are also increasing. The digital and intelligent operation and maintenance of natural gas storage and transportation facilities have gradually become a focus of attention in the industry. The digital construction of natural gas storage and transportation plays an important role in station and warehouse, long-distance pipeline transportation, dispatching, control, and safety management, it can effectively ensure the normal operation of the pipeline and provide timely production operation dynamic monitoring and management data for production commanders.

(1) The digital information system of storage and transportation management can monitor and manage the real-time dynamics of storage and transportation production, realize the management and dispatching of the quantity and quality of station and warehouse through digital control of the station and warehouse, provide automatic detection of oil depot production status, external transmission pipelines, real-time query of static and dynamic data of storage and transportation production, and automation management function of metering, providing timely, accurate, and highly integrated production operation information for production commanders.
(2) A large number of instruments are used in the storage and transportation process of oil and gas. Digital technology is used to carry out automatic control and management of these instruments, thereby enabling remote monitoring and dynamic management of the oil and gas storage and transportation process and improving the safety and operational efficiency of oil and gas storage.
(3) Long-distance oil and gas pipelines have many points, long lines, and wide areas, passing through regions with complex natural environments and social conditions, facing a large amount of uncertainty and risk. The digital construction operation and maintenance of oil and gas storage and transportation facilities can effectively prevent various risks and strengthen emergency response mechanisms. For example, by using the historical accident characteristics of long-distance oil and gas transportation to establish a dataset, research on accident emergency models can be conducted to improve the effectiveness of accident emergency responses and perfect the emergency plan system for oil and gas storage and transportation facilities.
(4) Based on the interconnected natural gas pipeline system, the use of the Internet of Things and advanced digital technologies strongly supports the centralized control and management model of natural gas, achieving natural gas dispatching and peak shaving, alleviating supply and demand contradictions, and improving the operational level and efficiency of natural gas storage and transportation.

3. Developments in the digitalization of the natural gas consumption market

As the marketization reform of the natural gas industry continues to advance, competition in the natural gas market is intensifying. Natural gas sales companies need faster market response speeds and more refined customer marketing services. Facing different industries and types of users, their gas usage characteristics and patterns vary greatly. The application of digital technology provides more possibilities for the expansion of the natural gas consumption market, and research on the digital transformation of the natural gas market has become an important topic.

(1) The urban gas industry is an important part of the downstream of the natural gas industry chain. The gas industry involves many links, such as production (pressure regulation), distribution, storage, metering, utilization, and marketing after the gas pipeline network is established. Its transportation network is large, and the user range is wide, resulting in low-quality management, monitoring, and customer service. With the introduction of the Internet of Things and digital and information technology applications, the smart gas transmission and distribution network uses IoT technology to transform the original gas pipeline network, helping gas companies improve their business capabilities.
(2) The gas wireless automatic meter reading system and sales system, with the help of wireless remote transmission meters and their controllers and other Internet of Things technologies, realize bank collection and online gas purchase, mobile phone gas purchase and other functions, providing customers with a good user experience. At the same time, by mining the value of data and applying it to natural gas sales and procurement, remote metering and charging, etc., the flexibility and scientific nature of natural gas distribution supply is realized.
(3) Through major monitoring points such as gate stations and pressure regulation metering stations, combined with key area video surveillance and gas leak monitoring, remote wireless monitoring of the pipeline network is realized. This effectively improves the efficiency of gas pipeline network monitoring and management, improves the operation efficiency of the gas pipeline network, and achieves scientific management.
(4) Trends in digital transformation in the natural gas industry.

Overall, the digitalization level of the oil and gas industry is low, and the returns brought by digitalization are not yet obvious. To unleash the huge value hidden in digitalization, it is still necessary to accelerate the research and application of digital technology. From the current theoretical research and typical application practices, the digital transformation of the natural gas industry mainly revolves around the deep application of artificial intelligence in the field of exploration and development, the deep empowerment of integrated intelligent technology in production and operation, and the deep development of natural gas market marketing data resources.

1. Deep application of artificial intelligence in natural gas exploration and development

With the continuous growth of data scale in the natural gas industry, artificial intelligence technology based on large-scale analysis will play a more important role. The application of artificial intelligence in the field of exploration and development will use big data analysis technology to process and analyze massive data from various links to extract valuable insights and knowledge; use real-time connections with many different data sources to effectively reduce manual processing of large amounts of address and seismic data inaccuracies, successfully carry out construction, fault, sequence interpretation, and can also be used for pre-stack and post-stack seismic data analysis, well logging data and other multi-dimensional data processing and analysis, to obtain three-dimensional data bodies that can directly predict oil and gas, greatly reducing manual workload and improving interpretation accuracy; industrial internet, internet of things, wearable devices, drones, robots, end-to-end edge computing and other technologies related to artificial intelligence are widely used in geophysical exploration, well logging, drilling and completion, oil extraction and oil and gas field surface engineering, etc., for all-round real-time perception and remote control, transforming production and operation methods, reducing operation costs and risks.

Through big data intelligent modeling of oil and gas operation scenarios and the application of artificial intelligence technology in various links such as drilling, sweet spot identification of oil and gas reservoirs, description and simulation of oil and gas reservoirs, production optimization, diagnosis of downhole and surface faults in oil wells, allocation of surface resources and risk warning, etc., with advanced analysis solutions, meet prediction needs, optimize production plans, improve safety and environmental sustainability, etc.

2. Deep empowerment of integrated intelligent technology for production and operations

Using technologies such as big data, digital twins, and artificial intelligence, an intelligent and efficient production planning execution system is established, a visual dispatch command platform is built, and the optimal plan scheduling and integrated collaborative management of global benefits are realized. Natural gas companies are mostly heavy asset companies, owning many assets such as urban pipelines and gas stations. With the development of digital technology and healthy market competition, the platform economy connecting all major stakeholders in the natural gas industry may become a reality. By building an integrated production and operation platform, the collaborative formulation of production plans, dynamic analysis and adjustment of business plans, and integration of production, supply, and sales are realized.

In the context of integrated production and operation, big data intelligent control technology is constructed, such as introducing robots to replace humans in patrol operations and tasks in the well field, realizing the perception of the real production environment in the virtual production environment, realizing production status display, historical trend tracking, fault diagnosis, intelligent early warning and alarm, real-time working condition simulation, automatic operation of equipment self-stop and self-start, by establishing accurate digital models, including natural gas equipment, pipeline systems and production processes, combined with sensor data, real-time monitoring and large-scale analysis, reducing downtime, gradually realizing the dynamic operation status prediction of the entire industry chain equipment of oil and gas fields, providing more accurate and reliable references for operation decision-making and risk management.

3. Deep development of natural gas marketing data resources

Oil and gas trade, market sales and other businesses are based on digital platforms, integrating internal and external information data, strengthening data integration and sharing, and achieving data consistency across departments and enterprises. By establishing a data integration platform and standardized data model, different systems and departments can share and access related data. Through data sharing with supply chain partners and services, cooperation and innovation within the industry will be enhanced, jointly promoting the natural gas industry to achieve more intelligent and sustainable development, improving production efficiency, reducing operational risks, and providing customers with better services and experiences.

Driven by user demand data to improve user services, using information technology means, multi-faceted, multi-dimensional improvement of user service experience, improvement of user service quality and effectiveness; guided by business needs, comprehensive integration of various data resources, in promoting digital transformation, continuously expanding production data application scenarios, transforming traditional data analysis to full data analysis, further releasing data value, continuously strengthening data empowerment, strengthening online and offline multi-channel sales innovation, expanding business models; relying on data and digital technology-driven and digital platforms, building a digital industrial ecosystem of data sharing, technical collaboration, and element collaboration, and then achieving value co-creation.

Design of Digital Integration Scheme for the Natural Gas Industry

As the proportion of natural gas in China's primary energy is getting higher and higher, natural gas companies are developing rapidly, the scale of assets is getting larger and larger, the requirements and importance of asset management are getting higher and higher, the difficulty of asset management is also getting larger and larger, and the demand for asset management in the natural gas industry to transform from informatization to digitalization is becoming more and more urgent. The digital integration solution for the natural gas industry insists on data-driven as the core, builds data sharing capabilities, deep analysis capabilities, data visualization application scenarios, and realizes effective processing and multi-dimensional presentation of business management, business data, and business indicators on the platform.

(I) Overall approach

The digital integration solution for the natural gas industry integrates digital technology into the entire industry chain of natural gas, promotes the transformation of

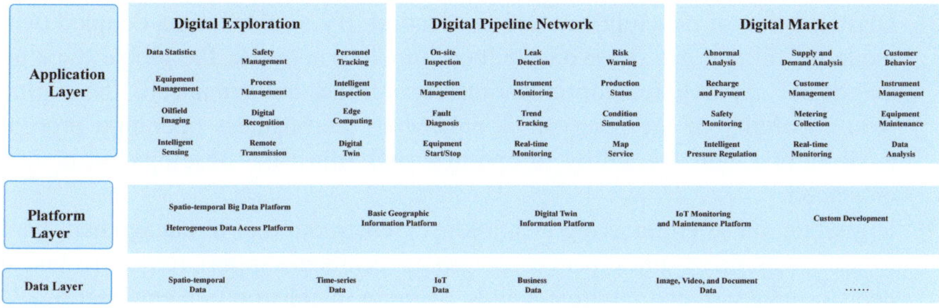

Fig. 11 Schematic diagram of digital integration scheme design for the natural gas industry

enterprise exploration and development digitalization, pipeline network digitalization, and market marketing digitalization, creates new capabilities such as personalized services, intelligent production, and networked collaboration, and creates a new industry based on users, data, and innovation-driven. The digital integration solution for the natural gas industry is a pioneering set of solutions that can effectively help enterprises achieve data exploration and digital transformation. By connecting internal and external IoT data, business data, and social data through big data, cloud computing, artificial intelligence algorithms, etc., the value chain of enterprise data is realized, helping enterprises to realize data decision-making or data-driven business models in different business scenarios, thereby promoting the transformation of intelligent operation. The schematic design of the digital integration solution for the natural gas industry is shown in Fig. 11.

(II) Business composition

1. Digitalization of exploration and development

Natural gas exploration and development is a complex process that requires the use of a large amount of data and tools. The application of digital technology is to fully realize the platformization of business and process management, realize business model innovation and management model transformation, can effectively improve exploration and development efficiency, reduce costs, improve the production environment, and ensure the sustainability and safety of exploration and development. Digitalization of exploration and development is the digitalization of all businesses and process management of natural gas exploration and development, from the operation area, and gas collection plant, gas field company to the overall digital system of the company.

(1) Digitalization of exploration. Establish seismic, wellbore, logging, experiment and other data cloud management and application, enhance data visualization function, through knowledge graph, automatic push to quickly meet application needs, improve the standardization, intelligent level of unconventional, new energy, new area exploration process.

(2) Digitalization of development and production. By establishing a coupled twin model of oil and gas reservoir-wellbore-pipeline network, the global tracking diagnosis and dynamic optimization capabilities are enhanced, the digital twin capability of the reservoir is enhanced, the business operation process is simplified, and the ability to optimize and adjust the development plan is enhanced.
(3) Digitalization of engineering technology. Construction of a management platform for remote on-site supervision, remote technical support, remote technical decision-making, and intelligent early warning in construction engineering.
(4) Digitalization of production operations. Strengthen the informatization support for production control processes and inspection task supervision, build effective single-well monitoring applications, and improve the intelligent perception and diagnostic capabilities of intelligent production allocation, injection, and the entire production cycle.
(5) Digitalization of collaborative research. Through the digitalization of business such as geophysical exploration, drilling, logging, geology, well logging, and comprehensive supporting operations, we can improve the collaboration between the front and rear, the integration of geology and engineering, and the collaborative research capabilities of multiple specialties, realize the underlying flow of data and processes, and increase the sharing of results and real-time feedback mechanisms.

2. Digitalization of the pipeline networks

(1) Digital twin of the pipeline. Efficient integration of static and dynamic data to achieve digitalization, integration, and visualization of the pipeline entity. By collecting and processing structured data from completion, construction, procurement, and operation and maintenance, and digitalizing various professional data from the station, the handover time at the station can be effectively reduced.
(2) Digitalization of pipeline operation and maintenance. Through usage frequency and usage plans, pre-judgments can be made on whether to repair or replace parts to assess the health of the equipment, transforming traditional "problem-occurring" passive maintenance into predictive maintenance. Advanced data analysis tools can identify, detect, and solve maintenance problems that occur, effectively reducing the probability of supply interruption and extending the life of the equipment.
(3) Digitalization of refining and storage transportation. Through functions such as production optimization, energy consumption, and intelligent equipment monitoring, ensure production quality, achieve visual and traceable management of the entire process, reduce manpower and physical consumption, reduce downtime due to failures, reduce maintenance and other costs, achieve optimization and low energy consumption of various production processes, improve the ability to detect hidden dangers and handle failures, and make production plans and supply chains more accurate.

3. Digitalization of the consumer market

(1) Digitalization of procurement and transportation. Using internet information technology to achieve fine management data, collaborative advancement of business, intelligent dispatching of vehicles, and panoramic insight into the process of natural gas supply chain services, providing customers with one-stop natural gas transportation and trade solutions.
(2) Digitalization of terminal market management. Based on natural gas sales volume and user number data, establish a consumption model and carry out real-time transportation dispatch and digital services. Establish systems such as natural gas revenue management, customer service calls, maintenance management, safety inspection management, and customer installation management. Provide users with refined gas usage analysis, leak active alarms, and real-time safety monitoring, and other multi-level, high-quality, personalized gas services, effectively enhancing users' sense of gain, happiness, and security.
(3) Digitalization of integrated energy services. Deep cultivation of users is conducive to accelerating the development of integrated energy services and value-added businesses. While leveraging its existing resources and experience, it uses the advantages of new technologies such as big data to promote the deep integration of new technologies and traditional businesses, integrating big data thinking and capabilities into integrated services, cultivating the habit of managing, deciding, and innovating with data, realizing the empowerment of new technologies to traditional businesses, and continuously expanding value-added services. For example, developing businesses related to natural gas, such as gas appliance sales, gas insurance sales, etc., based on providing better services to users.

Practice and Achievements of Digital Transformation in the Natural Gas Industry

The digital transformation of the natural gas industry is in full swing. For example, the National Oil and Gas Pipeline Network Group Corporation Limited has built and put into operation a data application platform for natural gas pipeline operations. This platform effectively enhances the capability for in-depth data mining and comprehensive analysis of natural gas pipeline operations, providing strong support for the intelligent control and regulation of operations. The Energy Index Database of the China Energy Index Research Center at Southwest Petroleum University, which is based on the construction of a natural gas industry prosperity index, is gradually evolving into a comprehensive information platform for energy market observation and energy data analysis.

(I) The Natural Gas Pipeline Operation Data Application Platform

The digitalization and utilization of natural gas pipeline assets are the key to intelligent pipelines, as well as the difficulty and focus of the digital transformation of the national pipeline network. The full and effective use of pipeline data is of great significance to the oil and gas regulation center as a control hub. The completion and operation of the Natural Gas Pipeline Operation Data Application Platform (hereinafter referred to as the "platform") mean that the data utilization of the natural gas pipeline network has made greater progress, effectively promoting the deep advancement of the digital transformation of the oil and gas regulation center. As the mileage of oil and gas pipelines in the national pipeline network becomes longer and longer, and there are more and more shippers and service objects, the oil and gas regulation center is implementing full remote control and advancing towards intelligent regulation. The more than 90,000 kilometers of pipelines in service, more than 1,000 stations, and more than 3,300 valve rooms have all been included in the centralized monitoring and surveillance scope of the center.

1. Institutional overview

China National Petroleum and Natural Gas Pipeline Group Co., Ltd. (hereinafter referred to as the "Group Company") was established on December 9, 2019, and is a key state-owned backbone enterprise supervised by the State-owned Assets Supervision and Administration Commission of the State Council. The National Pipeline Group is mainly engaged in the investment, construction, and operation of oil and gas trunk pipelines and gas storage and peak-shaving infrastructure, responsible for the interconnection and connection of trunk pipelines and social pipelines, as well as the operation and dispatch of the national oil and gas pipeline network, and regularly discloses the remaining pipeline transportation and storage capacity to the public, realizing the fair opening of infrastructure to users.

The main function of the Oil and Gas Regulation Center of the National Petroleum and Natural Gas Pipeline Network Group Co., Ltd. (hereinafter referred to as the "Oil and Gas Regulation Center") is to centrally dispatch and command, remotely monitor and operate, coordinate maintenance operations, optimize pipeline network operations, and train professional technical personnel for the oil and gas storage facilities owned by the National Pipeline Network Group. The Oil and Gas Regulation Center formulates pipeline operation plans according to the production and sales plans of the group company. It uniformly allocates oil and gas pipeline resources through seasonal coordination, monthly planning, weekly balancing, and daily designation, optimizing operations. At present, the Oil and Gas Regulation Center has become one of the control hubs for long-distance oil and gas pipelines with the most pipelines, the most comprehensive pipeline transport media, and the most complex operating environment in the world.

2. Background

As the control hub of the oil and gas pipeline network, the Oil and Gas Regulation Center controls more than 600 natural gas stations, connects nearly 50 gas source upload points, and collects 700,000 data points in real-time every day through SCADA and pipeline production management systems. With the rapid development of business, the scale of the pipeline network continues to expand, the system is becoming more complex, and the regulation management and operation face huge challenges. Against the background of the construction of a "smart pipeline network, intelligent pipeline and intelligent regulation", from the perspective of the safety, efficient operation and management of the natural gas pipeline network, a natural gas pipeline network operation data application platform is established to quickly extract, analyze and display data, mine data value to meet the needs of different levels for production operation data, strengthen the support for regulation business, and the value is huge.

As the business develops rapidly, the scale of the pipeline network continues to expand, the system is becoming more complex, and the regulation management and operation face huge challenges. Against the background of the construction of a "smart pipeline network, intelligent pipeline and intelligent regulation", from the perspective of the safety, efficient operation and management of the natural gas pipeline network, a natural gas pipeline network operation data application platform is established to quickly extract, analyze and display data, mine data value to meet the needs of different levels for production operation data, strengthen the support for regulation business, and the value is huge.

3. Platform architecture and function

The platform helps regulation personnel to grasp the production situation more directly and quickly, enhances the deep mining and comprehensive analysis ability of natural gas pipeline network operation data, promotes the acquisition and utilization of production operation data at different levels of the center, and provides strong support for the intelligent operation of regulation. The overall architecture of the platform is shown in Fig. 12.

(1) Data collection. Collect production operation data of various businesses such as pipeline network operation, resource intake, user distribution, unit status and

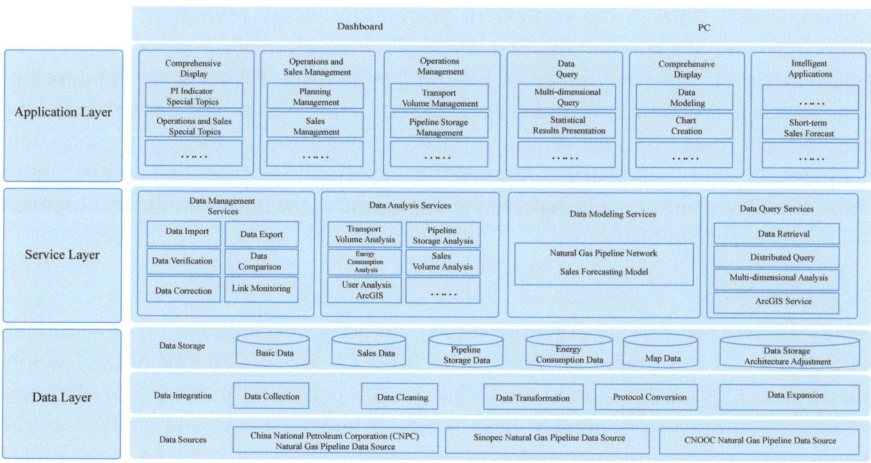

Fig. 12 Natural gas pipeline network operation data application platform architecture diagram

operation management. The main sources of data include Pipeline Production Management System (PPS), Intermediate Database (PI), and Online Simulation System, etc.

(2) Data integration and data governance. Integrate and integrate data from different system sources (Pipeline Production Management System, Intermediate Database, etc.) and store them in a database established according to business logic. Carry out data governance in accordance with the overall requirements of the group company's data governance work policies and standard norms, laying the foundation for later data utilization.
(3) Data analysis and application. Based on data integration and governance, associate data from various natural gas application systems (such as PPS, online simulation system, etc.) to form an overall data model, use industrial data analysis and machine learning technology to carry out related predictions, enhance the deep mining and comprehensive analysis ability of natural gas pipeline network operation data, and provide strong support for the intelligent operation of regulation.
(4) Data visualization. Improve the level of data visualization interaction according to the business needs of the regulation operation. Use charts (line charts, bar charts, pie charts, maps, scatter plots and other graphics) to display diversified data, fully reflect the trend and difference of data, and use data platform PC page and large screen system display.

(II) Energy index database of the China Energy Index Research Center, Southwest Petroleum University

1. Basic information

The China Energy Index Research Center of Southwest Petroleum University, relying on the discipline advantage of the natural gas industry of Southwest Petroleum University, was born in the context of promoting energy revolution and supply-side reform and has become the first domestic research institution focusing on the measurement of prosperity index of natural gas industry, focusing on the development research of natural gas industry. In recent years, the center has extended the measurement research of the industry prosperity index to gas, hydrogen energy and other specific energy industries. Against the background of profound adjustment of China's energy structure, it is expected to contribute to the high-quality development of China's energy industry.

2. Background

Energy is the core driving force of socio-economic development. With the rapid development of the global energy market, factors such as energy prices, supply and demand relationships, and policy changes are becoming increasingly complex. Businesses and investors need to have a deeper understanding and analysis of the energy market. The rapid expansion of emerging energy markets and the complex and changing energy investment environment pose higher requirements for corporate management strategies and development strategy decisions. The construction of the Energy Index Database of the China Energy Index Research Center of Southwest Petroleum University (hereinafter referred to as the "Index Database") stems from the demand for energy market development and the importance of energy data analysis, aiming to serve national development strategies, government economic decisions, domestic and foreign academic research, and corporate data needs. The Index Database can provide comprehensive, accurate, and timely energy data, helping businesses and investors better grasp market dynamics, serving the energy market, policy formulation, and the development of the energy industry.

3. Main functions

In its early stages, the Index Database mainly served the expert survey, data collection, data sorting, index calculation, and report writing work for the prosperity index indicators of the natural gas industry, providing support services for the research of the prosperity index of the Chinese natural gas industry. With the continuous development of digital and intelligent technologies, the ability to obtain, process, and analyze energy data is also constantly improving. The Index Database is constantly expanding its application scenarios. It has been widely used in index reports, scientific research topics, etc., providing an information platform for leaders, investors, and scholars to meet the needs of energy market development and energy data analysis. The schematic diagram of the Index Database platform architecture is shown in Fig. 13. The schematic diagram of the Energy Index Database monitoring screen is shown in Fig. 14 (the data in the figure is test data).

(1) Centralized data management. The platform provides a system for centralized management of energy data, which can store a large amount of data and classify, tag, and organize it as needed. Researchers can easily search and access. The data provides classified retrieval, currently mainly including industry information, economic information, industrial policies, exploration and development, oil and gas pipelines, oil product sales, and other columns.
(2) Data sharing and collaboration. The platform can easily share data and research results, accelerating the research process. It can also save various parameters

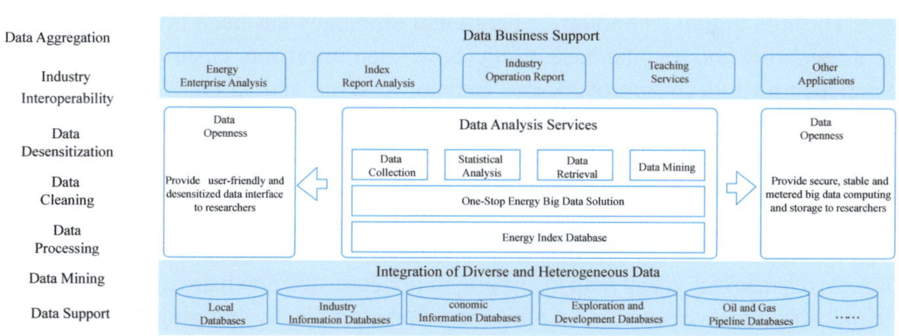

Fig. 13 Schematic diagram of energy index database platform

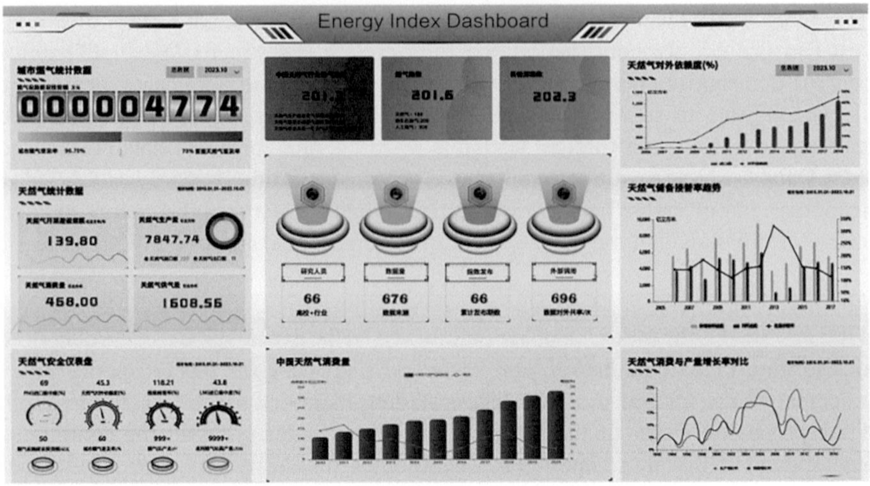

Fig. 14 Schematic diagram of energy index database dashboard

and data in the data query process of researchers, adjust retrieval conditions and parameters in time, and improve data utilization.

(3) Data analysis and mining. The platform provides basic data analysis and mining functions, which can perform statistics, trend analysis, chart display, etc., based on selected data. This helps researchers discover patterns, extract useful information from large amounts of data, and assist scientific research.
(4) Monitoring and statistics. The platform provides key business indicator extraction, supports the effective transmission of information and decision-making support in the actual application of the monitoring screen, and meets the generation and export of charts for various research topics.

(III) Main achievements of digital transformation in the natural gas industry

The effect of digital transformation practices in the natural gas industry is significant. Through digital technology, operational efficiency has been improved, equipment reliability and maintenance effects have been improved, customer service and experience have been improved, and decision-making and strategic planning have been improved. Digital transformation will bring more opportunities and competitive advantages, and companies can achieve more efficient, intelligent, and sustainable development, creating greater value and returns.

1. Effectively improved corporate operational efficiency

Digital transformation can help upstream, midstream, and downstream companies in the industry improve operational efficiency. For example, by applying real-time data monitoring and analysis technology, gas companies can monitor and optimize the production process, equipment operation and maintenance in real time. Real-time data-driven decision-making and operations can reduce delays in operations, reduce downtime, and improve resource utilization efficiency.

2. Effectively improved equipment reliability and asset maintenance results

Digital transformation allows gas companies to use advanced equipment monitoring and predictive maintenance technologies. Through sensors and data collection equipment, companies can monitor the operating conditions and performance parameters of equipment in real time. Combined with large-scale analysis and machine learning technology, fault prediction and maintenance optimization can be carried out, potential equipment faults and abnormalities can be discovered in advance, and corresponding maintenance measures can be taken. This helps to improve the reliability of equipment, extend the life of equipment, and reduce maintenance costs.

3. Effectively improved customer service and user experience

Digital transformation can help companies provide better customer service and user experience. For example, at the consumer market end, through digital channels and platforms, customers can conveniently query and manage their gas accounts and obtain real-time energy usage information and bill details. Companies can use data analysis technology to provide personalized services and customized energy solutions based on customer needs and consumption habits. It can also realize intelligent metering and remote monitoring, reduce the workload of manual meter reading and fault troubleshooting, effectively improve customer satisfaction and loyalty, and increase customer retention rate.

4. Effectively improved management decision-making and strategic planning

Digital transformation provides businesses with a more accurate and comprehensive data foundation to support decision-making and strategic planning. Through big data analysis and predictive modeling, market trend analysis, demand forecasting, and supply chain optimization can be carried out, thereby formulating more accurate production plans and resource allocation. It can also help businesses conduct risk assessment and management, predicting and responding to potential operational risks.

Suggestions for Digital Transformation in the Natural Gas Industry

The development and progress of digitalization in the natural gas industry cannot be separated from policy support and the joint efforts of various related industries and enterprises. The government plays an important role in the digital transformation of the natural gas industry. It can promote the digital transformation of the natural gas industry and promote industry innovation and sustainable development by striving to improve policy measures for digital transformation, actively building cooperation platforms and ecosystems around digital transformation, and increasing guidance for talent cultivation in the digital transformation of the natural gas industry. In the digital transformation of the natural gas industry, the enterprise is key. Only by fully exerting the subjective initiative of the enterprise, scientifically planning the digital strategic transformation, optimizing the process of management systems, and innovating talent management, can the enterprise choose a digital transformation path suitable for itself, help the enterprise improve production and operation efficiency, and bring richer application scenarios and higher value to the enterprise. Enterprises should keep pace with the times, actively embrace digital transformation, and lay a solid foundation for their long-term development.

(I) Strive to improve policy measures for digital transformation

The government should formulate clear policies and regulations to encourage the digital transformation of the natural gas industry. This includes but is not limited to providing financial support, tax reductions, and incentives to attract enterprises to invest and adopt digital technologies. Ensure the stability and continuity of policies and regulations to enhance the confidence of enterprises in digital transformation. The government can formulate policies and plans to support sustainable energy development and the application of digital technologies, encouraging the natural gas industry to develop in a cleaner, more efficient, and intelligent direction.

(II) Actively build cooperation platforms and ecosystems around digital transformation

The government establishes cooperation platforms and ecosystems within and outside the industry to promote cooperation and exchange in digital transformation. Promote cooperation and collaborative innovation among enterprises, technology providers, research institutions, and government departments, promote cross-industry and cross-domain cooperation, and encourage cooperation between the natural gas industry and enterprises and institutions in the energy, technology, finance, and other fields to jointly explore new digital technology applications and business models.

(III) Increase guidance for talent cultivation in the digital transformation of the natural gas industry

Strengthen education and training, and cultivate and reserve talents needed for the digital transformation of the natural gas industry. The government can cooperate with educational institutions and enterprises to carry out related training courses and seminars, providing financial and resource support. This includes industry digital talents, data scientists, artificial intelligence experts, Internet of Things engineers, and other professional talents to meet the talent needs of the natural gas industry for digital transformation.

(IV) Enterprises should make good digital strategic transformation planning

Under the new pattern of natural gas industry development, enterprises must also carry out digital strategic transformation planning according to their core competitiveness and promote digital transformation through top-down thinking. At present, large Chinese oil and gas enterprises have proposed to create innovative driving models, new business models based on user data, and proposed to achieve full industry digitalization, networking, and intelligent development construction, etc. The leading enterprises in all walks of life have recognized the changes in the market economic development environment and actively carried out digital transformation planning and put it into practice.

(V) Enterprises should optimize the process of management systems

The traditional internal approval decision-making process in the operation management process of enterprises is cumbersome, which seriously slows down the operation efficiency and increases the burden on people. Through digital transformation, the procedures in the management process can be effectively reduced. Then the overall organizational structure of the enterprise can be effectively innovated, unnecessary management positions can be cut, a sound organizational structure suitable for digital transformation can be established, repetitive and repetitive work can be avoided, intermediate management links can be reduced, work and production operation efficiency can be improved, and process optimization can be achieved.

(VI) Enterprises should innovate digital talent management models

The natural gas industry is a complex and large system with many sub-industries and a complex industrial chain. To do a good job in the digital transformation of the industry and enterprises, a team of talents who understand the business and have digital capabilities is needed. Enterprises should formulate a series of digital talent training plans according to the needs of digital transformation, including recruitment, training, etc., and actively build an enterprise-level talent management platform. In the process of human resource management, reasonable arrangements should be made for the positions of relevant personnel according to different construction needs and talent development directions. At the same time, enterprises should pay attention to talent cultivation, strengthen the ability of relevant personnel to use the platform, cooperate with universities, establish internship and training bases, enhance the digital capabilities of enterprise talents, and further improve the operation efficiency of enterprises and the utilization rate of talents.

Special Topic 4: Full Life Cycle Planning of Natural Gas Development Projects[2]

Overview of Full Life Cycle Management of Natural Gas Development Projects

(I) Natural gas development projects

1. Definition

Natural gas development projects are a typical form of oil and gas engineering, mainly based on exploration results and necessary productive development tests. On the basis of comprehensive research, they plan and formulate reasonable production plans and development schemes for natural gas resources with industrial value. They also construct and put into production oil and gas fields, enabling them to produce for a long term with predetermined production capacity and economic effects until the end of development. This is a comprehensive management process.

2. Characteristics of natural gas development projects

(1) High technology and knowledge intensity.
(2) High investment and high risk.
(3) Possessing life cycle characteristics.

(II) Connotation of full life cycle management of natural gas development projects

1. Definition

The essence of full life cycle management of natural gas development projects is to manage the natural gas development project as the object, with the fundamental goal of optimizing the overall economic benefits of the full life cycle of the natural gas development project. Based on modern project management theory and full life cycle management thinking, it is a system engineering to achieve sustainable development of natural gas development projects through certain organizational forms, full process, full functional area, and full-level integrated management. This definition needs to be understood from the following aspects:

(1) Emphasis on integrated thinking

Full life cycle management of natural gas development projects is a new management model. It integrates the relatively independent stages in the life cycle of natural gas development projects, the various participants in the project organization, the various functions of management content, and the individual goals derived from the overall project goals.

---

[2] This section is excerpted from the research projects "Exploration and Practice of the Whole Life Cycle Management Model for Oil and Gas Development Projects" and "Research on the Complete Project Management Model for Oil and Gas Fields," which were presided over and completed by Liu Xianfeng and Shen Xilin.

(2) Comprehensive, full-process, full-field management control

Full life cycle of natural gas development projects emphasizes full-process control management, that is, full-staff, all-round, dynamic feedback, full-process, comprehensive management benefits, and in the full-process control management, it specifically achieves refined stage goals, clarifies work focus, and standardizes management.

(3) Integrated project management system of responsibility, power, and rights, and a perfect assessment and incentive system

Set up a natural gas development project management leadership team and operation area management organization, respectively, and be fully responsible for the complete project management of natural gas development. Establish a long-term and phased goal assessment system in project management to ensure the achievement of project construction and production operation goals.

(4) Emphasis on minimizing investment costs

The complete project management of natural gas development projects should consider the benefits of investment costs, which can be specifically expressed in three aspects. One is the integrated consideration of investment costs, the essence of investment is cost; the second is the overall consideration of the operating life cycle of natural gas development projects, that is, the investment costs of the planning, construction, operation, and end stages of natural gas development projects are considered as a whole; the third is the coordination and optimization of investment and costs of natural gas development projects.

2. Guiding thought

The guiding thought of complete project management of natural gas development projects is to adhere to the scientific development concept, follow the basic principles of complete project management "complete process, clear goals, and scientific methods", establish an efficient management structure, form an integrated complete project management model of responsibility, power, and rights, achieve the complete unity of project decision-making, operation, and benefits, promote the effective development of natural gas development projects, and promote the transformation of the development mode of oil companies through pilot promotion and leading demonstration.

3. Basic principles

Some basic principles need to be adhered to in the process of implementing full life cycle management of natural gas development projects.

(1) Goal-oriented

The fundamental goal of natural gas development projects is to maximize economic benefits. The goals of each management subject, each management stage, and each management function of natural gas development projects need to be determined and decomposed around the fundamental goal to ensure that the decomposed goal system is consistent with the fundamental goal.

(2) Node control

The characteristics of each stage of natural gas development projects are different, and the methods, means, and strategies of their management are also different. It is necessary to formulate different management strategies according to the characteristics of each stage, correctly execute and tightly control the management of each stage, and ensure the effectiveness of complete project management of natural gas development projects.

(3) Dynamic adjustment

In the life cycle of natural gas development projects, the internal and external environment will constantly change. It is necessary to adjust and change the countermeasures of complete project management of natural gas development projects in a timely and reasonable manner according to the characteristics of each stage of natural gas development projects and the real-time dynamics of internal and external environmental changes to ensure the maximization of economic benefits of natural gas development projects.

(4) Comprehensiveness

The full life cycle management of natural gas development projects should run through all stages of project operation, cover all functions of project management, and integrate all project stakeholders.

(5) Materiality

The goal of complete project management for natural gas development projects involves multiple levels and multiple management functions. It is necessary to determine the core management goals and formulate a management goal system around the core goals.

(6) Consistency

Although the investment scale, life cycle, and output of natural gas development projects vary, the methods, means, evaluation indicators, and calculation calibers used in investment decision-making, planning, and evaluation management activities should be consistent.

(7) Objectivity

The goals, management content, and management information of complete project management for natural gas development projects should be objective, taking into account the practical feasibility and long-term development needs of various businesses.

4. Strategic goals

According to the aforementioned definition of complete project management for natural gas development projects, the fundamental goal of the project is to pursue the optimal benefits of the entire life cycle of the natural gas development project under the premise of safety and efficiency. Specifically, it can be divided into five dimensions: economy, management, technology, safety and environmental protection, and social responsibility. Table 8 lists the important indicators of these five dimensions.

In Table 16, the fundamental goal of natural gas development projects is listed, as well as important first-level indicators reflecting the comprehensive performance of natural gas development projects by dimension. For natural gas development projects to achieve the fundamental goal of natural gas development projects, this topic believes that the most critical goal indicators are output, operating costs, investment, and economic life cycle. Therefore, how to apply the full life cycle management idea to realize the scientific decision-making and management of output, operating costs, and investment is the core content of complete project management for natural gas development projects.

Content.

(III) Full life cycle management stages of natural gas development projects

The full life cycle of a natural gas development project is divided into four phases, namely the project planning phase, the project construction phase, the project operation phase, and the project decommissioning phase.

The project planning phase begins with project conceptualization and initiation, and ends with the approval of the feasibility study report or the formulation of the project development plan.

The project construction phase starts with land acquisition and demolition, and construction preparation, and ends when the project reaches steady production.

The project operation phase begins when steady production is achieved and continues until the annual production declines to the point of economic shutdown. It mainly includes the steady production period and the decline period, which can last for several years.

The project decommissioning phase occurs when the annual production reaches the shutdown level, where the production volume no longer generates economic

Table 8 List of important indicators for life cycle management of oil and gas development projects

Dimension	Important indicators
Economy	Output, operating costs, return on investment, asset preservation and appreciation
Management	Performance, quality, schedule, risk, innovation
Technology	Reserves, capacity, production efficiency, technological innovation
HSE	Health, safety, environmental protection
Social responsibility	Local fiscal revenue, employee income, employment

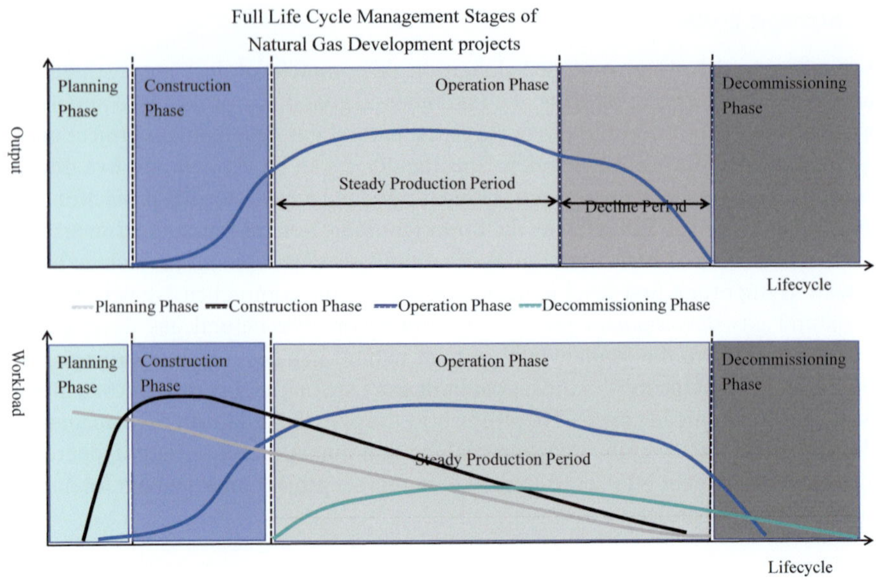

Fig. 15 Schematic diagram of the life cycle of a natural gas development project

benefits. This phase involves the closure of the gas field, cessation of production operations, asset transfer, scrapping, and dissolution of organizational structures.

As shown in Fig. 15, it illustrates the changes in the work of each phase of the natural gas development project throughout the entire life cycle.

(IV) Full life cycle management model of natural gas development projects

1. Three-dimensional conceptual model

Drawing on the three-dimensional structure system model of American systems engineering scholar Hall, a three-dimensional conceptual model of full life cycle management of natural gas development projects is proposed, with the full life cycle of the project, all participants of the project, and project management content as dimensions, as shown in Fig. 16.

2. Logical structure

The logical structure diagram of the full life cycle management of natural gas development projects is shown in Fig. 17.

As shown in Fig. 17, the full life cycle management of natural gas development projects is a complex systems engineering. It is supported by project management theory and full life cycle thinking, based on internal and external environmental conditions, guided by goal planning, and grasped by organizational management and evaluation and assessment systems, to achieve the maximization of overall project benefits.

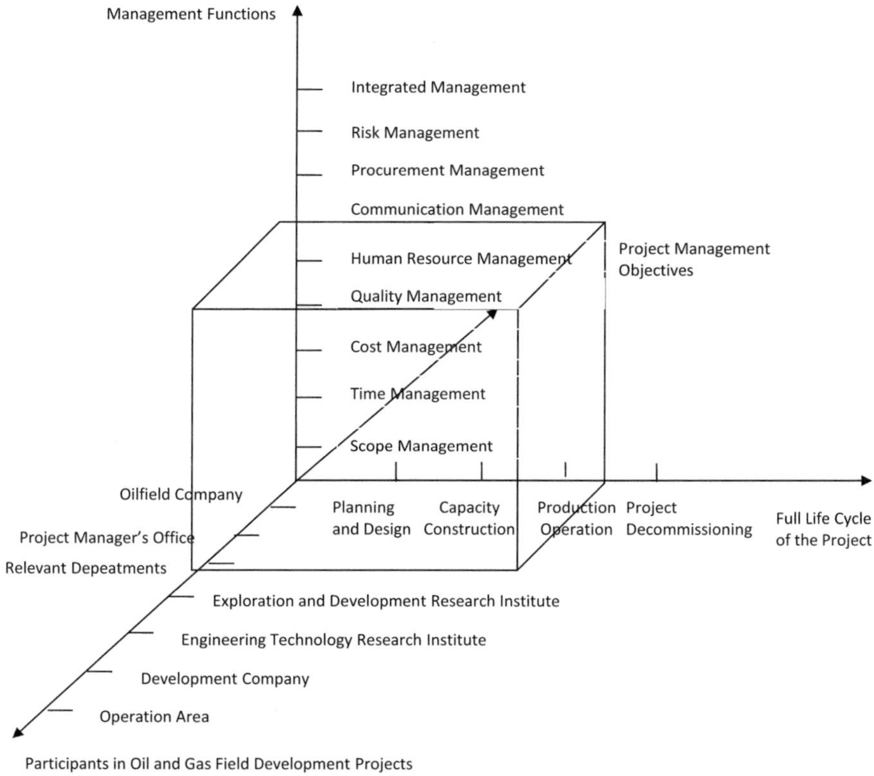

Fig. 16 Three-dimensional conceptual model of the full life cycle management mode for oil and gas development projects

Planning Framework for the Full Life Cycle Management of Natural Gas Development Projects

(I) Definition

Project planning is the process of predicting the future, determining the goals to be achieved, estimating the problems that will be encountered, and proposing effective plans, policies, measures, and means to achieve the goals and solve the problems. Project planning is thinking, imagining, and planning from reality, and then determining, deciding, and arranging the various activities and work results necessary to achieve the project goals.

The characteristics of the full life cycle management planning of natural gas development projects are manifested as:

(1) The planning content involves the entire life cycle of the project;
(2) The planning content covers all functions of project management;

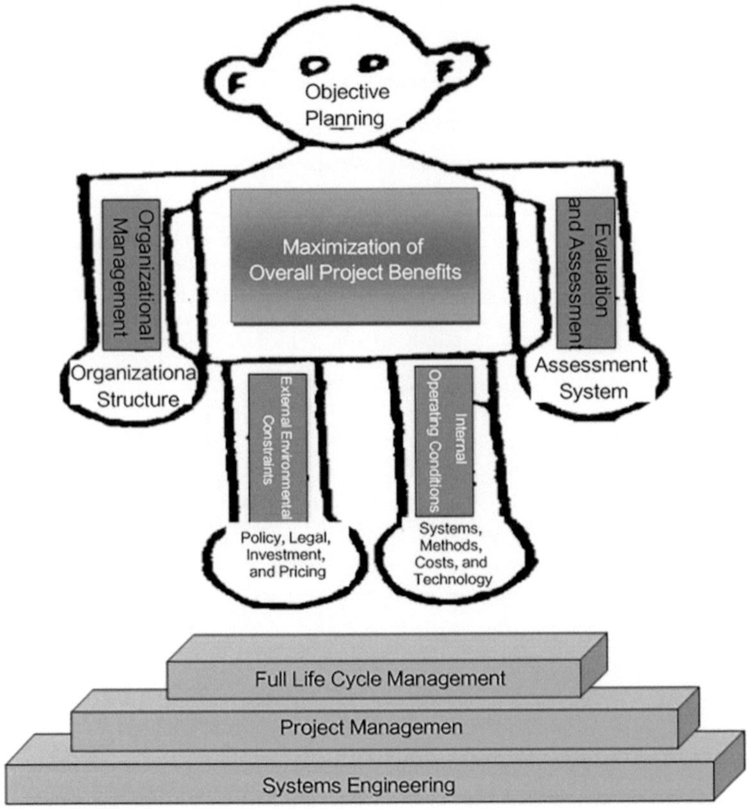

Fig. 17 Logical structure diagram of the full life cycle management of natural gas development projects

(3) The planning work requires the full participation of all personnel related to the project;
(4) Planning focuses on managing and grasping the project as a whole, and efficiently planning, organizing, and controlling each link of the project;
(5) Planning itself is also a stage in the full life cycle of the project.

(II) Principles of planning

In the process of project planning formulation, the following six principles should generally be followed:

(1) Purposefulness

Every project has one or several definite goals, to achieve specific functions, roles, and tasks, and the formulation of any project plan is centered around the realization of project goals. When formulating a plan, the first thing to do is to analyze the goals and clarify the tasks. Therefore, project planning is purposeful.

(2) Systematic

Project planning itself is a system, composed of a series of sub-plans. Each sub-plan does not exist in isolation, they are relatively independent of each other, yet closely related, thus making the project plan also have the basic characteristics of systematic purpose, relevance, hierarchy, adaptability, and integrity, making the project plan form an organically coordinated whole.

(3) Coordination

The formulation and implementation of project planning are not based on the organizational structure within a certain organization or department, nor on their own interests and requirements, but on the overall functions of the project and project management, involving all departments and institutions of project management.

(4) Economy

The goal of project planning not only requires the project to have high efficiency but also high benefits. Therefore, multiple schemes must be proposed for optimization analysis in the planning.

(5) Relevance

Project planning is a systematic whole. Any changes in the sub-plans that make up the project planning will affect the formulation and execution of other sub-plans, and ultimately affect the normal implementation of project planning. The formulation of project planning should fully consider the relevance between sub-plans.

(6) Dynamism

This is determined by the life cycle of the project. The life cycle of a project is short for several months and long for several years. During this period, the project environment is often changing, causing the implementation of the plan to deviate from the project baseline plan. Therefore, project planning needs to be constantly adjusted and modified with the changes in the environment and conditions to ensure the completion of project goals. This requires project planning to be dynamic to adapt to the changing environment.

(III) Planning Content

The planning content mainly includes:

(1) Basic information collection;
(2) Logical framework construction;
(3) Work structure decomposition;
(4) Scope planning;
(5) Schedule planning;
(6) Cost planning;

(7) Quality planning;
(8) Charter writing;
(9) Risk analysis;
(10) Planning scheme evaluation.

(IV) Planning process

The full life cycle management planning process includes: project planning basic information collection, analysis, processing; establishing the logical framework matrix of project planning; work structure decomposition of project planning; CSSQ planning of the project; planning scheme evaluation. The project planning process is shown in Fig. 18.

(V) Natural gas development project full life cycle management planning model

1. Planning Model Goals

The strategic goals of the full life cycle management planning of natural gas development projects involve multiple dimensions such as economy, management, technology, HSE, social responsibility, etc. The goals corresponding to each dimension have different impacts on the success or failure of the project. The key goals are mainly cycle, investment, output, operating cost, and benefit. Specific goals include:

(1) According to the life cycle change law of resources (reserves), combined with the current mining technology, determine the production capacity construction scheme, and then determine the production capacity scale during the stable production period within the life cycle;
(2) According to the investment law of natural gas development project capacity construction, considering the feasibility of project financing, determine the annual investment within the life cycle;
(3) According to the change law of construction and operation stage costs, determine the annual unit cost within the life cycle;
(4) Combining the first three items, draft several feasible cycle, output, investment, and cost combination planning schemes, calculate the income indicators of each scheme, and then determine the optimal scheme according to the goal of maximizing income to get the optimal cycle, output, investment, and cost combination.

2. Variable assumption

Determine the full life cycle variables of natural gas development projects as follows:

(1) Cycle (T)—the full life cycle of the project, t is the annual identifier;
(2) Profit Π—the profit of the project;
(3) Price (P_t)—the natural gas price in the t year;
(4) Output (Q_t)—the output in the t year;
(5) Cost (C_t)—the cost in the t cost of the year;
(6) Investment (I_t)—Investment of the t th year;

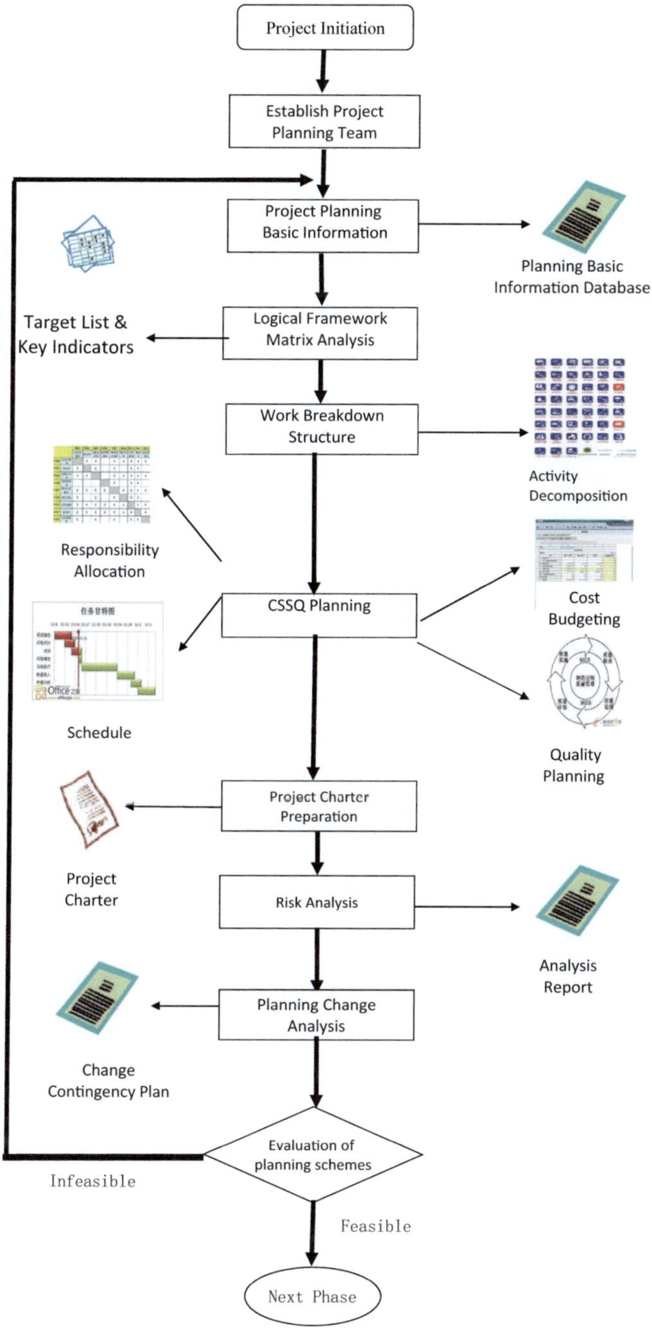

Fig. 18 Logical structure diagram of the full life cycle management of natural gas development projects

(7) Capacity (X_t)—Capacity of the t th year;
(8) Discount rate r;
(9) Net Present Value (NPV_t)—Net cash flow of the project in the t th year.

3. Decision variables

According to the objectives of the full life cycle management plan of the natural gas development project, the decision variables include: annual natural gas production, investment, cost, and economic life during the life cycle.

4. Objective function

(1) Quantitative objective.

The decision-making objective of the natural gas development project is to maximize the economic benefits of the project's full life cycle. Considering the time value, the net present value of the project's full life cycle is used as the quantitative objective.

(2) Expression of the objective function.

$$Max\Pi = \sum_{t=1}^{T} NPV_t = \sum_{t=1}^{T} \frac{P_t Q_t - I_t - C_t}{(1+r)^{t-1}} \tag{1}$$

5. Constraints

(1) Production constraint.

$$Q_t \leq Q_{t0} \tag{2}$$

Indicates that the planned annual production is greater than the target production.

(2) Investment constraint.

$$I_t \leq I_{t0} \tag{3}$$

Indicates that the planned annual investment is less than the funds available for investment.

(3) Cost constraint.

$$C_t \leq C_{t0} \tag{4}$$

Indicates that the annual production cost used for operation is less than the target cost.

(4) Non-negativity constraint.

$$Q_t > 0; I_t > 0; C_t > 0 \quad (5)$$

Indicates that the planned annual production, investment, and cost cannot be negative.

Full Life Cycle Management CSSQ Planning of Natural Gas Development Project

The core element of the full life cycle planning for natural gas development projects is CSSQ planning, which, in logical sequence, includes scope planning, schedule planning, cost planning, and quality planning. The following section will focus on introducing the main content and methods of CSSQ planning.

(I) Scope planning

1. Scope management process

The project scope management process is shown in Fig. 19.

2. Definition of project scope planning

Project scope planning involves all the work and activities to achieve a goal. It mainly focuses on the identification interface of work or activities, such as what activities or work need to be done and what do not need to be done. The main purpose of project scope management is to ensure that project stakeholders reach a consensus on the final deliverables of the project. The series of work carried out to complete the deliverables through five steps: scope planning, scope definition, creation of Work Breakdown Structure (WBS), scope verification, and scope control (scope change).

Project scope planning refers to the further formation of various documents to provide a basis for future project decisions. These documents include standards to measure whether a project or project phase has been successfully completed. As the output of the scope planning process, the project team needs to develop a scope statement and a scope management plan. The scope statement should at least explain the project demonstration, project product, project deliverables, and project objectives. The scope management plan describes how to manage the project scope and how the project scope changes to be consistent with the project requirements.

3. Content of scope planning

The content of project scope management planning includes: defining the deliverables of the project; preparing a detailed project scope statement based on the preliminary project scope statement (i.e., tender documents, contracts); creating a Work Breakdown Structure based on the scope statement; establishing related processes and methods to control changes.

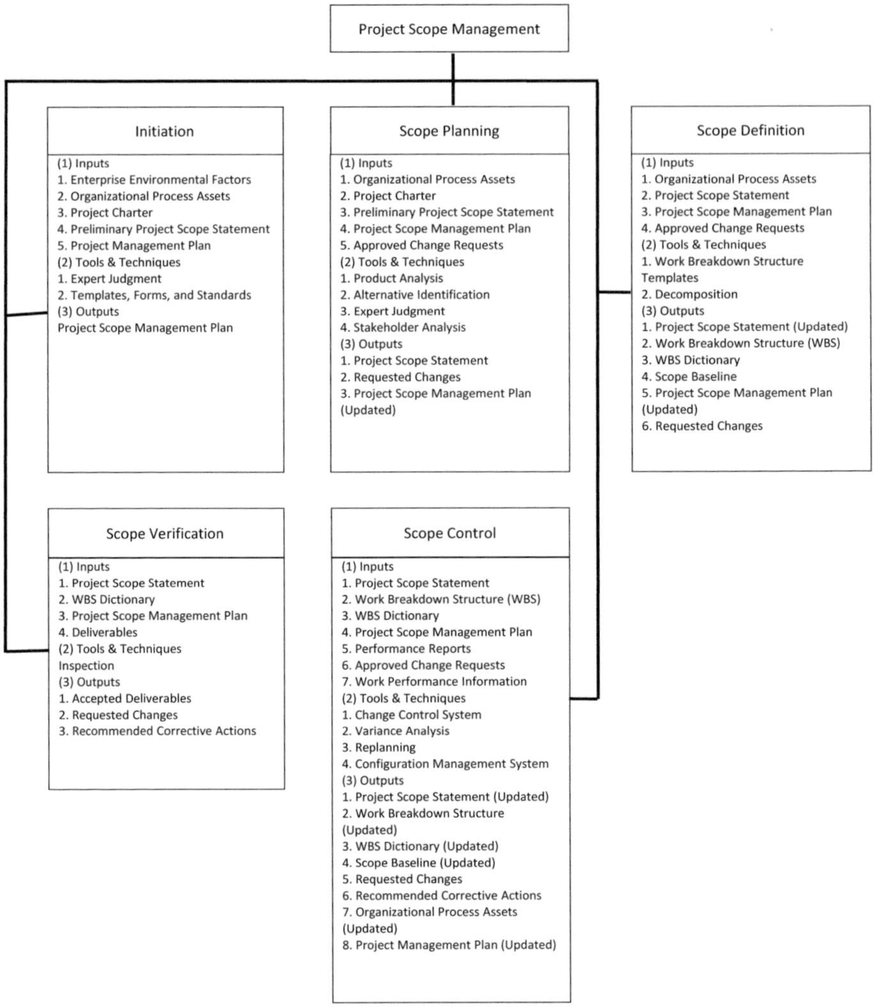

Fig. 19 Logical structure diagram of the full life cycle management of natural gas development projects

Developing a project scope management plan is a gradual process that needs to be updated in a timely manner in conjunction with the project's own characteristics and progress. From the start of a project, planning for the project's scope, time, and cost is necessary, but generally, only rough estimates and planning are carried out. Since the project scope directly affects the three elements of project time, cost, and quality, it is necessary to quantify the project scope at the start of the project, with verifiable, traceable, and measurable planning. Scope planning is a plan for the management of project scope and the management of project scope changes. The plan stipulates how to determine, verify, and control the project scope, as well as how

Table 9 Definition of responsibility classification and code

Type of responsibility	Code
Primary responsibility	P (Primary)
Secondary responsibility	S (Secondary)
Involvement	I (Involvement)
Monitoring and approval	M (Monitoring)

to develop and define the work breakdown structure (WBS). The content involved in the project scope management plan may not necessarily be detailed or quantified; it could also be general or informal. Therefore, at this stage, the project team needs to fully understand the objective conditions and relevant background of the project, and formulate a relatively detailed plan based on the project charter.

The allocation of responsibilities for each activity is determined based on the work breakdown structure of the project. The written description of the project scope is the result document of project planning. The content of the scope description should include:

(1) The division of activities at each stage of the oilfield enterprise that the project aims to solve;
(2) The objectives of the project, how to achieve them, who will achieve them, what the ultimate goal of the project is, the list of project deliverables, delivery time, and also the deliverables not included in the project scope;
(3) The project manager must clearly define what is within the project scope and what is not because the less clear the boundary between the two, the more difficult it will be to control the project later, and the details of what is and is not within the scope are key inputs to complete the project schedule planning;
(4) Key success factors (generally cost, schedule, and quality metrics) determine whether a project is successful.

The project responsibility classification is shown in Table 9.

This way, the responsibility breakdown of the natural gas development project life cycle stage based on project activities can be obtained, as shown in Table 10.

4. Task breakdown for each stage

(1) Planning phase

The task breakdown for the planning stage is shown in Table 11.

(2) Construction phase

The task breakdown for the construction stage is shown in Table 12.

(3) Operation phase

The task breakdown for the operation stage is shown in Table 13.

Table 10 Full life cycle responsibility matrix for oil and gas development projects

Serial Number	Work breakdown project activities	Responsible units			
		Project team	Project manager	Outsourcing unit	Head of outsourcing unit
Project planning stage					
1	Analysis of project environmental conditions Oil and gas reservoir engineering design and planning Drilling engineering design and planning Surface engineering design and planning Oil and gas extraction engineering Design and Planning CSSQ planning HSE planning Risk planning Project evaluation analysis Organizational structure planning	M	P	I	
2	Feasibility study	P	S	I	I
Project construction stage					
1	Implementation of drilling and completion engineering	M	M	I	P
2	Well completion project management	M	M	P	S
3	Ground construction project implementation	M	M	I	P

(continued)

Table 10 (continued)

Serial Number	Work breakdown project activities	Responsible units			
		Project team	Project manager	Outsourcing unit	Head of outsourcing unit
4	Ground construction project management	M	M	P	S
Project operation phase					
1	Oil and gas project implementation	M	M	P	
2	Operation management	M	P	I	
Project end phase					
1	Asset evaluation	M	P	I	I
2	Asset disposal	M	P	I	I
3	Environmental restoration	M	P	I	
4	Employee resettlement	M	P	I	
5	Post-project evaluation	M	P	I	
6	Archiving work	M	P	I	

(4) Termination phase

The task breakdown for the termination stage is shown in Table 14.

5. Lower-level task planning

Plan the scope of the journey for the bottom-level tasks obtained from the task breakdown of each stage, as shown in Table 15.

In this way, the output results of the natural gas development project scope planning are formed.

(II) Schedule planning

1. Definition of project schedule planning

Schedule planning is a plan that expresses the order of work in the project, the start and completion time, and the interconnection relationship. Through the preparation of the schedule plan, the project implementation forms an organic whole. Schedule planning is the basis for schedule control and management. Project schedule planning will directly affect the allocation of various resources in the project, and the quality of project schedule planning will directly affect various aspects of project human resource management, schedule management, cost and quality management.

Table 11 Task breakdown in the planning phase

Level 1 project activities	Level 2 project activities	Level 3 project activities
Task scope planning in the planning phase	Project environmental condition analysis	Geological condition analysis
		Engineering condition analysis
		Economic condition analysis
		Environmental condition analysis
	Oil and gas reservoir engineering design and planning	Oil and gas reservoir engineering design
	Drilling engineering design and planning	Drilling engineering design
	Ground engineering design and planning	Ground engineering design
	Oil and gas engineering design and planning	Oil and gas engineering design
	CSSQ planning	Scope planning
		Quality planning
		Schedule planning
		Cost planning
	HSE Planning	Health management
		Safety management
		Environmental management
	Risk planning	Reservoir geological risk
		Reservoir engineering risk
		Personnel safety risk
	Project evaluation analysis	Technical evaluation analysis
		Economic evaluation analysis
	Organizational structure planning	Planning stage organizational structure planning
		Construction stage organizational structure planning
		Operation stage organizational structure planning
		End stage organizational structure planning

Table 12 Task breakdown in the construction phase

Level 1 project activities	Level 2 project activities	Level 3 project activities	Level 4 project activities
Tasks of the construction stage scope planning	Well completion engineering	Pre-drilling tasks	Land lease
			road construction
			Clear water pool
			Sewage pool
			Sewage disposal engineering
			Blowout pool
		Drilling tasks	Drilling labor
			Owner-supplied materials
			Mud labor
			Cementing labor
			Directional well labor
			Post-drilling treatment
		Logging tasks	Logging labor
			Data processing and interpretation
		Well logging tasks	Well logging
			Special well logging project analysis
		Well completion tasks	Gas testing operations
			Fracturing and acidizing
			Perforation
	Surface engineering	Oil and gas station construction	Oil and gas station construction
		Oil and gas station construction	Oil and gas station construction
		Gathering pipeline construction	Gathering pipeline construction
		Oil and gas pipeline	Oil and gas pipeline
		Road engineering	Road engineering
		Gathering engineering	Gathering engineering
		Water injection engineering	Water injection engineering
		Wastewater treatment engineering	Wastewater treatment engineering
		Power supply engineering	Power supply engineering

(continued)

Table 12 (continued)

Level 1 project activities	Level 2 project activities	Level 3 project activities	Level 4 project activities
		Communication engineering	Communication engineering
Tasks of the construction phase scope planning	Management work	Project scope management	Project scope management
		Project schedule management	Project schedule management
		Project cost management	Project cost management
		Project quality management	Project quality management
		Project risk management	Project risk management

Table 13 Task breakdown in the operation phase

Level 1 project activities	Level 2 project activities	Level 3 project activities
Tasks of the operation phase Scope planning	Production activities	Well completion
		Perforation
		Reservoir protection
		Oil and gas testing
		Water injection
		Enhanced recovery measures
		Anti-corrosion
		Downhole operations
		Oil and gas extraction
		Oil and gas gathering
	Management activities	Planning management
		Cost management
		Quality management
		Safety management

2. Content of project schedule planning

The main content of project schedule planning is shown in Table 16.

(III) Cost planning

1. Definition of project cost planning

Project cost planning is the process of planning costs through cost estimation. Project cost estimation is a project cost management work that estimates and determines the

Table 14 Task breakdown in the termination phase

Level 1 project activities	Level 2 project activities	Level 3 project activities
Task scope planning of the termination phase	Asset assessment	Asset assessment
	Asset disposal	Asset write-off
		Asset transfer
	Environmental restoration	Soil and water restoration
		Ecological restoration
	Employee resettlement	Employee resettlement
	Post-project evaluation	Post-project evaluation
	Archiving work	Paper document archiving
		Electronic document archiving

Table 15 Scope planning table for natural gas development project (Illustrative)

Task Code	Task Name	Superior Task	Achieve Goal	Limiting Conditions	Existing Risks	Milestone Plan	Submit Results	Responsible Department	Responsible Position

Table 16 Main contents of project schedule planning (Illustrative)

Basic Task	Task Code	Duration (Week/Year)	Earliest Start Time	Latest Start Time	Earliest End Time	Latest End Time	Flex Time

cost of various activities in the project and the total cost of the project based on the project's resource requirements and plans, as well as the price information of various project resources. Cost estimation is an estimate and plan of the costs required to complete the project and is an important part of the planning plan.

2. Main factors affecting project cost

    (1) The quantity and price of resources consumed and occupied by the project
    (2) Project duration
    (3) Project quality
    (4) Project scope

3. Main basis for project cost estimation

    (1) Work breakdown structure
    (2) Project scope statement
    (3) Resource requirements

(4) Resource unit price
(5) Activity duration estimation
(6) Historical information
(7) Ledger

4. Composition of project cost

The specific composition of the project cost includes:

(1) Material cost
(2) Fuel power cost
(3) Depreciation of fixed assets
(4) Equipment maintenance cost
(5) Employee compensation
(6) Management fee
(7) Others

5. Content of project cost planning

The main content of project cost planning is shown in Table 17.

(IV) Quality planning

1. Definition of project quality planning

Project quality planning is part of project quality management, dedicated to setting quality objectives and specifying necessary operating processes and related resources to achieve project quality objectives. The project quality objectives refer to the goals pursued by project quality management. Project quality planning is the process of determining the specific quality objectives to be achieved by a project and how to achieve these objectives based on relevant requirements.

Project quality planning includes: identifying and determining necessary work processes, configuring required human and material resources, and the process of careful consideration and overall arrangement to ensure the achievement of expected quality objectives. Project quality planning is one of the processes that ensure project success.

Table 17 Main contents of project cost planning (Illustrative)

Underlying Task	Material Cost	Fuel Power Cost	Fixed Assets Depreciation Cost	Equipment Maintenance Cost	Employee Salary	Management Fee	Others	Subtotal
Total								

Table 18 Main contents of project quality planning (Illustrative)

Task Code	Task Name	Superior Task	Quality Standard	Quality Control Measures	Responsible Department	Responsible Position

2. Procedure of project quality planning

The process of project quality planning generally includes four aspects of work:

(1) Data collection

That is, to clarify and collect the data and information needed for project quality planning.

(2) Understand the project implementation organization or project quality policy

Understand the quality policy of the project implementation organization or project client and the assumptions, premises, and constraints of the project. The results to be created by project quality planning largely depend on the resources available for the project quality team. When the project quality management team plans for project quality, it should consider the following content:

① Product quality planning involved in the project;
② Project quality management and operation planning;
③ Preparation of quality plan.

(3) Learn and utilize project quality planning

Learning and utilizing the methods, tools, techniques, knowledge, and experience of project quality planning. These methods, tools, techniques, knowledge, and experience are collectively referred to as "tools and techniques". Nowadays, when doing project quality planning, it is necessary to be meticulous, and it is impossible to rely solely on brainstorming without necessary tools and techniques.

(4) Write project quality planning documents and related auxiliary files

3. Content of project quality planning

The main content of project quality planning is shown in Table 18.

Appendix
Major Events in China's Natural Gas Exploration and Development in 2023

Major Events in Natural Gas Exploration

(I) Major Events in CNPC Exploration

1. Significant strategic breakthrough in oil and gas exploration in the Sichuan Basin of Southwest Oil and Gas Field

On April 8, the shale gas well Ziyang 201, deployed by Southwest Oil and Gas Field Company in Neijiang, tested a stable daily gas production of 738,800 cubic meters. This is the first time in China that a high-yield industrial gas flow with commercial development value has been drilled in the ancient shale strata of the Cambrian System, which dates back 540 million years.

On October 14, after obtaining a daily gas production of 421,600 cubic meters from the Deng'er Formation, Well Datan 1 of PetroChina Southwest Oil and Gas Field Company again achieved a high-yield industrial gas flow of 818,500 cubic meters per day from the Dengsi Formation. This is the first high-yield industrial gas well in the platform margin zone of the Dengying Formation on the west side of the De'yang-An'ue rift basin in the Sichuan Basin. The exploration has made another significant strategic breakthrough, opening up a new block for large-scale reserve increases.

2. Significant breakthrough in oil exploration in the complex fault region of the Western Ordos Basin

On December 1, Changqing Oilfield of PetroChina announced that after two years of exploration, PetroChina's Changqing Oilfield discovered a geological reserve of over 100 million tons in the Hongde area of Huan County, Gansu Province, opening up a new field for oil exploration and development in the western part of the Ordos Basin. So far, Changqing Oilfield has submitted a proven oil reserve of 50.24 million tons in the Hongde area, and it is predicted that there is still an oil reserve of 56.20 million tons, and a large oilfield of over 100 million tons has finally surfaced. This discovery marks a major breakthrough in oil exploration in the complex fault zone

© Petroleum Industry Press 2025

L. Zhang et al., *Annual Report on Operation of China's Natural Gas Industry: 2023*,
https://doi.org/10.1007/978-981-95-2371-9

in the western part of the Ordos Basin, and opens up a new field for oil exploration and development in the western part of the basin.

3. China's first 10,000-m scientific exploration well drilling depth breaks through 8,900 m

On December 18, in the heart of the Taklamakan Desert, China's first 10,000-m scientific exploration well—China National Petroleum Corporation's Deep Earth Taklamakan Well 1—achieved a drilling depth of over 8,900 m, fully entering the stage of drilling into highly challenging strata. Over the past year, China National Petroleum Corporation has intensified theoretical and technological breakthroughs and accelerated continuous innovation in drilling and completion, reservoir modification, geophysics, and other engineering technologies, making the foundation for major exploration discoveries more solid.

(II) Major events in Sinopec's exploration

1. Major breakthrough in exploration of ultra-deep condensate oil and gas in the central part of Shunbei, Tarim Basin

Accelerating the construction of the "Deep Earth No. 1" Shunbei Oil and Gas Field Base, wells were concentrated in the oil-rich central area of Shunbei. Thirteen high-yielding wells were successively drilled, and two hundred-million-ton-level and four ten-million-ton-level oil-rich fracture zones were efficiently discovered. The new field between the fractures in the risk exploration area yielded a thousand-ton high production. The central area of Shunbei successfully created a large scene of "overall enrichment and continuous oil-bearing areas." A total of 276 million tons of oil and 209.3 billion cubic meters of natural gas geological reserves have been proven, and a production capacity base of 3 million tons of oil equivalent has been built. New breakthroughs have been made in the efficient drilling and completion engineering technology for ultra-deep layers, and Well Yuejin 3-3XC set a new record for ultra-deep large displacement drilling, further consolidating the ability to "advance into the ten-thousand-meter deep earth."

2. Major breakthrough in shale oil exploration in the New Area of Lijin Sag of the Jiyang Depression

Efforts were continuously strengthened to build the national demonstration zone for Jiyang continental shale oil. In 2023, after fracturing, Well Lijin 101 in the Lijin Sag of the Dongying Depression produced 201 tons of oil per day. This achieved a major breakthrough in the exploration of new shale oil areas and a new type of clayey mixed lithofacies. A hundred-million-ton-level resource base was newly identified. So far, Jiyang continental shale oil has achieved comprehensive breakthroughs in "five sags, three lithofacies, two stratigraphic systems, and multiple types." Twenty-two wells have produced over ten thousand tons of oil in total, with three wells producing over 30,000 tons, realizing rapid growth in reserves and production.

3. Major breakthrough in deep coalbed methane exploration in Daniudi Gas Field, Ordos Basin

Efforts to deepen the evaluation of deep coalbed methane exploration areas in the Ordos Basin were intensified. The deep coal seams of the Taiyuan Formation in the transitional marine-continental phase were chosen as the strategic breakthrough direction. In the Daniudi Gas Field, the risk exploration well Yangmei 1HF was deployed. After fracturing, the No. 8 coal seam of the Taiyuan Formation in this well produced 104,000 cubic meters of gas per day, achieving a major breakthrough in deep coalbed methane at a depth of 2,800 m. It was confirmed that the deep coalbed methane had a high gas content, with free gas accounting for 20% to 25% of the total gas content. New predicted reserves amounted to 122.6 billion cubic meters. The well has been in production for 254 days, with a cumulative gas production of 17.48 million cubic meters. This has significant implications for the exploration and development of China Petrochemical Corporation's deep coalbed methane resources, which exceed one trillion cubic meters.

4. Major breakthrough in continental tight oil and gas exploration in the northern part of the Sichuan Basin

A systematic evaluation of the tight sandstone in the Triassic-Jurassic formations of the northern Sichuan Basin was carried out. Well Bazhong 1HF first achieved a stable oil flow of over 100 tons per day through fracturing in the Jurassic channel sandstone in the Sichuan Basin. The evaluation confirmed that the resource potential of the Lianggaoshan Formation channel sandstone in the northern Sichuan Basin exceeded 100 million tons, thus opening up a new area for large-scale reserve increases. Well Yuanlu 2HF was deployed to conduct a production enhancement pilot in the porous reservoirs of the Xusi Formation. Through fracturing, it achieved a daily gas production of 500,000 cubic meters. This marked a major breakthrough in the production capacity of the matrix-type reservoirs of the Xujiahe Formation, and the ability to develop tight oil and gas resources was lifted to a new level.

5. Major breakthrough in deep marine shale gas exploration in the Dalong Formation of the Permian in the Puguang area of the Sichuan Basin

Actively implementing the strategic concept of shale gas exploration "stepping out of the Silurian System," the risk exploration well Leiyue 1 Well, deployed in the Dalong Formation of the Permian in the Puguang area, was fractured and tested to achieve a daily gas production of 426,000 cubic meters. This marked a significant breakthrough in the exploration of deep marine shale gas in the Permian in the Sichuan Basin, with a confirmed resource potential of 17.27 billion cubic meters, which is expected to cultivate a new field to support the stable production of the Puguang gas field.

6. Major Breakthrough in shale oil exploration in the Fu 2 Section of the Subei Basin

The exploration scene of shale oil in the Subei Basin continued to expand. In 2023, several major breakthroughs were achieved in the Haian, Zhentong, and Gaoyou depressions. In the Haian depression, risk exploration in the new area with low to medium evolution degree of shale oil was successful. Well Haiye 1HF was fractured and tested to achieve a daily oil production of 50.7 tons, with an additional predicted reserve of 25.51 million tons. In the Zhentong depression, different well-type experiments were carried out in different areas. Five single wells all produced more than 10,000 tons of oil, with an additional controlled reserve of 30.03 million tons. In the Gaoyou depression, the exploration of shale oil was extended to deeper, medium–high maturity and new types of shale oil. Well Huaye 7HF was fractured in the Fu 2 section and tested to achieve a daily oil production of 52.3 tons and a daily gas production of 10,400 cubic meters, with an additional controlled and predicted reserve of 52.12 million tons.

7. Major breakthrough in the exploration of the Fengcheng Formation in the Heshan Foreland Belt of the Junggar Basin

By tackling the structural modeling technology of the Heshan foreland belt on the northwest edge of the Junggar Basin, a new understanding of "multiple periods of thrusting, progressive superposition" of the thrust structure was achieved, leading to the discovery of the Mahu Northeast Subsidence-Heshan Depression. The Heshan 5 well in the Fengcheng Formation produced 95.8 cubic meters of oil and 6869 cubic meters of gas per day after fracturing, marking the first major breakthrough in the foreland thrust belt of the Junggar Basin, adding a predicted reserve of over 30.63 million tons, and potentially opening up a new billion-ton reserve base.

8. Major breakthrough in the exploration of Permian Shale Gas in the Hongxing Area of Western Hubei-Eastern Chongqing

Focusing on the geological study of deep-water continental shelf shale, efforts were intensified to tackle the exploration technology of new layers and types of shale gas. The Hongye Mao 1HF and Hongye Mao 2HF wells in the Mao Fourth Section produced 64,000 and 60,000 cubic meters of gas per day respectively after fracturing, marking the first major breakthrough in the exploration of shale gas in the Maokou Formation, with an estimated resource volume of over 500 billion cubic meters. The Hongye 7HF and Hongye 3–2 wells produced over 200,000 cubic meters of gas per day after fracturing in the Wu Second Section, breaking through the commercial production capacity barrier at depths below 4500 m. Currently, six wells in the Hongxing area have reached commercial development standards, forming a new exploration field in the Permian Wu Second Section and Mao Fourth Section of Eastern Sichuan.

9. Major breakthrough in the exploration of a new type of tight gas in the Xu 2 Section of the Hexing Field in Western Sichuan

In response to the challenges of the widespread gas content but low porosity and permeability, and large differences in single well production capacity in the Xujiahe Formation of the Sichuan Basin, a new enrichment model of "fracture bodies" was proposed. The exploration strategy shifted from "occupying high points, finding fractures" to "overlapping ancient and modern high points, fracture-matrix enrichment". High-yield gas flows were obtained from wells such as Fenggu 110, Fenggu 113, and Xinsheng 202 in the Hexing Field-Fenggu area, revealing the enrichment rule of "early east–west fractures are gas-rich, late north–south fractures are optimal, and areas with developed matrix and micro-fractures can also produce high yields". Proven reserves of 133 billion cubic meters were efficiently implemented, promoting the large-scale increase in reserves and production of tight gas in the Xujiahe Formation in Western Sichuan.

10. Major breakthrough in exploration in the New Area of the Changling Fault in the Songliao Basin

Based on the exploration of the new oil and gas-rich fault zone in the southern part of the Songliao Basin, the exploration strategy was changed and actively explored. The Changling Fault was re-analyzed, and a new model of "detachment sliding controls the formation of the basin" was innovatively proposed. The risk exploration well Dashen 1 achieved a breakthrough, expanding the range of hydrocarbon-generating depressions and adding a natural gas resource volume of 50 billion cubic meters. The North 12 and North 1201 wells deployed on the western slope of Changling discovered an atypical tensional lake basin controlled by volcanic topography, revealing for the first time the development of effective source rocks in the western part of the Changling Fault, implementing a natural gas resource volume of 180.9 billion cubic meters, and potentially cultivating a new reserve production base.

(III) Major events in CNOOC exploration

1. Major discovery of a hundred million-ton oil field in the deepwater and deep layers of the South China Sea at Kaiping South

The Kaiping Depression in the Pearl River Mouth Basin of the South China Sea has not achieved a breakthrough after two rounds of exploration over more than forty years. Since the "14th Five-Year Plan," systematic research on the structural and basin-forming mechanisms, hydrocarbon distribution and potential, and reservoir formation models and enrichment rules of the depression has significantly increased the resource potential of the depression. The oil and gas enrichment rules of the Kaiping South Belt were clarified, and a breakthrough was achieved in the exploration field in 2022. In 2023, the focus continued to be on the discovery of large and medium-sized oil and gas fields in deepwater and deep layers, focusing on the four major oil and gas convergence ridges, efficiently evaluating and expanding the scale of reserves. The Paleogene single well test produced over a thousand tons per day, setting a new record for deep layer testing in the eastern South China Sea, and is expected to build the first billion-ton light oil field in deepwater and deep layers domestically.

2. Major discovery of a 300 million-ton oil field in the shallow lithologic reservoir of the Bohai Sea

After the discovery of the Qinhuangdao 33–1 South Oilfield in 2011, the Shi Jiu Tuo uplift in the Bohai Bay Basin had not made any significant discoveries after more than ten years of exploration. In 2023, by liberating thought and innovatively proposing a shallow oil and gas enrichment model in the slope area, shallow-scale lithologic oil and gas exploration was carried out, achieving efficient evaluation of the Qinhuangdao 27–3 oilfield. The daily oil production of the Ming lower section test exceeded 400 cubic meters, making it the first hundred-million-ton lithologic oil and gas field in the central and northern Bohai Sea. The discovery of this oilfield provides a solid material guarantee for the Bohai Oilfield to produce and stabilize 40 million tons.

3. Major discovery of 200 million tons of oil and gas reserves at the concealed submerged hill of Bohai Bozhong 26–6

The western section of the Bohai South Low Uplift has been explored for 43 years, and the submarine mountain has not made any commercial discoveries. Since the "14th Five-Year Plan," through liberating thought, the mechanism of "ductile shear rock formation, differential sealing" was proposed, and the development model of "one mountain, multiple circles" for the submarine mountain was established, further clarifying the exploration direction. After making a breakthrough in 2022, an evaluation was carried out in 2023. The average thickness of the oil layer in the exploration wells exceeded 100 m. The highest daily oil production was 324.7 cubic meters, and gas was 324,000 cubic meters, becoming the world's largest metamorphic rock oilfield.

4. Major discovery of a large and medium-sized oilfield at Weizhou 11–6 in the South China Sea

The exploration degree of the middle and shallow layers in the Weizhou Southwest Depression is high and the potential is difficult to tap. The exploration degree of the deep layer is relatively low, but the difficulty of closure implementation and reservoir prediction is high, and the law of oil and gas enrichment is unclear. In 2023, through innovative thinking and technical tackling, the development model of "lake invasion, high-position sheet sand overlapping" structural-lithologic compound closure was proposed, breaking through the understanding limitations of the formation rules of the faulted lake basin. After integrated deployment drilling, Weizhou 11–6 achieved the largest integrated oilfield discovery in the Beibu Gulf Basin of the South China Sea, opening up a new exploration area in Weizhou Southwest, and also paving a new way for the development of other offshore faulted lake basin old oil areas.

5. Major discovery of a 100 billion cubic meters deep coalbed methane gas field in Shenfu area of the Ordos Basin

The onshore Shenfu coalbed methane field is located in the northeastern margin of the Ordos Basin. In 2023, by strengthening the research on the gas accumulation mechanism of deep coalbed methane, an innovative concept of "complementary" and "stereoscopic" exploration of tight gas and coalbed methane was proposed. The research on reservoir modification and drainage technology was also enhanced, which enabled the "renewed use of old wells" and "multiple uses of a single well".

This significantly reduced exploration costs and effectively accelerated the exploration process of deep coalbed methane. The highest daily output of a single well reached 29,000 cubic meters. The field successfully declared itself as China's first deep coalbed methane field with reserves of 100 billion cubic meters, demonstrating the prospect of exploration and development of deep coalbed methane reservoirs in the northeastern margin of the Ordos Basin.

6. Major discovery of a hundred million-ton oil and gas field in the deepwater and deep layers of Guyana, South America

In 2023, by applying reservoir description and hydrocarbon detection technologies to meticulously study deep targets and drive drilling operations, four wells were drilled, all of which yielded discoveries. Among them, two wells targeting the deep layers achieved daily production exceeding 1,000 cubic meters. The oil and gas discoveries in this ultra-deepwater and deep formation have opened up broader prospects for the exploration of subsequent blocks.

7. Major discovery of the Huizhou 26–6 north large and medium-sized oilfield in the deep layer of the Huizhou Depression in the Pearl River Mouth Basin

In the South China Sea's Pearl River Mouth Basin, the Huizhou Depression, a mature exploration area, has undergone systematic efforts to tackle challenges. These efforts have led to a breakthrough in the traditional understanding that the Enping Formation was unlikely to form large-scale reservoirs. The exploration has deepened the understanding of the reservoir formation theory and identified high-quality reservoirs in complex lithological belts. This has resulted in a significant exploration breakthrough in the large fan delta sedimentary system of the Enping Formation in the northern part of Huizhou 26–6. It has further expanded the exploration potential of the middle and deep layers and consolidated the scale of oil and gas reserves in the shallow deep layers of the Pearl River Mouth Basin.

8. Major discovery of the Panyu 10–6 large and medium-sized oilfield in the deep layer of the Panyu Depression in the Pearl River Mouth Basin of the South China Sea

In 2023, explorers broke traditional thinking and innovatively proposed a sealing model for the Enping Formation's sand-rich background, featuring regional lake flooding mudstone + local marine transgression calcareous sandstone. They boldly innovated and carefully verified, with the Enping Formation testing daily production of 516 cubic meters. This led to the first discovery of a large and medium-sized oilfield in the deep layer of the Panyu 4 Sag in the Pearl River Mouth Basin, at Panyu 10–6. This discovery has laid a solid foundation for the exploration and development of new stratigraphic systems in old oil areas.

9. Major breakthrough in the exploration of natural gas in the super-deep buried hills of the Mesozoic in the Bohai Sea at Bozhong 8–3 South

Breakthroughs have been achieved in the exploration of the Archean and Paleozoic strata in the Gulf Basin, but the Mesozoic reservoirs are complex and highly heterogeneous. Despite multiple rounds of exploration in earlier years, no large-scale oil and gas discoveries were made. By changing the approach and actively innovating geological understanding, explorers shifted their focus from the early uplift and slope areas to the proximal sag areas. In 2023, Bozhong 8–3 South was selected as a breakthrough point for the exploration of Mesozoic volcanic rocks in the Bohai Sea. Drilling achieved a significant breakthrough, successfully discovering a 125-m gas layer at a super-deep depth of 5,000 m. Testing yielded a daily production of 167,000 cubic meters of natural gas and 84 cubic meters of condensate oil. This major discovery in the exploration of Mesozoic buried hills in the Bozhong Sag has opened up a new situation for the exploration of Mesozoic buried hills in the Bohai Sea.

10. Major breakthrough in natural gas exploration in the ultra-deepwater and ultra-shallow layers of the Qiongdongnan Basin, South China Sea

In 2023, by further liberating thinking and innovating the large-scale gas accumulation model for ultra-deepwater and ultra-shallow layers, key technologies were developed and a rapid evaluation system for ultra-shallow operations was formed. This led scientific and efficient exploration deployment, and promoted the combination of open and closed-loop drilling of six wells in the area, all of which were successful and yielded high-saturation pure hydrocarbon gas layers. There is potential for the discovery of a large gas field, demonstrating the broad exploration prospects of the ultra-deepwater and ultra-shallow layers in the South China Sea.

Major Events in Natural Gas Development

(I) PetroChina

1. Changqing Oilfield

The development milestones of Changqing Oilfield in 2023 are as shown in Table A.1.

2. Tarim Oilfield

The development milestones of Tarim Oilfield in 2023 are shown in Table A.2.

3. Southwest Oil and Gas Field

The development milestones of Southwest Oil and Gas Field in 2023 are shown in Table A.3.

4. Daqing Oilfield

The development milestones of Daqing Oilfield in 2023 are shown in Table A.4.

(II) Sinopec

The natural gas development milestones of Sinopec in 2023 are shown in Table A.5.

Table A.1 Major events in the development of changqing oilfield in 2023

Date	Major events
February 20	Changqing Oilfield Gas Production Plant 1 has put into production 14 gas wells, with a daily gas production of 239,000 cubic meters
May 6	The first bauxite horizontal well in Changqing Oilfield tested a daily natural gas production of 3.534 million cubic meters, setting a new domestic record. This also marks a major breakthrough in China's bauxite gas reservoir
June 8	Changqing Oilfield Gas Production Plant 2 has produced a total of 40.14 billion cubic meters of natural gas this year, adding enough "gas" for the "hard overproduction" throughout the year
August 18	The cumulative gas production of the largest gas production area on land in China has exceeded 600 billion cubic meters, equivalent to replacing 720 million tons of standard coal, reducing carbon emissions by 1.657 billion tons
October 11	The natural gas production of PetroChina Changqing Oilfield this year has reached 400.93 billion cubic meters, and the annual natural gas production of Changqing Oilfield has exceeded 400 billion cubic meters
November 3	Changqing Oilfield's First Gas Production Plant has an accumulated annual natural gas production exceeding 90 billion cubic meters
November 13	GeoEast2023 officially released, marking a new step for China's geophysical core software and geophysical technology
December 17	China's largest onshore fully equipped tight gas field—Sulige Gas Field, with an annual natural gas production of 300.6 billion cubic meters, breaking through 300 billion cubic meters for two consecutive years
December 26	News from the Fourth Gas Production Plant of Changqing Oilfield, the plant's cumulative production of natural gas has reached 60 billion cubic meters, setting a new historical record

Data source Changqing Oilfield official website and official public account

Table A.2 Major events in the development of tarim oilfield in 2023

Date	Major events
February 9	China's first ultra-deep clastic rock gas field with an annual production of 10 billion cubic meters—Tarim Kedeep Gas Field's cumulative production of natural gas exceeded 700 billion cubic meters, reaching 701 billion cubic meters
March 9	Tarim Oilfield's Guole 3C well completed drilling to a depth of 9396 meters, setting a record for the deepest horizontal well in Asia
March 19	The cumulative production of natural gas in the Dabeiqi area exceeded 200 billion cubic meters, reaching 202.7 billion cubic meters. The annual gas production has increased for 8 consecutive years, providing solid guarantee for the gas supply along the West-East Gas Pipeline and for the people's livelihood in southern Xinjiang
May 1	The drilling of the deepest well in Asia, the Leap Forward 3-3XC well, deployed by Sinopec, began

(continued)

Table A.2 (continued)

Date	Major events
May 14	Tarim Oilfield has supplied more than 3300 billion cubic meters of gas to the West-East Gas Pipeline network, achieving more than 6800 days of continuous safe production
May 19	The Bozhi 1-2 well in Tarim Oilfield is undergoing drilling operations. The well uses air drilling technology in the three-open gravel layer section, drilling to a depth of 5068 meters, setting a new record for the deepest air drilling well in China with a 333.4 mm wellbore
May 30	China's first 10,000-meter scientific exploration well—Deep Earth Tako 1 well began drilling in the Tarim Basin in Xinjiang, marking the start of China's "10,000-meter era" of drilling capabilities into the deep earth
July 29	The cumulative oil and gas production equivalent of the Bozhi Gas Field exceeded 10 million tons, reaching 10.0065 million tons, of which the cumulative production of natural gas was 106.2 billion cubic meters
July 31	The Bozhi-Dabei natural gas pipeline project, with the largest transmission capacity, was successfully put into operation, increasing the maximum annual gas transmission capacity of the gas field from 50 billion cubic meters to 120 billion cubic meters, and constructing a large interconnected transmission artery in the southern part of the Tianshan Mountains
August 15	China's first 10,000-meter deep well—Deep Earth Tako 1 well, successfully completed the second open cementing operation. During the cementing casing operation, the well set several records in China, including the deepest large-diameter casing run and the largest tonnage of casing run
August 29	China's first ultra-deep gas field—Tarim Oilfield's Kedeep Gas Field, cumulative production of natural gas 753.11 billion cubic meters
September 4	The exploration and development of the Tarim Basin has fully entered the ultra-deep layer, and has built China's largest ultra-deep oil and gas production base
September 24	China's first 10,000-meter deep well—Deep Earth Tako 1 well safely drilled to the middle of the third opening, with a well depth of 7856 meters
October 15	China's largest ultra-deep condensate gas field—Bozi-Dabei gas field's annual processing capacity has crossed the 100 billion cubic meters mark
October 30	Tarim oilfield's west-to-east gas supply breaks through 340 billion cubic meters
November 5	The successful commissioning of the Fuyuan joint station in the Fuman oilfield located in Shaya County, Aksu region, Xinjiang Uygur Autonomous Region, marks the large-scale economic development of China's largest ultra-deep oilfield, Fuman oilfield, at a depth of 8000-9000 meters
December 1	The annual oil and gas production equivalent of the Bodao oil and gas management area of the Tarim oilfield broke through 7 million tons, reaching 7.02 million tons, an increase of 1.056 million tons compared to the same period in 2022. In 2023, a total of 8.12 billion cubic meters of natural gas has been exported, with a daily export volume exceeding 29.8 million cubic meters, setting a new historical record
December 13	The first gas storage facility of the Tarim oilfield—Kekoya gas storage facility officially started gas extraction, effectively improving the peak-shaving supply capacity of natural gas in southern Xinjiang
December 29	China's largest ultra-deep oilfield—Fuman oilfield's annual oil and gas production has changed its title three times in three years, breaking through the 4 million tons mark in 2023, climbing to a new peak of 4.036 million tons

Data source Tarim Oilfield official website and official public account

Table A.3 Major events in the development of southwest oil and gas field in 2023

Date	Major events
February 1	The Feixianguan formation gas reservoir development well Po 005-X4 in the Tieshanpo gas field in the northeastern Sichuan area of the Southwest Oil and Gas Field Company tested a daily gas production of 2.2219 million cubic meters, with an unimpeded flow rate of 3.6465 million cubic meters/day. This is the first large gas well in the Tieshanpo gas field with a daily production exceeding 2 million cubic meters, marking a new step in PetroChina's independent development of high-sulfur gas fields in northeastern Sichuan
February 13	PetroChina Southwest Oil and Gas Field Company's Pengshen 6 well in the Penglai gas area was successfully drilled and cemented, with the deepest well depth reaching 9026 meters, setting a new record for the deepest vertical well in Asia, and can be called the "underground Everest"
February 20	The ninth gas extraction period of the Xiangguosi gas storage of the Southwest Oil and Gas Field Company accumulated peak-shaving gas extraction of 2.033 billion cubic meters, breaking through 2 billion cubic meters for the first time in a single period, setting a new record for single-period gas extraction
February 26	The Feixianguan formation of the Tieshanpo gas field of the Southwest Oil and Gas Field Company's Po 002-H5 well tested a daily gas production of 1.62 million cubic meters and an unimpeded flow rate of 2.78 million cubic meters, becoming the third high-yield well with a daily gas production exceeding 1 million cubic meters since February, following the Po 005-X4 well and the Gaoshi 018-1-H2 well
March 9	PetroChina deployed 6 production wells in the Tieshanpo gas field, perfectly ending with a total daily gas production of over 10 million cubic meters, an average daily gas production of 1.71 million cubic meters per well, and an average unimpeded flow rate of 3 million cubic meters/day, marking a major progress in PetroChina's independent development of high-sulfur gas fields
April 16	PetroChina Southwest Oil and Gas Field Company's Gaoshi 018-4-H1 well in the pilot test of the Neidengsi gas reservoir in Gaoshi Terrace tested a high-yield industrial gas flow of 1.1227 million cubic meters/day and an unimpeded flow rate of 1.8823 million cubic meters/day, setting a new record for the test production and unimpeded flow rate of the Neidengsi gas reservoir in the area, marking a major new progress in the development evaluation of the Neidengsi gas reservoir in Gaoshi Terrace
May 12	Southwest Oil and Gas Field Longgang 83-C1 well initially tested a daily gas production of 537,800 cubic meters
May 23	The well with the highest pressure in the Sichuan Basin—Southwest Oil and Gas Field Company's Jiaotan 1 well has been safely producing for 312 days, with a cumulative production of natural gas exceeding 100 million cubic meters
June 3	The Gao Shi 045-H1 well in the Gao Shi Ladder Block of the Anyue Gas Field of the Southwest Oil and Gas Field Company tested and produced 1.3672 million cubic meters of high-yield industrial gas per day, with an unimpeded flow rate of 2.9333 million cubic meters per day
June 6	China Petroleum's first independently developed ultra-high sulfur gas field—Tieshanpo gas field has reached full production, with a daily production of natural gas of 4 million cubic meters, marking a new breakthrough in the key core technology of China's ultra-high sulfur gas field development

(continued)

Table A.3 (continued)

Date	Major events
June 8	The acidizing construction of Pengshen 6 well, the deepest vertical well in Asia, has fully completed all design engineering indicators, achieving a major breakthrough in the acidizing transformation of ultra-deep carbonate rocks
June 29	In the first half of the year, the Southwest Oil and Gas Field Company completed 16 conventional gas development wells, with an average test production of 1.282 million cubic meters per day and an average unimpeded flow rate of 2.368 million cubic meters per day, setting a new historical record
July 3	The Pengshen 16 well, a daily fee pilot well implemented by the Southwest Oil and Gas Field Company, drilled a 333.4 mm borehole to a medium completion depth of 2786 meters, and entered the medium completion operation. The Pengshen 17 well drilled a 241.3 mm borehole to the top of the Leikoupo Group at 3620 meters, and is currently drilling normally. With the completion of the third opening of the two wells, significant results have been achieved in speeding up and improving efficiency, marking the achievement of phased results in the daily fee pilot
July 20	The world's first 10,000-meter scientific exploration well with the most complex geological conditions and the highest drilling difficulty—Shendi Chuan Science 1 well started drilling in the Sichuan Basin
August 5	The Southwest Oil and Gas Field Company's Anyue Gas Field rolling exploration and development well—Moxi 031-H4 well Qixia Group tested and produced 1.4266 million cubic meters of high-yield industrial gas per day
August 24	It was learned from the ISO International Standardization Organization that the international standard ISO 7055: 2023 "Test Method for Slippery Water Drag Reduction Performance in the Upstream Field of Natural Gas", led by the Natural Gas Research Institute of the Southwest Oil and Gas Field Company, was officially released. This is China's first breakthrough in the field of shale gas international standards
August 25	The Southwest Oil and Gas Field Company's Chuan Dongbei high-sulfur gas field Dukouhe—Qili North Block Feixianguan Group tested a daily gas production of 2.0508 million cubic meters, this is the first well in the Dukouhe—Qili North Block to test over 2 million cubic meters of high-yield gas
August 30	The first gas storage cluster in Sichuan Province—Moujiaping, Laowenchang gas storage cluster has a cumulative injection volume exceeding 300 million cubic meters
September 4	The PDC drill bit designed by the company achieved a new breakthrough in the Maokou-Qixia Group of Pengshen 102 well, drilling from the top of Maokou to Liangshan Group in one run, with a progress of 329 meters, a mechanical drilling speed of 4.44 meters per hour, and a speed increase of 72% compared to the adjacent well
September 4	The cumulative production of natural gas by China Petroleum Southwest Oil and Gas Field Company has exceeded 600 billion cubic meters
September 4	The first gas storage in the southwest region—Xiangguosi gas storage has a cumulative injection volume of 20.13 billion cubic meters in the eleventh injection period, an increase of 0.32 billion cubic meters compared to the same period in 2022, with a single-day highest injection volume of 16.13 million cubic meters and a storage volume exceeding 40 billion cubic meters

(continued)

Table A.3 (continued)

Date	Major events
September 12	The enterprise standard "Operation Procedure for Wireline Testing Grease Injection System in Gas Wells" (Q/SYXN 0632—2023) drafted by the Southwest Oil and Gas Field Company of China Petroleum Natural Gas Co., Ltd. was officially released, filling the domestic gap in the technical standard for wireline testing grease sealing
September 22	The deepest well in Asia—China Petroleum Southwest Oil and Gas Field Company's Pengshen 6 well Maokou Group test produces 1.46 million cubic meters of gas per day, achieving significant exploration results
September 25	China Petroleum Southwest Oil and Gas Field Company's Pengyang 6 well Maokou Group test produces 2.4 million cubic meters of gas per day, setting a new record for the Maokou Group gas reservoir test in the Sichuan Basin
September 27	Southwest Oil and Gas Field Company's Anyue Gas Field Longnu Temple Block Moxi 039-H2 well Maokou Group test produces 1.6666 million cubic meters of gas per day
October 4	Southwest Oil and Gas Field Company's Tieshanpo Gas Field has been safely producing for 130 consecutive days, with a total production of 508 million cubic meters of natural gas, demonstrating the "four highest" industry level of Tieshanpo's high-sulfur gas field
October 12	Southwest Oil and Gas Field Company's shale gas production this year reached 10.051 billion cubic meters, an increase of 11.75% year-on-year, and the production exceeded 10 billion cubic meters for four consecutive years
October 13	From January to September, the Southwest Oil and Gas Field Company focused on tapping the potential of old gas fields, implemented the shutdown well recovery project, and increased the production of natural gas by 740 million cubic meters
October 15	Southwest Oil and Gas Field Company's Tianfu Gas Field tight gas daily production volume is 10.01 million cubic meters, breaking through 10 million cubic meters for the first time, achieving a leap in tight gas production
October 18	The first gas storage facility in the southwest region—Xiangguo Temple Gas Storage completed the eleventh cycle of gas injection work, with a total injection of 2.346 billion cubic meters in this injection cycle, the upper limit pressure increased by 2 trillion pascals, the working gas volume increased by 300 million cubic meters, it is the first gas storage facility in China to reach capacity, and all production indicators in a single period have achieved historical breakthroughs
October 21	The largest shale gas production base in China—China Petroleum Chuannan Shale Gas Field has produced a total of 70.025 billion cubic meters of gas, becoming the first shale gas field in China to break through 70 billion cubic meters of cumulative gas production
October 22	Southwest Oil and Gas Field's key reserve well Xuanxing 1 well was drilled to a depth of 6080 meters, with a drilling cycle of 140.56 days, speeding up by 101.19 days compared to the neighboring well Xuantan 1 well
November 1	Southwest Oil and Gas Field Company's Tianfu Gas Field Jianyang Block Yongqian 3 Platform 11 pilot test wells are fully operational, with a total daily production of over 1 million cubic meters, marking the completion of the first million cubic meter tight gas production platform in the Sichuan Basin by the Southwest Oil and Gas Field Company

(continued)

Table A.3 (continued)

Date	Major events
November 3	The Lu 203H91 platform, which was finely cultivated through geological engineering integration, successfully completed the reservoir transformation construction, achieving 100% design compliance and wellbore integrity
November 4	Southwest Oil and Gas Field Company learned that Weihan 1 well Ordovician Baota Group test produced 90.2 thousand cubic meters of gas per day
November 7	The 11 wells of Yongqian 3 platform are fully operational, marking the successful completion of the first tight gas hydraulic fracturing mine test in China
November 8	China Petroleum Southwest Oil and Gas Field Company's Shunan Gas Mine Moujiaping, Laowenchang Gas Storage Group's daily gas injection capacity reached 1.3 million cubic meters, and the first gas storage group pilot test project in Sichuan Province was fully operational
November 9	Yang 101H13-4 well was successfully drilled to 5920 meters, with a single run drilling of 1930 meters in the build-up + horizontal section, marking an important progress in the first deep shale gas oil-based drilling fluid downhole cooling test in China
November 14	The company's pipeline internal detection technology team successfully completed the magnetic leakage internal detection of the 813mm large-diameter pipeline at the Chongqing Changshou Duzhou New Station Gas Transmission Station Southwest Oil and Gas Field Company's innovative research and development of continuous tubing completion and gas extraction integrated technology was applied in the Zhongqian 203-6-H2 and Zhongqian 203-9-H2 wells, with a daily gas production exceeding 100,000 cubic meters. It has been stably produced for 15 consecutive days, marking the successful first application of the company's continuous tubing completion and gas extraction integrated technology in tight gas
November 16	The first new capacity block in the gas area of China National Petroleum Corporation to adopt the "oil company" model, the Southwest Oil and Gas Field Company's Central Sichuan Northern Gas Management Office, has produced a total of 10 billion cubic meters of natural gas
November 16	The Lux LNG storage station of Huayou Company, a subsidiary of Southwest Oil and Gas Field Company, was officially completed and put into operation
November 22	The work of comparing the measurement of standard substances for natural gas composition analysis was completed for the first time in China
November 24	The annual oil and gas equivalent production is 30.06 million tons, including 37.52 billion cubic meters of natural gas and 112,000 tons of crude oil, an increase of 2.836 million tons compared to the same period in 2022
November 24	The CX group of the Pengshen 10 well of the Southwest Oil and Gas Field Company achieved a daily gas production of over 2 million cubic meters in the straight well test, setting a new record for the CX group gas reservoir test in the Sichuan Basin
November 27	A new natural gas purification plant and two single-well desulfurization stations were built, with a daily increase in natural gas processing capacity of nearly 10 million cubic meters, always leading domestically. The company's daily natural gas purification capacity has exceeded 70 million cubic meters

(continued)

Table A.3 (continued)

Date	Major events
November 30	The cumulative production of shale gas from the Changning Shale Gas Field of China National Petroleum Corporation's Southwest Oil and Gas Field Company has exceeded 30 billion cubic meters
December 6	Two invention patents applied for by the Southwest Oil and Gas Field, "Fracturing Backflow Fluid Treatment Device and Method" and "Temporary Plugging Ball Performance Determination Method and Device", have been granted
December 11	The four horizontal wells of the Tianfu Gas Field Yanting Block in the tight gas of Zhongqian 210 platform have achieved a test daily production of over 2 million cubic meters and an unobstructed flow rate of over 4 million cubic meters, setting a platform test record for the SXM group 9 sand group of Zhongqian 1 well area
December 14	The annual natural gas production of China National Petroleum Corporation's Southwest Oil and Gas Field Company has exceeded 40 billion cubic meters, reaching 40.001 billion cubic meters, marking the official completion of the first 40 billion cubic meters large gas area in Southwest China
December 15	During the drilling process of the inclined section to the horizontal section of the 215.9 mm wellbore Longmaxi Group of the Ning 209H73-5 well of the Southwest Oil and Gas Field Company, a single day completed a footage of 1039 meters, breaking the 1000 meters mark for the first time in daily drilling footage
December 17	The Luzhou 213 well tested and obtained a high-yield industrial gas flow, making it another high-yield well in the deep shale gas of the central Luzhou block. The horizontal section of this well is 1800 meters long, with a platinum casing encounter rate of 95.4% and a wellbore integrity of 100%
December 19	All wells of the Luzhou 208H2 platform have been drilled. The average drilling cycle of a single well is 62.2 days, the average drilling cycle of four openings is 21.4 days, and the average platinum casing encounter rate is 99.1%, setting multiple records such as the drilling cycle of a single well and the average platinum casing encounter rate for deep shale gas platforms
December 20	The annual natural gas production of the Sichuan Central Oil and Gas Mine of the Southwest Oil and Gas Field Company exceeded 12.42 billion cubic meters
December 23	The annual shale gas production of China National Petroleum Corporation's Southwest Oil and Gas Field Company reached 12.58 billion cubic meters, and the oil and gas equivalent broke through the 10 million tons mark for the first time
December 25	The three-dimensional physical exploration project of the Zi 201 well area—the first three-dimensional physical exploration project for the new stratigraphic collection of shale gas in China was successfully completed
December 26	Southwest Oil and Gas Field Company successfully completed the three-opening cementing construction of the deep Sichuan Science 1 well, setting multiple world records including the deepest drilling of a 593.73 mm borehole, the deepest setting of 492.13 mm + 485.78 mm casing, and the heaviest, among others

Data source Southwest Oil and Gas Field official website and official public account

Table A.4 Major events in the development of daqing oilfield in 2023

Date	Major events
May 16	Songshen 9-Ping 5 well tested gas and achieved a daily production of over one million cubic meters of high-yield gas flow, achieving a new breakthrough in the production capacity of tight gas in the deep Shahezi Formation of Daqing
June 5	The Drilling Engineering Company successfully completed the drilling of the exploratory directional well—Qingyu 1 well, with a depth of 8738.29 meters, setting a record for the deepest well drilled in Daqing Oilfield
June 26	Daqing Oilfield Petroleum Engineering Research Institute successfully connected the first experimental well Xu Shen 1-Slant 202 well "injection dissolution" integrated automatic injection device with the digital gas well decision-making platform through the oilfield VPDN dedicated network, achieving the first digital remote measurement and control of Daqing Oilfield gas wells
June 27	Daqing Drilling International Business Division successfully completed the cementing acoustic amplitude operation of the Tamuchage block Tower 19-67 well—Flat 15 well in Mongolia, with a total drilling cycle of only 18.98 days, breaking the record for the fastest construction of horizontal wells in the block
July 5	In the first half of the year, Daqing Oilfield produced a total of 15.0913 million tons of crude oil, exceeding the plan and achieving "half the time, half the output"; it produced 2.969 billion cubic meters of natural gas, exceeding the plan by 0.19 billion cubic meters
July 24	Drilling Engineering Company's drilling—the company's Saudi DQ050 drilling team successfully started drilling the BRRI-548 well in the offshore artificial island construction of JUBAIL (Jubai) in Saudi Arabia. This is the first shallow sea artificial island drilling project undertaken by the Drilling Engineering Company's overseas projects
August 8	The all-metal packer independently developed by Daqing Oilfield Petroleum Engineering Research Institute has filled the domestic technical gap in metal sealing packers under high temperature, high pressure, and corrosive environments
August 18	Daqing Oilfield used its independently developed portable load displacement field detection device to complete 420 well times of online comparison, with an accuracy rate of 100%, filling the technical gap in the field of digital instrument calibration of Daqing Oilfield's mechanical mining well force parameters
September 25	The independently developed "Qing Spear" brand 127 type ultra-deep penetration perforator successfully completed the on-site inspection of the American Petroleum Institute (API), with an average perforation depth of 2593.6 mm, refreshing the world record again
November 2	Daqing Oilfield's 1205 drilling team is intensifying the construction of the GY25-Q9-H6 well in the Daqing Gulung continental shale oil national demonstration area. The well drilling applied the currently domestically produced automation level highest one-button human-machine interaction 7000 meters automated drilling machine, marking a new progress in the construction of Daqing Oilfield's first digital drilling team
November 10	Drilling Engineering Company's drilling—the company's DQ050 drilling team completed all construction work on the first well on an artificial island in the offshore of Saudi Arabia, marking the successful "first show" of Daqing Drilling's offshore drilling construction

Data source Daqing Oilfield official website and official public account

Table A.5 Major events in Sinopec's natural gas development in 2023

Date	Major events
March 14	"Deep Earth No.1·Shunbei Oil and Gas Field Base", Shunbei 84 inclined well test achieved high-yield industrial oil and gas flow, with a daily oil and gas equivalent of 1017 tons, becoming the 22nd thousand-ton well in Shunbei Oil and Gas Field
March 22	Jingwei Company's self-developed rotary geological steering drilling system achieved excellent results in the Fuling Shale Gas Field Jiaoye 18-S11HF well, with 261 hours of trouble-free operation, a single drilling advance of 3210 meters, and a 100% encounter rate of high-quality reservoirs. This is the first time that the single drilling advance of the self-developed rotary geological steering drilling system has exceeded 3000 meters
March 31	The welding operation of the 270,000 cubic meter large-scale liquefied natural gas storage tank at the Qingdao LNG receiving station of the Natural Gas Branch Company was successfully completed. This storage tank is currently the fastest built, largest area, and longest diameter of the inner and outer tank bottoms in China
April 11	"Deep Earth Engineering·Sichuan-Chongqing Natural Gas Base" in the shale gas exploration well Leiye 1 in Dazhou City, Sichuan Province, achieved a high-yield shale gas flow of 426,600 cubic meters per day. The shale layer is buried more than 4000 meters deep, which is China's first breakthrough in marine deep shale gas exploration in the Permian Dalong Formation
May 1	The "Deep Earth No.1" Yuejin 3-3XC well in the Tarim Basin started drilling. The designed depth of the well is 9472 meters, which will break the record for the deepest well in Asia
May 2	"Deep Earth No. 1" achieved another breakthrough in the new fault zone, and the first exploration well deployed in the Shunbei Oil and Gas Field No. 10 belt—Shunbei 10X well test achieved a high-yield oil and gas flow, with a daily gas production of 600,000 cubic meters
May 25	The new deep 105 well of Sinopec Southwest Oil and Gas Branch Company achieved a high-yield industrial gas flow of 1.056 million cubic meters per day and an unobstructed flow rate of 2.49 million cubic meters after segmented acid fracturing test transformation
June 26	The risk exploration well Pingdi 1HF deployed by the East China Oil and Gas Branch in Wulong, Chongqing, has produced more than 8 million cubic meters of gas, with a stable daily gas production of 46,000 cubic meters, and the preliminary evaluation of the recoverable reserves of a single well is 40 million cubic meters
July 12	The world's first over 7000 meters ultra-deep high-sulfur bio-reef large gas field—Yuanba Gas Field, has produced more than 30 billion cubic meters of natural gas and reduced carbon dioxide emissions equivalent to 40 million tons
August 8	From January to July, Henan Oilfield produced 654,800 tons of crude oil and 45.85 million cubic meters of natural gas, both exceeding the planned operation and achieving steady growth
August 21	The Southwest Oil and Gas Branch Company put into production more than 70 new wells, with new production capacity exceeding 600 million cubic meters, and a total production of 5.697 billion cubic meters of natural gas, an increase of 6% year-on-year

(continued)

Table A.5 (continued)

Date	Major events
August 22	The Southwest Oil and Gas Branch Company has produced more than 100 billion cubic meters of natural gas since it was put into production. Currently, the company's annual natural gas production capacity exceeds 10 billion cubic meters, and the proven reserves of natural gas exceed 1 trillion cubic meters
November 2	China's first 270,000 cubic meter liquefied natural gas storage tank was officially put into use at Sinopec's Qingdao LNG receiving station. This is the largest liquefied natural gas storage tank in China. After the storage tank is put into use, the Qingdao LNG receiving station will increase its gas storage capacity by 165 million cubic meters
November 7	The deep-layer tight sandstone key exploration well—Yuanlu 2HF well in the Yuanba Gas Field in Guangyuan City, Sichuan Province, achieved a high-yield industrial gas flow of 220,600 cubic meters per day and an unobstructed flow rate of 503,900 cubic meters at a burial depth of nearly 4500 meters in the Xujiahe Formation
November 8	The industry standard "Technical Requirements for the Compilation of Shale Gas Development Adjustment Plans" led by Jianghan Oilfield was approved and released by the National Energy Administration. This standard is China's first industry standard for three-dimensional development of shale gas
November 15	The "Deep Earth Project · Shunbei Oil and Gas Field Base" deployed by Sinopec in the Tarim Basin has achieved high oil and gas flow in the 3-3XC well test, with a daily output of 200 tons of crude oil and 50,000 cubic meters of natural gas. The completed well depth reached 9432 meters, setting a new record for the deepest well on land in Asia

Data source Sinopec official website and official public account

Bibliography

Beihai365. (2023). LNG 三期扩建报批, 再建四座 LNG 储罐 [The approval process for the Phase III expansion of the LNG project is underway, which includes the construction of four additional LNG storage tanks]. Retrieved from https://mp.weixin.qq.com/s/gWZUYm07p_-bCMzCwt6o0A

Chang, H., Zhou, L., Luo, Q., et al. (2022). 中国天然气国际标准化发展与思考 [Development and reflections on the international standardization of natural gas in China]. 天然气工业, *42*(10), 1–10.

China National Radio. (2024). 创历史新高!"十四五"以来我国已累计新建管道超 1 万公里 [A new historical high! Since the "14th five-year plan," China has built more than 10,000 kilometers of new pipelines]. Retrieved from http://news.cnr.cn/native/gd/20240126/t20240126_526572467.shtml

China Securities Network. (2024). 创历史新高!"十四五"以来我国已累计新建管道超1万公里 [Creating a historical high! Since the "14th five-year plan", China has built more than 10,000 kilometers of new pipelines]. Retrieved from https://news.cnstock.com/news/bwkx-202401-5183156.htm

Chongqing Petroleum and Natural Gas Trading Center. (2023). 跨省天然气管道分区定价的影响与下一步改革建议 [Cross-provincial natural gas pipeline zonal pricing and reform suggestions]. Retrieved from https://mp.weixin.qq.com/s/CgAd3OoHLW-pMyEb8BNHNw

Dai, B., Wang, D., Cao, Y., et al. (2024). 2023 年中国能源行业回顾及 2024年展望 [Review of China's energy industry in 2023 and outlook for 2024]. 当代石油石化, *32*(1), 1–7.

Dan, M., Zhang, H., Che, C., et al. (2024). 大能源视角下中国能源安全形势及中央企业作用研究 [Research on China's energy security situation and the role of central enterprises from a macro-energy perspective]. 中国矿业, *33*(2), 1–9.

Development and Reform Commission, Energy Bureau, Ministry of Industry and Information Technology. (2016). 关于推进"互联网 +"智慧能源发展的指导意见[Guiding opinions on promoting the development of "Internet +" smart energy]. Retrieved from https://www.gov.cn/gongbao/content/2016/content_5082989.htm

Fang, G. (2023). "双碳"目标下的天然气行业经济发展探索 [Exploration of the economic development of the natural gas industry under the "dual-carbon" goals]. 商业*2.0, 2023*(13), 90–91.

Feng, Q., Wang, C., He, J., et al. (2023). 油气储运企业生产运维流程开发探索与实践 [Exploration and practice of production operation process development in oil and gas storage and transportation enterprises]. 油气储运, *42*(1), 9–15.

Gao, F., Liu, B., Li, M., et al. (2022). 全球天然气发展趋势与启示 [Global trends in natural gas development and their implications]. 中国石油勘探, *27*(6), 13–21.

© Petroleum Industry Press 2025

L. Zhang et al., *Annual Report on Operation of China's Natural Gas Industry: 2023*,
https://doi.org/10.1007/978-981-95-2371-9

Gao, Z., He, M., Yang, F., et al. (2023). "双碳"目标下中国天然气发展的分析与建议 [Analysis and suggestions on the development of China's natural gas under the "dual carbon" goals]. 油气与新能源, *35*(4), 7–11.

Gao, Y., Wang, B., Hu, Y., et al. (2024). 2023 年中国天然气发展述评及2024年展望 [Review and outlook of China's natural gas development in 2023]. 天然气工业, *44*(2), 166–177.

General Administration of Customs of the People's Republic of China. (n.d.). 统计月 [*Statistical Monthly Report*]. Retrieved from http://www.customs.gov.cn/customs/302249/zfxxgk/2799825/302274/302277/4899681/index.html

He, D., Jia, C., Wei, Y., et al. (2022). 世界天然气产业形势与发展趋势 [The situation and development trend of the world natural gas industry]. 天然气工业, *42*(11), 1–12.

He, C., Duan, Y., Zhang, Y., et al. (2023). 中国天然气价格改革理论研究进展及其应用回顾与展望 [Review and outlook of theoretical research progress and application of natural gas price reform in China]. 天然气工业, *43*(12), 121–129.

He, C. (2021). 天然气热值计量发展趋势及展望 [Development trend and outlook of natural gas calorific value measurement]. 天然气与石油, *39*(1), 140–144.

Huaqi Energy Headhunter. (n.d.). 广西 LNG 三期扩建项目配套码头工程开工, 计划 2025 年完工 [The supporting wharf project of Guangxi LNG phase III expansion project started, and it is planned to be completed in 2025]. Retrieved from https://mp.weixin.qq.com/s/llUPrbR14rrvYvIUyoye4g

Huang, H., Xu, W., & Dai, X. (2021). 国际市场天然气价格大涨的原因, 影响及对策 [The reasons, impacts and countermeasures of the sharp rise in international natural gas prices]. 价格理论与实践, *2021*(10), 5–8.

Huaxia Climate. (n.d.). 数说中国能源 [Data on China's energy 2023/2024]. Retrieved from https://mp.weixin.qq.com/s/EYLQK9U9nglreLh1P5OTsw

Jiang, S., Du, H., Xu, M. (2024). "双碳"目标下能源与产业双重结构转型 [Dual-carbon goals and the dual structural transformation of energy and industry]. 数量经济技术经济研究, *41*(2), 109–130.

Jingchu Network. (2023). 西气东输三线湖北段试运行 [Trial operation of the Hubei section of the West-East gas pipeline III]. Retrieved from http://news.cnhubei.com/content/2023-11/21/content_16974143.html

Li, L., Li, Q., Ai, S., et al. (2021). 传统化石与新型生物质能源重整制氢研究现状 [Current research on hydrogen production by reforming of traditional fossil fuels and new biomass energy]. 化学与生物工程, *38*(11), 1–6.

Li, H., Yu, M., Zheng, Y., et al. (2023). 基于情景的 2050 年世界能源供需展望分析——基于《BP世界能源展望(2023 年版)》[Scenario-based analysis of world energy supply and demand outlook for 2050: Based on "BP World Energy Outlook (2023 Edition)"]. 天然气与石油, *41*(5), 131–137.

Liang, H., Wang, Y., & Jia, W. (2023). 基于机器学习的天然气钢质管道缺陷检测方法研究 [Research on defect detection method of natural gas steel pipeline based on machine learning]. 安全与环境学报, *23*(10), 3528–3537.

Liu, J., & Liu, Y. (2022). 中国地下储气库建设的发展现状及展望 [Development and outlook of underground gas storage construction in China]. 应用化工, *51*(4), 1136–1140, 1145.

LNG Receiving Station Builder. (2023). 惠州LNG二期规划建设3座27万立方米的 LNG 储罐及配套设施, 已启动项目报批前期工作![Huizhou LNG phase II plans to build 3 270,000 cubic meters LNG storage tanks and supporting facilities, and has started the pre-approval work of the project!] Retrieved from https://mp.weixin.qq.com/s/eefk3cVm1m_VDPJL4cNFAw

Lock Oil Treasure Subscription Account. (2024). 我国能源行业盘点与展望——低碳转型加速韧性持续增强 [Review and outlook of China's energy industry — Low carbon transition accelerates, resilience continues to strengthen]. Retrieved from https://mp.weixin.qq.com/s/byYdWkhp3Ywuzz2_G-gkSQ

Ministry of Industry and Information Technology, Development and Reform Commission, Ministry of Science and Technology, et al. (2022). 关于"十四五"推动石化化工行业高质 量发展的指导意见 [Guiding opinions on promoting high-quality development of petrochemical industry

during the “14th five-year plan”]. Retrieved from https://www.gov.cn/gongbao/content/2022/content_5696258.htm
Ministry of Industry and Information Technology. (2024).《石化化工行业数字化转型实施指南》解读材料 [Interpretation materials of “Implementation guide for digital transformation of petrochemical industry”]. Retrieved from https://www.miit.gov.cn/jgsj/ycls/gzdt/art/2024/art_8abb0cb8eb5d4129b2beef3892fb3aec.html
Ministry of Industry and Information Technology. (2022). “十四五”信息化和工业化深度融合发展规划 [“14th Five-Year Plan” for the deep integration of informatization and industrialization.] Retrieved from https://www.miit.gov.cn/jgsj/ghs/zlygh/art/2022/art_21ab63dacb6a49b4b6072498abf3ecfc.html
National Bureau of Statistics of the People’s Republic of China. (2023). 中国统计年鉴 *2023* [*China statistical yearbook 2023*]. 中国统计出版社.
National Energy Administration. (2023). 《中国天然气发展报告 (2023)》[*China natural gas development report (2023)]. 天然气勘探与开发, 46*(3), 32.
National Energy Administration. (2023). 中国天然气发展报告 *(2023)* [*China natural gas development report (2023)*]. 石油工业出版社.
National Pipeline Network Group Oil and Gas Control Center. (2023). 数字化转型|天然气管网运行数据应用平台 (二期) 建成试用 [Digital transformation: trial use of natural gas pipeline operation data application platform (Phase II)]. Retrieved from https://mp.weixin.qq.com/s/SkM7u_qFoOtVri-ABDDDBw
Natural Gas Industry Information. (2023). 国家管网山东龙口南山 LNG:建设进度已达 89.01%![National pipeline network Shandong Longkou Nanshan LNG: Construction progress has reached 89.01%!] Retrieved from https://mp.weixin.qq.com/s/P7rrebrwohR6t41AIwoa4Q
Polar Star Energy Network. (2023). 国家管网潜江—韶关输气管道广西支干线工程顺利投产 [The Guangxi branch line project of the national pipeline network Qianjiang-Shaoguan gas pipeline was successfully put into operation]. Retrieved from https://news.bjx.com.cn/html/20230906/1330415.shtml
Ren, Z., Chen, X., & Xu, Q. (2022). 油气田勘探数字化转型现状及展望 [Current status and prospects of digital transformation in oil and gas exploration]. 油气与新能源, *34*(1), 67–73.
Ruili River Network. (2023).“一带一路”十周年|中缅油气管道管网人:为国输油气 为民保供应 [The 10th anniversary of the Belt and Road Intiative: pipeline workers of the China-Myanmar oil and gas pipeline have been transporting oil and gas for the country and ensuring supply for the people]. Retrieved from https://www.ruili.gov.cn/tzrl/qy/content-169-10465-1.html
Shen, X., Zhou, S., Wu, C. (2023). 中国天然气市场发展 2022 年回顾与 2023 年展望 [Review and outlook of China’s natural gas market development in 2022 and 2023]. 国际石油经济, *31*(4), 25–36.
Shi, N. (2021). 油气田企业数字化业务管理研究 [Research on digital business management of oil and gas field enterprises]. 中国管理信息化, *24*(18), 107–108.
Southern Energy Observation. (n.d.). “双碳”目标下的天然气市场深化改革的建议 [Suggestions on deepening the natural gas market reform under the “dual carbon” goals]. Retrieved from https://mp.weixin.qq.com/s/AjAxUKIRicrx9ThdVuuugA
Su, K., Zhou, H., Xin, J., et al. (2023). 2023—2024 年供暖季中国天然气市场供需预测 [Forecast of China’s natural gas market supply and demand in the heating season of 2023–2024]. 国际石油经济, *31*(9), 61–69.
Sun, M. (2023). “双碳”背景下天然气行业数字化转型策略研究 [Research on digital transformation strategy of natural gas industry under the background of the “dual carbon” goals]. 中国市场, *15*, 81–84.
The State Council of the People’s Republic of China. (2021). 中华人民共和国国民经济和社会发展第十四个五年规划和2035年远景目标纲要 [The 14th Five-Year Plan for National Economic and Social Development and the Outline of the Long-Term Goals for 2035 of the People’s Republic of China]. Retrieved from https://www.gov.cn/xinwen/2021-03/13/content_5592681.htm

Wang, Z., & Du, X. (2014). 焦炉煤气制天然气产业的发展探讨 [Discussion on the development of natural gas industry from coke oven gas]. 煤化工, *43*, 7–11.
Wang, K., Yang, T., Liang, K., et al. (2023). 碳达峰碳中和战略下天然气的能源优势与发展建议 [Energy advantages and development suggestions for natural gas under the carbon peak and carbon neutrality strategy]. 能源与节能, *2023*(3), 27–31.
Wang, L., Li, L., Luo, X., et al. (2023). "双碳"目标下天然气行业影响因素分析及建议——以川渝地区为例 [Analysis of the influencing factors of the natural gas industry under the "dual carbon" goals and recommendations: Taking the Sichuan and Chongqing region as an example]. 天然气技术与经济, *17*(5), 53–61.
Wei, H., Ding, G., & Wang, C. (2023). 国内外地下储气库建设现状及规模研究 [Research on the current situation and scale of underground gas storage construction at home and abroad]. 中国石油学会天然气专业委员会. 第 *33* 届全国天然气学术年会 *(2023)*论文集 *(06* 综合*)* [*Proceedings of the 33rd National Natural Gas Academic Annual Meeting (2023) (06 Comprehensive)*]. 中国石油勘探开发研究院, 中国石油天然气集团公司油气地下储库工程重点实验室, 国家管网集团北方管道责任有限公司沈阳输油气分公司.
Wu, Y., Li, R. (2023). 中国东南沿海地区天然气市场竞争态势分析及发展策略建议 [Analysis of the competitive situation of the natural gas market and development strategy recommendations in the southeast coastal region of China]. 国际石油经济, *31*(11), 47–54.
Xiao, X., Sun, R. (2023). 中国能源可持续安全:理念塑造, 现状解析与路径构建 [China's sustainable energy security: Concept shaping, current situation analysis, and path construction]. 太平洋学报, *31*(10), 56–68.
Xiao, J., & Wang, X. (2019). 中国液化天然气现货价格的传导机制 [The transmission mechanism of China's liquefied natural gas spot prices]. 天然气工业, *39*(11), 9.
Xu, Q., Li, W. (2023). 全球碳中和背景下的中国能源安全方略 [China's energy security strategy under the backdrop of global carbon neutrality]. 南通大学学报*(*社会科学版*)*, *39*(1), 117–129.
Yang, X. (2023). 中国地下储气库建设需求与关键技术发展方向 [Demand and key technology development direction of underground gas storage construction in China]. 油气储运, *42*(10), 1100–1106.
Yin, W. (2024). 当前我国能源行业形势, 走势及对策研究 [Current situation, trends, and countermeasures of China's energy industry]. 中国物价, *2024*(2), 17–20.
Yuan, Q. (2023)."双碳"背景下天然气地下储气库的建设思考 [Thoughts on the construction of natural gas underground storage under the "dual carbon" background]. 中国石油和化工标准与质量, *43*(14), 86–88.
Zhang, K., Hu, J., Hu, Y., et al. (2017). 石油库储运数字化生产管理信息系统设计与实现 [Design and implementation of digital production management information system for oil storage and transportation]. 中国石油石化, *9*, 1–2.
Zhang, K., Miao, M., Zhang, L. (2022). "双碳"目标与中国能源转型思考(二)——能源转型中的化石能源 [Dual-carbon goals and reflections on China's energy transition (II) — Fossil fuels in the energy transition]. 中外能源, *27*(4), 1–7.
Zhang, K., Meng, F., Zhang, L. (2023). 21 世纪初期世界天然气格局变化及启示 [Changes in the world natural gas pattern in the early 21st century and their implications]. 世界石油工业, *30*(1), 20–29.
Zhang, Y., He, C., Du, B., et al. (2023). 对中国天然气价格改革的回顾与政策建议 [Review and policy suggestions on China's natural gas price reform]. 天然气技术与经济, *17*(3), 1–9.
Zhang, G., Yang, F., Zhao, H., et al. (2023). 西南油气田开发储量管理数字化转型探索与实践 [Exploration and practice of digital transformation in reserves management of southwest oil and gas field development]. 中国管理信息化, *26*(7), 102–108.
Zhu, X., Bai, H., Fan, H., et al. (2022). 天然气在中国构建新型能源体系中的作用及保障供应安全的建议 [The role of natural gas in building a new energy system in China and suggestions for ensuring supply security]. 煤炭经济研究, *42*(9), 12–18.
Zou, C., Lin, M., Ma, F., et al. (2024). 碳中和目标下中国天然气工业进展, 挑战及对策 [Progress, challenges, and countermeasures of China's natural gas industry under the carbon neutrality target]. 石油勘探与开发, *2024*(1), 1–18.

Zou, C., Xiong, B., Li, S., et al. (2024). 碳中和背景下世界能源转型与中国式现代化能源革命 [World energy transformation and Chinese-style modern energy revolution under the background of carbon neutrality]. 石油科技论坛, *43*(1), 1–17.

Zuo, K. (2023).“双碳”背景下, 天然气行业发展前景探索 [“Dual-carbon” goals and the development prospects of the natural gas industry]. *2023* 年中国城市燃气协会标准工作委员会年会暨燃气安全运营和智慧建设研讨会论文集 [*Proceedings of the 2023 China City Gas Association Standards Work Committee Annual Conference and the Seminar on Gas Safety Operation and Intelligent Construction]*, 258–261.

MIX
Papier aus verantwortungsvollen Quellen
Paper from responsible sources
FSC® C105338

If you have any concerns about our products,
you can contact us on
ProductSafety@springernature.com

In case Publisher is established outside the EU,
the EU authorized representative is:
Springer Nature Customer Service Center GmbH
Europaplatz 3, 69115 Heidelberg, Germany

Printed by Libri Plureos GmbH
in Hamburg, Germany